AF602031

Medicinal Plants

Cultivation and Uses

The Authors

Dr. R.N. Nakar is currently working as Assistant Professor in Department of Botany, Sheth PT Arts and Science College, Godhra (Under Gujarat University), Gujarat, India. Earlier he worked as Research Associate at Central Arid Zone Research Institute, Regional Research Station, Kukma, Bhuj, Gujarat,India also worked as Senior Research Fellow in Department of Plant Physiology, Directorate of Groundnut Research, ICAR Institute. He did his Ph.D. in Botany on subject entitled "Studies on phenological patterns of Girnar Reseve Forest, near Junagadh- Gujarat", has published 9 papers in Internationals Journals of repute, presented papers in conferences, symposia and seminars of various levels. He was University first during his M.Sc. He has recently got International Travel Grant from Science and Engineering Board, India to present his research paper in International Conference at Colombo, Sri Lanka. His areas of Interest include Phenology, Plant physiology, Floristic diversity, Forest ecology, along with MAPs.

Dr. H.L. Dhaduk is currently working as Associate Professor in Department of Genetics and Plant Breeding, B. A. College of Agriculture, Anand Agricultural University, Anand Gujarat. He has vast experience of research and teaching of 15 years, has published 20 research papers in reputed Journals of International and National levels. He is also associated with genetic improvement in Guava and Custard Apple. He has guided 5 M.Sc. students and currently, guiding two M.Sc. and two Ph.D. students. His areas of interest are Plant breeding, Genetics, MAPs, Conservation of Biodiversity, is one of the well known members of many National scientific bodies.

Dr. V.P. Chovatia is currently working as Principal, College of Agriculture, Amreli (Junagadh Agricultural University), has vast experience of more than 20 years of research and teaching. He has worked as Principal Investigator in 8 different projects, also worked as team leader in Bt-cotton trials in Saurashtra region, Gujarat during 2004 to 2007. He has authored 4 books, published 24 papers in repute Journals of International and National level. He has guided 12 MSc, 3 Mphil students, and currently guiding 2 PhD students. His areas of interest include Oil seed crops, sugarcane, and biodiversity along with MAPs. He was recipient of ICAR Junior Research Fellowship in 1982, is member of many National scientific bodies.

Medicinal Plants

Cultivation and Uses

Dr. R.N. Nakar

Dr. H.L. Dhaduk

Dr. V.P. Chovatia

2016

Daya Publishing House®

A Division of

Astral International Pvt. Ltd.

New Delhi – 110 002

Cataloging in Publication Data--DK
Courtesy: D.K. Agencies (P) Ltd. <docinfo@dkagencies.com>

Nakar, R. N., author.
Medicinal plants : cultivation and uses / Dr. R.N. Nakar, Dr. H.L. Dhaduk, Dr. V.P. Chovatia.
pages cm
Includes bibliographical references and index.

ISBN 978-93-5130-991-8 (International Edition)

1. Medicinal plants--India--Gujarat. 2. Materia medica, Vegetable--India--Gujarat. I. Dhaduk, H. L., author. II. Chovatia, V. P., author. III. Title.

QK99.I4N35 2016 DDC 581.634095475 23

Published by : **Daya Publishing House®**
A Division of
Astral International Pvt. Ltd.
– ISO 9001:2008 Certified Company –
4760-61/23, Ansari Road, Darya Ganj
New Delhi-110 002
Ph. 011-43549197, 23278134
E-mail: info@astralint.com
Website: www.astralint.com

Laser Typesetting : **Classic Computer Services**, Delhi - 110 035

Printed at : **Replika Press Pvt. Ltd.**

Dedicated

to

All

Environment Lovers…

Acknowledgement

Authors wish to thank deeply to Dr. B.K. Kikani and Dr. N.C. Patel (Ex-Vice Chancellor, JAU), Dr. J.M. Parekhiya (Director of Extension Education, JAU), Dr. Butani (Principal, Collge of Agriculture, JAU) for their encouragement. We convey our sincere thanks to Dr. I.U. Dhruj (Associate Director of Research, JAU) for providing moral support for compiling this book. We also thank current Vice Chancellor, Dr. A.R. Pathak for giving inspiration. We convey our sincere thanks to Dr. Maltiben Chauhan (Professor, L.M. College of Pharmacy, Ahmedabad) and Dr. Padmanabh S. Nagar (Associate Professor, M.S.University, Baroda) for their technical help whenever required.

Dr. R.N. Nakar

Dr. H.L. Dhaduk

Dr. V.P. Chovatia

Dr. A.R. Pathak
Vice Chancellor

Junagadh Agricultural University
Junagadh – 362 001, Gujarat

Foreword

India is considered to be hub of medicinal plants. The use of these plants for curing various diseases has been given in *"Charaka Samhita"* and in many Indian Veda and Strata. In recent lime allopathic medicine are being used for curing diseases. However, these medicine has their own side effect. Due to awareness of the effect of allopathic, use of *Ayurveda* medicine is increasing not only in our country but abroad too. The demand for such medicine is increasing and farmers can grow important medicinal plants and earn good profit however, importance of medicinal plants for specific diseases and its scientific cultivation practices are not known to many. In view of this, the authors have put their efforts to compile details from various authentic sources in the current book *"Medicinal Plants: Cultivation and Uses"*.

The book *"Medicinal Plants: Cultivation and Uses"* not only provides cultivation details and uses, but also includes various local names in different languages of India. I congratulate authors for their sincere efforts, compiling gigantic information. I am sure this book will be useful to all researchers, students and farmers, and finally to whole society.

Dr. A.R. Pathak
Vice Chancellor
Junagadh Agricultural University
Junagadh – 362 001
Gujarat

Preface

Plants are main source of food on earth. They are the only plants with help of them, world is existed. Plants not only provide food, fodder and economics but also are highly useful for medicines since ancient period of Atharva Veda. In earlier literature of Susruta- samhita, description of 700 medicinal plants is given however all of them do not belong to India. But about 1,500 Indian native plants are believed to be contributed for different treatments in different diseases. Rigveda reported plants such as Semal, Palash, Pithvan and Pipal with high significance. However after these veda there is no clear and precise information regarding these medicinal plants.

According to one estimation, there are about 2,000 medicinal plants or drugs utilized for various ailments in India but only about 200 are considered of animal origin and almost same number is of mineral origin. Hence about 1,500 are of plants origin. India has wide range of plants ranging from Himalayas to Western Ghats but still we are adopting only few plants with medicinal properties. Gujarat is in the Western Side of India and considered wide importance of floristic diversity. Areas such as Saurashtra, Kutch, North Gujarat and South Gujarat harbors great source of plant diversity. There are mountains such as Aravalli in North side of Gujarat, Girnar and Bardo in Saurashtra, Kalo Dungar in Kutch provide good research opportunities for researchers as well as farmers for farming of medicinal plants. Geography and atmosphere of Gujarat region favors growth of important medicinal plants.

People have stated different notes describing about various uses of plants from different areas but still literature availability is not much sufficient for readers and users. Most of literature is explaining on medicinal uses but very little literature has given details on their agricultural cultivation at farmer's level which is one of the very important aspects of the current time in Gujarat and country. As we know that there is increasing demand of medicinal trees, herbs, shrubs and climbers as these medicines don't have side effects at all!!. However, still most of people relay on new medicinal systems of Allopathy and surgery as there is no systematic information on production of medicinal plants and their marketing. There is lot of confusions on local names of plants as it varies with place to place but here names are given from authentic sources along with their scientific names and botanical family. Here arrangement of chapters in current book is alphabetical by botanical names of plants. For the convenience of readers, latest names have been written but some older names, one or more synonyms have been used too.

In this book, short description with figure along with identification using botanical keys, cultivation process and harvesting have been compiled of 164 medicinal plant. Moreover, their local names in most of languages including Gujarati, Hindi, Sanskrit, Tamil and Telugu are also depicted which are of immense use. Information which is generated in this book has been taken from authentic sources and publications as well as from ethnobotanic guides which have been recognized in British Pharmaceutical Codex and/or United States Dispensatory or which are experimentally proved. However, still many medicinal preparations based on plants are prepared with proper proportions of different ingredients, processing and doses which can be used in ailment treatment and hence we request that nobody of his own self would try to use any plant mentioned in this book because different conditions need different treatments and specific prescriptions. Intention of bringing out this compilation is to provide proper cultivation techniques along with its scientific detail and medicinal uses of important medicinal plants to farmers, researchers and students, particularly of Gujarat. We hope that this book would be useful guide to researchers, students, pharmacists, farmers and all those who are related with farming and uses of medicinal plants. We also hope this book will be useful to the society welfare and people would move towards traditionally useful plant.

Dr. R.N. Nakar

Dr. H.L. Dhaduk

Dr. V.P. Chovatia

Contents

Introduction

Just after human existence many natural materials were identified and plant was one of them. Medicinal plants have deep history. In Rig-Veda around 400 plants have been shown containing medicinal values. Atharva-veda also showed plants with medicinal value. There have been references of medicinal plants in ancient literatures like "Charak Samhita" and Upveda. Dhanvantri is well known person who is considered as God of Ayurvedic Medicines. Thus from the time immemorial, medicinal plants and their derivatives have been main source of affordable health.

Medicinal Plants in different System of Medicines

In most of the known system of medicines, like Ayurveda, Siddha and Unani, medicinal plants are widely used. Ayurveda is ancient system of medicine which is even older that traditional Chinese Medicine. Ayurveda is derived from Indian words "Ayur" and "Veda". Ayur means life and Veda means knowledge. Thus, ultimately leads meaning as 'The Science of Life'. Ayurveda is similar to Galenical medicine which is based on bodily humours (doses) and inner life force (Prana) which also known to maintain digestion and mental activity. Six tests very important in Ayurveda, are Sweet (Madhura), Sour (Amla), Salt (Lavana), Bitter (Tikta), Pangent (Katu) and astringent (Kasaya). These most of the properties are provided by medicinal plants to cure diseases.

Medicinal Plants and Diseases

Plants have been main source of the major raw materials for treating various diseases and ailments of human being. In the world population is increasing at rapid rate and deseases are also encountering to people's daily lives. Because of side effects and wide spread toxicity synthetic drugs have lost their importance as that were in earlier days, people are now converting towards natural drugs which are mainly in the form of medicinal plants. Some important medicinal plants are

Adhatoda vasica, Achyranthus aspera, Boerhavia diffusa, Acorus culamus, Cassia tora, Cassia fistula, Commiphora mukul etc. *Adhatoda vasica* is known as expectorant and antiasthmatic, while *Achyranthus aspera* and *Boerhavia diffusa* are diuretic, *Cassia fistula* has catharitic property.

Medicinal Plants Cultivation

Forests are the main source of medicinal plants but wild plants are now being destroyed at rapid pace by different activities like deforestation, grazing, for making houses, buildings etc. Some of plants are now a day's hard to find as they are very few in numbers. Government has set up some policies to prevent cutting down some of important trees and conserve important natural plants. Important plants have been put by government into different categories such as extinct, endangered, rare and vulnerable So, these kinds of plants should be grown at commercial level. Farmers should be encouraged to cultivate medicinal plants. This becomes helpful to farmers as well as countries' economy. Some institutions are working for this purpose are CIMAP (Central Institute on Medicinal and Aromatic Plants-Lackhnow), NBRI (National Botanical Research Institute, Lackhnaw), RRL(Regional Research Laboratories under CSIR), CCRAS (Central Council for Research on Ayurveda and Siddha), CCRIMH (Central Council for Research on Indian Medicine and Homeopathy), Directorate of Medicinal and Aromatic Plants (National Research Centre on Medicinal and Aromatic Plants, Boriavi, Gujarat), NBPGR (National Beuro of plant Genetic Resources-New Delhi), BARC (Bhabha Atomic Research Centre, Bombay), (IIA) International Institute of Ayurveda, Coimbatore, (SAUs) State Agricultural Universities and Private Organizations.

National Medicinal Plants Board (NMPB) is well known institute or national body working on development sector of medicinal plants. Board was established by Government of India on 24th November, 2000 under chairmanship of Union Health and Family Welfare Ministry. Its main function is to offer financial and technical support for all factors of medicinal plants cultivation. It also works in co-ordination of all matters related to medicinal plants like drawing up of policies and strategies for conservation, cost effective cultivation, providing financial assistance to promotional schemes and projects and sustainable development of medicinally useful plants.

Conservation of Medicinal Plants

The high demands of herbal drugs and medicinal plants have been resulted in over exploited leading to rarity and endangerment of many valued plant species. International Union for Conservation of Nature (IUCN) states that around 200 plant species of medicinal plants have become endangered and rare. This is actually warning for society and researchers to take immediate steps for conservation of medicinal plants. Government has taken some steps and prepared policies to conserve important plants. Some important points related with this policy are:

(1) *In situ* Conservation

In situ conservation allows populations of plant species to be maintained in their natural and agricultural habitat, thus permitting the evolutionary processes that shape the genetic diversity and adoptability of plant populations to operate continuously.

(2) *Ex situ* Conservation

When germplasm conservation is attempted outside the natural habitat, it is known as ex-situ conservation. It includes setting of ethnomedicinal plants garden and gene banks. Gardens have been established by many governmental and non-governmental organizations, universities.

(3) Legislation

Some important steps are taken by government to prevent important plants. For example, in 1994, Indian Government banned around 56 medicinal plants for export.

(4) Rules for Proper Identification and Botanical Names

Priority is given by preparing herbaria. Institutions like National Botanical Research Institute, Lacknow and Tropical Botanical Garden Research Institute, Palode are doing such works.

(5) Public Awareness

People should be given knowledge about indigenous plant resources for health care.

(6) Promote Cultivation

Need is increased due to over progress of industry so, only wild collections are not only option, but instead farmers should be trained to cultivate medicinal plants in their farms. Proper market ultra structure and linking must be developed at district, state and natural level.

Local people and farmers can play vital role in conservation of medicinal plants of local region however rules set up by government bodies are also necessary. Accurate knowledge regarding medicinal plants local names, scientific names, and their cultivation along with their uses enrich people with vast utilization of these potentially useful plants.

Description of
Medicinal Plants

1

Abelmoschus manihot Linn. (Jangli Bhindo)

Botanical Name: *Abelmoschus manihot* Linn.

Family: Malvaceae

Local Names

- ☆ **Gujarati:** Bhindo, Jangali Bhindo, Kantalo Bhindo
- ☆ **Hindi**: Bhindi, Jangali Bhindi
- ☆ **English:** Sunset musk mallow

Introduction

It is Perennial herb growing upto two meter at a fast rate. It flowers from July to September, and the seeds are ripen from August to October. The flowers are hermaphrodite (have both male and female organs) and are pollinated by Insects. The plant prefers light (sandy), medium (loamy) and heavy (clay) soils and well-drained soil. The plant preferes acid, neutral and basic (alkaline) soils. It cannot grow in the shade. It requires moist soil. A perennial plant, it is generally tender in the temperate zone but can be grown outdoors as an annual, flowering well in its first year and setting seeds. It grows well in an ornamental vegetable garden.

Origin and Distribution

This plant is distributed in East India to South eastern Asia and to North Australia. Particularly in weste lands and humid areas this plant is found to be distributed. Plant is native of South eastern Asia.

Cultivation

This plant can be grown easily in any well drained soil. Plants can tolerate occasional short-lived lows down to about -5°C so long as they are in a very well-drained soil. Seed are sown during March in a warm greenhouse.

Seed propagation is good and fast method. The seeds are germinated with two weeks, when it is large enough to handle pick it out into individual pots and plant out after the last expected frosts. Seeds can be sown *in situ* in summer also.

Harvesting

Fruit sets continuously from October to April. As mature pods remain opened and shatter their seed. Harvesting starts when most pods begin to turn from green to brown and just start to open. Pods are generally picked when three-quarters of their body has turned blackish-brown; Seeds are removed manually. Picking is difficult task because plants, including the pods, possess hairs that cause itching. In India harvesting has often stopped by the end of February, as later harvesting rounds yield too little to be economical. Pods are dried in the shed.

Yield

Average seed yield obtained in India is 0.8-1 t/ha.

Handling after Harvest

After drying in the shade, the pods are threshed by being beaten with sticks. The husk is then removed by winnowing. Steam distillation of whole seed give ambrette seed oil, while distillation of ground seed and hydrocarbon extraction of ground seed produces ambrette seed concrete, largely consisting of palmitic and myristic acids. Ambrette seed absolute is prepared from the seed concrete either by neutralization and subsequent elimination of fatty acids or by steam distillation of the concrete followed by washing with alcohol.

Medicinal and Economic Importance

- Roots provide mucilage which is used in sizing paper.
- Seeds yield fatty oil.
- Plant also provides fiber which has convolutions like cotton fibre and resembles jute in colour, dry twist and microscopic structure.
- Leaves are cooked and eaten. Bark of plant is used in emmegagogue.

2

Abelmoschus moschatus Linn. (Musk Dana)

Botanical Name: *Abelmoschus moschatus*

Family: Malvaceae

Local Names

- **Gujarati:** Muskdana, Lata kasturi
- **Hindi:** Latakasturi,
- **Telugu:** Kasturibendavittu
- **English:** Ambrette plant

Introduction

The plant is cultivated through out India. It is erect herb which is annual or biennial. It gets 0.6 to 1.8 m height. Leaves are generally simple type but they vary in shapes, they are 3 to 7 lobed and lobes are narrow-acute, oblong-ovate, crenate, serrate or irregularly toothed. They are hairy on both surfaces. Flower are bright-yellow and large. They are usually solitary axillary. Sometimes in few flower racemes are found. Fruits are hairy and capsular. Capsule is 6.5 to7.8 cm long. Seeds are many subreniform, black in color, musk scented.

Origin and Distribution

The genus belongs to trees, herbs and also shrubs. It is native of Tropical Africa and Asia. Generally 8 species occur in India. Andhra Pradesh, Karnataka, Himachal Pradesh, Kerala, Punjab and Gujarat are some of the states of in India where musk dana is grown.

Cultivation

It grows on rich and well drained soil, sandy and clayey soils are not suitable for musk dana. It is rainy season crop, and sown in the land ploughed 3 to 4 times and leveled thereafter by planking during June-July which is generally rainy season. Before soaking seeds are soaked for 1 day and night (24 hrs), then seeds are sown in furrows 45 cm apart by dibbling one or two seeds about 2.5 cm deep and 30 cm apart.

Harvesting

Hand picking is the only method for harvesting. Harvesting of the plant is very arduous. The fruits posses irritating hair causing itching. When fruits attain blackish color, they are plucked to prevent scattering of seeds. Seeds are similar to that of ladies finger.

Yield

A normal crop may give a yield of 7.5 to 10 quintal seeds per hectare.

Storage

Dried seeds are stored in polythin bags then they are taken to market.

Diseases and Prevention

After rainy season insects and pest attack on to the stem. It is advantageous to spray pesticides like Indosulfon and Macrocraptophos to prevent diseases. Domestic ash, fly ash, neem ash are useful. A solution is useful which is made from 6 liters of neem oil and one sunlight soap cake dissolved in 100 liters water. This can be sprayed in one hectare of field. In the whole life cycle of plant 4 times spray is useful to prevent disease. In the evening smoke of gugal is also helpful against diseases. Paste made form neem leaves and cow urine is also useful in prevention of stem diseases.

Medicinal and Economic Importance

- ✰ Generally seeds, leaves and root are used. Seeds are aphrodisiac, carminative, demulcent.
- ✰ Seeds are used in dryness of throat and in inhalation hoarseness.
- ✰ Seeds made to paste with milk are used to cure scabies.
- ✰ Seeds are used to protect woolen garments again moth, they are pounded with cosmetics and used to scent hair.
- ✰ Its oil is rich, sweet, floral musky distinctly wine or brandy type smell with bouquet and roundness which is very rarely found in other perfume material.
- ✰ It can blend with rose, naroli, methylionones, sandalwood oil *etc.* That's why it is very expensive and used in specific perfumes

3

Abrus precatorius Linn. (Chanothi)

Botanical Name: *Abrus precatorius* Linn.

Family: Laguminosea- Fabaceae

Local Names

- ☆ **Gujarati:** Chanothi
- ☆ **Sanskrit:** Gunja
- ☆ **Hindi:** Gunchi, Rati
- ☆ **Tamil:** Kuntumani
- ☆ **English:** Indian liqueorice, Jequirity
- ☆ **Telugu:** Guruginja

Introduction

It is deciduous climbery coines with tough branches. It is perennial plant. Leaves are abruptly pinnate with many pairs of leaflets are found. Spine is there at the end of raches. Leaflets are oblong, rounded at both ends, thinly membranous leaflets are seen. Flowers are pink in colour; they are clustered on tubercles arranged along the rachis of one sided pedunculate raceme. Fruits are pods type, turgid with a sharp deflexed beak. Seeds are usually scarlet with a black spot or sometimes pure white.

Origin and Distribution

Abrus is the genus of shruyby climbers distributed throughout the tropics. Two species are found in India. It grows up to a height of 1200 m and above the

sea level. It is also found on hedgews and bushes in exposed areas. The species is native to tropics.

Cultivation

Generally cultivation of this plant is done by seeds and from cutting of firm shoots. Sandy loamy soil is useful for plants cultivation. Land is prepared with cow dung, compost mixture of leaf mould, cow dung and sand. During dry season plants are watered freely. Seeds are directly sown in the soil at 2 x 1m spacing, in the first week of june before the onset of monsoon. Germination of the seeds will be complete in 15-20 days.These plants are slowly trained to grow up on the pandal. Ten grams of Urea is applied per plant after the planting. The plant develops good root system and aerial growth in the first year. In the second year also, one weeding along with soil working and manuring are carried out as in the first year. From 3rd year and onwards, manuring the plants should be continued for regular good yield of seeds.

Harvesting

Plant start flowering and fruiting in the second year. When the seeds are ripen, the pods start dehiscing harvesting is done. Then the pods are collected and seeds are separated at the time of maturity.

Yield

It is estimated that a production of about 1500 kg of seeds can be obtained from one hactare.

Chemical Constitution

Roots and leaves contain glycyrrhizin, precol, abrol, abrasine and precasine from roots. Gallic acid, abrine, gypaphorine, alanine, serine, valine, choline, trigonelline, precatorine, and methyl ester, 5B-cholanic acid, abrin A and abtrin B from seeds.

Medicinal and Economic Importance

- The seeds, leaves and roots are useful in the medicinal treatment.
- Seeds are used in piles. They are laxative, expectorant and also used in chronic cyatitis, gleet and gonorrhea.
- Leaves are demulcent, locally applied to boils and ulcers
- Decoction of plant is used in toothache and tender gums. It is also used for inflammation bladder.
- Infusion or root and leaves are demulcent and diuretic which is applied in fever, chest affection and urethritis.
- Bark is astringent and diuretic. Leaf juice is sweet in taste and used to treat hoar senses. The leaf juice mixed with oil is applied on painful swellings of the body.
- Seeds are used as beads for rosaries. Roots are used as a substitute of liquorices.
- In Punjab fronds are given with pepper as a febrifuge pounded with honey, they are administered in the catarrhal affections.
- It has also proved to be good hair tonic.

4

Abutilon indicum Linn. (Atibala)

Botanical Name: *Abuliton indicum* Linn.

Family: Malvaceae

Local Names

- ☆ **Gujarati:** Atibala
- ☆ **Hindi:** Kanghi
- ☆ **English:** Indian mallow, country mallow
- ☆ **Telugu:** Tuttur benda
- ☆ **Tamil:** paniyartutti

Introduction

It is hairy under shrub of 75 to 200 cm height growing throughout India. It grows quickly on all types of soil but generally it grows well on to sandy loam soil. It has good resistant power and can grow on poor soil as well as saline soil. Leaves are generally 3 lobbed with hairy petiole. Leaves are broadly ovate or sub orbicular. Flowers are bright yellow in color. They are solitary and axillary. Fruits are 1 to 2.5 cm length, apprised hairy, ripe carpels 15-20 in number and almost 1 to 2 cm long. Seeds are reniform, blackish brown in color.

Origin and Distribution

It is still not clear that what is the main native of atibala but it is grown in India, Pakistan, Sri Lanka, Burma, Bangladesh, and Africa like countries. In India it is grown in most of the states for its economic and medicinal importance. In India it is grown in the Gujarat, Maharastra, Uttar Pradesh, Punjab, etc like states. In Gujarat atibala is found in the junagadh, Dang, Val sad, Katchh, and Jamnagar like districts.

Cultivation

Sandy-loamy soil is quite useful for growth of Atibala but it can grow in all kind of soils. Generally it is propagated by seed. In the beginning of the rainy season, seeds are directly being sown in the land in line spacing of 1 meter distance. By proper care plants grow easily. After second year plant is ready for harvesting. It is free from diseases.

Harvesting

Harvesting is done generally after 2 to 2.5 years when plants bear fruit.

Yield

It yields different items for village level medical treatment. Although it has not commercial status.

Medicinal and Economic Importance

- ☆ Whole plant is medicinally useful. The leaves are rich in mucilage and are used as demulcent tonic.
- ☆ Decoction of the herb can be given in bronchitis, catarrh and biliousness.
- ☆ Drug in the form of paste in combination with coconut oil has proved to be useful in allaying irritation to the skin and reducing swelling and pain.
- ☆ Seeds are used in curing piles, in killing thread worms when the rectum of the effected child is exposed to the smoke of the powdered seeds.
- ☆ The decoction can also be used for mouth wash for tooth ache.
- ☆ Bark of the plant gives good fibers.

5

Acacia catechu (L.f) Wild (Khair)

Botanical Name: *Acacia catechu* (L.f) Wild

Family: Mimosaceae

Local Names

- ☆ **Gujarati:** Khair
- ☆ **English:** Black cutch, Catechu
- ☆ **Hindi:** Khair, Khadira
- ☆ **Telugu:** Tella Tumma, Sundra

Origin and Distribution

Khair is tree type of plant which is grown in most of the states in India but in Gujarat and Utter Pradesh khair is grown as highly profitable crop. Other states are Himachal Pradesh, Madhya Pradesh, Bihar, Maharastra, and Karnataka which grow khair. Nepal, Burma and Pakistan also grow khair. Khair is found in most of the districts of Gujarat like Junagadh, Jamnagar, Dang, Rajkot, Porbandar etc.

Cultivation

Generally there is not proper timing of sowing seeds but it can be sown in any season. Seeds are sown by dribble method in rows 45 cm apart. Seeds are buried by a small layer of soil over them. Manure of cow dung is also applied in the field. Approximately after 7 to 10 years trees are ready to give fruits and all other product.

Harvesting

Seeds are harvested in April-May. When seeds are matured they are collected by shaking branches or pulling down beans with the help of stick or bamboo. Seeds are dried for 15 days under sun light. Then these seeds are spread on a clean cloth in and dry and airy place for 4-9 days.

Storage

Dry seeds are filled in the holed bags and then they are stored. For storing polythin bags or poly bags are not used. These are to be stored on a clean place little above ground at 25 c moisture free temperature. If we store seeds for longer time sprouting percentage is declined so they are to be stored in cold storage. Once seeds are put under strong sun, their edges automatically start cracking. When seeds get matured they look light brown in color.

Medicinal and Economic Importance

- Bark of the tree gives kaththa which is very useful product. Liquid extract gives to fraction when concentrated and cooled. A less soluble fraction is separated when it is cooled. This gives White Kaththa.
- The highly soluble portion of water extract, when it is evaporated to dryness, gives Dark Kaththa which is also known as Cutch of Commerce. Cutch of Commerce contains water soluble tannins whose building units are catechins.
- Acacia catechu is complex mixture of catechins, catechu tannic acid, catechuic acid, catechu red, tennis, gums, quercetin and ash etc.
- Kaththa is used against tooth decay, gum troubles, urinary dis-orders, leprosy etc.
- Bark is used as astringent and digestive which is useful in diarrhea and cough; it is also applied in eruption on skin.
- Cutch is good economic product which is widely used in building construction, bridge construction, in making of agricultural equipments. It is also useful in making of knees of boats and good quality tool handles, good charcoal is also obtained from cutch.

- ☆ It acts as host for lac insects. Cutch which is obtained from heart wood serves as useful ingredient of paan chewed in India.
- ☆ It is useful to make ply wood, dyening canvas, fishing nets and ropes *etc.*

6

Acacia nilotica Delile. (Baval)

Botanical Name: *Acacia nilotica* Delile.

Synonym: *Acacia aerabica* Willd.

Family: Mimosaceae

Local Names

- ☆ **Gujarati:** Baval
- ☆ **English:** Indian gum, Arabic tree.
- ☆ **Hindi:** Babul
- ☆ **Tamil:** Karuvael
- ☆ **Sanskrit:** Babula
- ☆ **Telugu:** Nallatumma

Introduction

There are about seven hundred tropical and subtropical species under this genus. Nearly twenty five species are native to India. These are trees or shrubs, erect or climbing in habit of growth and are usually armed. Many species yield tannin, gum and timber. This is medium sized tree with a short trunk, usually attaining a height of 15 m, bark is almost black to dark brown, deeply cracked or longitudinally fissured. Leaves are bipinnate with spinescent stipules 2.5 to 5 cm long. Flowers are crowded in long-peduncled globose heads, forming axillary clusters of 2-5 heads, fragrant, golden-yellow. Pods are white in colour. They are flat containing length of 7-15cm. Each pod contains 8-12 seeds.

Origin and Distribution

It is distributed in tropic and sub tropic areas of the world. This species occur throughout in drier region of India. It is distributed in most of the states like Gujarat, Rajasthan, Bengal, Madhya Pradesh, Orrisa *etc.* This pant is native from Egypt south to Mozambique and Natal.

Cultivation

It is cultivated in dry and semi dry regions where sun light is in proper amount. It is dry evergreen weedy perennial plant. Generally sandy loam soil is ideal for plant cultivation. Seed propagation is good method although it can be propagated by cuttings of half ripened shoots. Direct seeding is the common practice. Stored seed may require scarification. Young seedlings are said to "require full sun and frequent weeding".

Harvesting

Harvesting is carried out when fruits are metured. Generally fruits are taken by hands or they are allowed to fall on land. Gum is obtained from the bark of the tree. After some year whole plant is cut by axe. Although there are other sources of gum arabic, trees are still tapped for the gum by removing a bit of bark 5–7.5 cm wide and bruising the surrounding bark with mallet or hammer. The gum harvest from the various species lasts about five weeks. About the middle of November, after the rainy season, it exudes spontaneously from the trunk and principal branches. The resulting reddish gum, almost completely soluble and tasteless, is formed into balls.

Chemical Constitution

Bark yields several polyphenolic compounds, catechin, epicatechin, cpigallocatechin, quercetin, gallic acid, leucocyanidin, gallate, surcrose and tannin, m-Digallic acid and chlorogenic acid. Gum contains galactose, L-rhamnose, L-arabinose and its derivatives along with four aldobiouronic acids. Seeds contain amino acids, fatty acids and ascorbic acid along with tannin as the major constituent.

Medicinal and Economic Importance

- ☆ Infusion of tender leaves is used as an astringent and remedy for diarrhea and dysentery.
- ☆ Decoction of bark is used as a gargle insures throat and toothache. Dry powder applied externally in ulcers.
- ☆ Gum is astringent and styptic.
- ☆ Pods are good fodder for cattle.
- ☆ Wood is used by poor people for making their houses and making furniture.
- ☆ Extensively used, *e.g.* in India, for firewood and charcoal, this species has been used in locomotives and steamships as well as industry balers.
- ☆ The calorific value of the sapwood is 4,800 kcal/kg of the heartwood 4,950. The species does nodulate and fix nitrogen.

7

Acanthospermum hispidum DC. (Acanthospermum)

Botanical Name: *Acanthospermum hispidum* DC.

Family: Acanthaceae

Local Name

☆ **English:** Acanthospermum

Introduction

This plant is an annual, much branched, erect herb. Leaves are ovate, sessile, membranous, sub-hispid, sub-dentate. Base of the leaf is cuneate. Flowers are heterogamous. Many flowered heads are found. In the female flowers ray florets, seen in one series. Disc florets are tubular in male. Involucral bracts are found in one series; generally they are five in numbers. The achenes are tightly covered with prickly bracts of paleae. The florets are pale yellow in colour. Generally flowering time is August to December.

Distribution

This plant is found in the moist spots the whole year. This is seen in dry and moist forest areas particularly near the river banks and dams. In Gujarat this plant is widely seen in the areas near the dams and river banks in Girnar forest. Origin of this plant is believed to be central and southern Amrica.

Medicinal Uses

☆ This plant is used in the treatment of leprosy.

8

Aconitum ferox Wall.ex.Ser (Indian Aconite)

Botanical Name: *Aconitum ferox* Wall.ex.Ser

Family: Ranunculaceae

Local Names

- ☆ **Gujarati:** Vachh
- ☆ **English:** Indian aconite
- ☆ **Hindi:** Bacchanag, Bish
- ☆ **Tamil:** Vashanabi
- ☆ **Sanskrit:** Vatsanava
- ☆ **Telugu:** Ativasanabhi

Introduction

Aconitum is a large genus consisting of about hundred species. Thirty species are native to India in Himayalayan region. The tuberous roots of several Aconitum species are used in medicine and number of them is known to contain highly toxic alkaloids. This plant is perennial herb. Roots are tuberous, paired. Daughter tuber is ellipsoid to ovoid oblong, 2.5-4.0 cm in length, with filiform root fibers. Leaves are scattered, orbicular-cordate to reniform, palmately 5-lobed, resembling those of melon leaves. Petiole and blade are pubescent. Straight peduncle is also there which bears flowers on both the sides. Flowers are pale dirty blue and they are borne in dense terminal raceme. Flowers are 10-25 cm long, helmet-vaulted with a short sharp beak, resembling a pea flower. Follicles are oblong; Seeds are long, obpyramidal to obovoid, winged along the raphe.

Origin and Distribution

This plant is distributed in sub alpine and alpine zone of Himalayas up to 3,600m. Origin of this plant is India. Particularly hilly areas are suitable for its growth. This plant is Shrubberies and forest clearings; It is distributed from Central Nepal to Bhutan.

Cultivation

This plant can be cultivated by two common methods (1) Seed propagation and (2.) Vegetative propagation. This plant can thrive well in most soils and in the light shade of trees. It grows well in heavy clay soils, moist soil in sun or semi-shade, calcareous soils. It also grows well in open woodlands A greedy plant, inhibiting the growth of nearby species, especially legumes. It is closely related to A. napellus.

Seed propagation is best method for cultivation of the plant. Seed - best sown as soon as it is ripe in a cold frame. The seed can be stratified and sown in spring but will then be slow to germinate. When large enough to handle, prick the seedlings out into individual pots and grow them on in a cold frame for their first winter. Plant them out in late spring or early summer. Division - best done in spring but it can also be done in autumn. Another report says that division is best carried out in the autumn or late winter because the plants come into growth very early in the year.

Harvesting and Yield

The plants are uprooted and the root stock is separated and dried in the sun for 4-5 days. The the roots/tubers are cleaned and sent to market. It is estimated that each plant will produce about 200 grams of dry root. Thus, 500kg. of dry roots would be available every year. The market rate of aconite roots is around Rs.100/kg

Chemical Composition

It contains alkaloids pseudaconitine, chasmaconitine, indaconitine and bikhaconitine recently two new alkaloids, veratroyl pseudaconitine and diacetyl pseudaconitine.

Medicinal and Economic Importance

- ✰ The root and underground stems are highly toxic. But the toxicity may be reduced by suitable processing.
- ✰ In small doses they are beneficial in nasal catarrh, uvula hypertrophy, sore throat, gibbous, paralysis and chronic fever. In large doses it acts as narcotic poison and powerful sedative.
- ✰ Internally, the tincture of root is used in combination with other drugs for the treatment of fever and rheumatism.
- ✰ The root is considered to be cardiac stimulant, hypoglycemic, diaphoretic and antiphlogistic.
- ✰ Powdered roots in the form of liniment or paste are spread over the skin in case of arthritis and in scabies.

9

Achyranthes aspera L. (Anghedo)

Botanical Name: *Achyranthes aspera* L.

Family: Amaranthaceae

Local Names

- ✰ **Gujarati:** Anghedo
- ✰ **Sanskrit:** Apamarga
- ✰ **Hindi:** Chirchira, Chichitta
- ✰ **Tamil:** Chirukadaladi, Nayurivi
- ✰ **English:** Prickly Chaff Flower, Rough chaff
- ✰ **Telugu:** Apamarganu, Uttareeni

Introduction

There are about fifty five species under this genus and four are native to India. It is an erect herb or undershrub which attains a maximum height up to one meter. Stem is stiff and erect, pubescent, swollen at the nodes. Leaves are opposite, short-petioled, margins undulate. Flowers are numerous, stiffly deflected against the pubescent rachis in elongate terminal spike, they are 20-30 in cm. and long. Utricle oblong-cylindrical, enclosed in the hardened perianth, brown. Seeds are oblong to ovoid. Flowering is found in July to September.

Origin and Distribution

Achyranthes aspera is native to India. It grows throughout in India. This species is distributed in different parts of India, tropical Asia, Africa, Australia, Ceylon and Baluchistan.

Cultivation

This is wild plant. It grows any where. It is naturally produced on weste lands or it can be propagated from seeds also. Seeds are dipped into soil in rows. After some days they start to germinate. It grows quickly on land where there is loam and sunny condition present. The plant prefers light (sandy), medium (loamy) and heavy (clay) soils. The plant prefers acid, neutral and basic (alkaline) soils. It can grow in semi-shade (light woodland) or no shade.

It is propagated by seed. It is a problematic weed in many parts of India. This species is rarely cultivated, as it is available in nature in abundance. The seeds are collected in November/December from the plants before they fell on the ground and kept well preserved till next monsoon. The area is ploughed well and continuous furrows are made at 20 cm apart. At the beginning of the rainy-season, the seeds are directly sown in the furrows continuously. The seeds germinate soon and the plants get established very easily with little or no care. The plants can be harvested in the month of November/December.

Harvesting and Yield

Various parts of this species are put to different medicinal purposes. Therefore, the management of the crop depends mainly on the purpose for which it is grown.

Chemical Composition

Roots contain triterpenoid saponins, betaine, achyranthine, hentriacontane, ecdysterone and two glycosides of oleanolic acid have been reported. The dried dehusked seeds contain amino acids.

Medicinal and Economic Importance

- Whole plant is useful as medicinal drug.
- Roots are useful as anti-inflemmatory and uterine stimulant activity, are prescribed in the rheumatism, lumbago, osteodynia, dysuria, post-partum haematometra and dysmenorrheal.
- The seeds are nutritious when cooked with milk and is a potential source of food. Seed powder is used beneficially in the treatment of bleeding piles.
- Roots are astringent, their paste is applied to clear opacity of cornea, and to wounds as a haemostatic.
- It is reported to be useful in cancer.
- A decoction of the roots is used for stomach troubles and an aqueous extract for stones in the bladder.
- Decoction of plant is diuretic and used in renal dropsy and generalized anasarca.
- The leaf is used as a remedy for boil and abscess. Leaf juice is useful in stomachache, bowel complaints, piles and skin eruptions.
- Paste of leaf is used to treat bites of poisonous insects, wasps and bee. Decoction of whole plant is given in the painful delivery.
- Roots are used in toothache. In menstrual disorders extracts of the roots are used.
- Seeds are used as brain tonic.
- The ash from the burnt plant, often mixed with mustard oil and a pinch of salt, is used as a tooth powder for cleaning teeth
- The ash of the burnt plant is a rich source of potash. It is used for washing clothes.

10

Adansonia digitata L. (Gorakh Ambali)

Botanical Name: *Adansonia digitata* L.

Family: Bombacaceae-Malvaceae

Local Names

- ☆ **Gujarati:** Gorakh ambali, Rukhdo
- ☆ **Hindi:** Gorak Ambali, Gorakh imli
- ☆ **English:** Baobab tree, monkey bread tree
- ☆ **Sanskrit:** Gopali, Gorakshi, Panchparnika
- ☆ **Tamil:** Anaippuli, Paparapulia, Puria
- ☆ **Telugu:** Anaipuliamaram, Brahmaamlika

Introduction

The name Adansonia was given in the memory of *Michel Adanson* (1727-1806), the species name digitata meaning hand like, is in reference to the shape of the leaves. This deciduous tree is referred as largest succulent plant in the world; the baobab tree is steeped in a wealth of mystique, legend and superstition wherever it occurs in Africa. Main stem of large tree gets 28 m diameter in girth. This tree gets height of 25 m. It has cylindrical trunk which gives rise to thick tapering branches resembling a root system. This tree is also known as grotesque. Bark is present on the stem which is 50-100 mm thick. The bark is grayish brown and smooth but can be variously folded and seamed from years of growth. Leaves are hand sized and divided into 5-7 fingers like leaflets. Flowers are large, white and sweetly scented. Flowers are pendulous type. They emerge in the late afternoon from large rounds

buds on long drooping stalks from October to December. Flowers fall within 24 hours turning brown and smelling quite unpleasant. Pollination is done by bats in the night. The fruits are large and egg shaped capsule is found. Yellowish brown hairs are seen. Fruit consists of a hard, woody outer shell with a dry, powdery substance inside that covers the hard, black, kidney shaped seeds. The off white, powdery substance is apparently rich in ascorbic acid. All though tree is slow growing. So many legends are attached to the plant. For example it is believed that an elephant frightened the maternal ancestor of the baobab. In some parts the baobab is worshiped as symbol of fertility. There is another belief that water in which seeds have been soaked will may attract crocodiles. It is also believed that a man who drinks an infusion of the bark will become strong.

Origin and Distribution

This plant is native to tropical Africa, one of the largest and long lifed tree. The baobab tree is found distributed in South Africa, Botswana, Namibia, Mozambique and other tropical African countries it is very common. It grows easily on hot, dry, woodland on stony, well drained soils, in frost-free areas that receive low rainfall. In South Africa it is found only in warm parts. In India this tree found in tropical and sub tropical areas. It is distributed in Gujarat, Maharastra, and Madhya Pradesh like states. It doesn't like higher rainfall areas where they are frost free and don't experience cold winters.

Cultivation

Baobas are quite easily grown from seed although they are seldom available in nurseries. Seeds can be collected from dry fruits by cracking the fruit open and washing away they dry powdery coating. The dark brown to black, kidney-shaped seeds should be soaked in container of hot water and allowed to cool, they may then be sown after soaking for 24 hours. Seeds are best sown in spring and summer in a well drained seedling mixture containing one third sand. Seeds are covered with sand to a depth of 4-6 mm, place the trays in a warm semi shaded position and water regularly until the seeds have all germinated. Germination occurs in two to more weeks. Seedlings are transplated when they become 50 mm in height into sandy soil with some well rotted compost and bone meal.

Diseases and Prevention

During growth of plant in nursery, damping off fungal disease can attack on the plant. This can be prevented with fungicidal drench.

Medicinal and Economic Importance

- ✰ Trees with hollow stems are used for various purposes including houses, prisons, pubs, storage barns etc. Rain water often collects in the clefts of the large branches. It has been recorded that in some cases the centre of the tree is purposely hollowed out to serve as a reservoir for water. A hole is drilled in the trunk and a plug is inserted so that water can be easily retrieved by removing the plug.
- ✰ The leaves are said to be rich in vitamin C, sugar, potassium tartrate, and calcium. They are cooked fresh as vegetable or dried and crushed for later use by local people. Leaves are used locally for a variety of inflammatory conditions, including insect bites and guineaworm sores.
- ✰ Internally leaves are given as an astringent, sudorific, tonic and febrifuge. Leaves are also used as a lotion in earache and opthalmia.

- ☆ The root of very young trees is also edible. Seeds are also edible and can be roasted for use as a coffee substitute.
- ☆ Fruits are edible, acidic pulp of fruit has been used as a rubber coagulant, and has been used to curdle milk. Seeds are eaten as nuts. Powder of seeds is used as a manure.
- ☆ When wood is chewed it provides vital moisture to relieve thirst, humans as well as certain animals eat it in times drought.
- ☆ Bark is used as substitute of cinkona bark. It yields a large quantity of semifluid, white gum, like tragacanth, wich turns reddish brown on ageing.

11

Adenanthera pavonina L. (Coral Wood Tree)

Botanical Name: *Adenanthera pavonina* L.

Family: Mimosaceae

Local Names

- ☆ **Gujarati:** Baragunci, Rakt chandan
- ☆ **English:** Coral wood tree, Red wood tree
- ☆ **Hindi:** Baragunci
- ☆ **Tamil:** Vanai Kuntumani
- ☆ **Sanskrit:** Kucandanah, Tamarakah
- ☆ **Telugu:** Bandigurvina, Mansenikotta

Introduction

This is medium sized, unarmed deciduous tree which can get hight of around 20 m or more with a clear bole of 6 meter. Bark is grayish brown in colour and has longitudinal fissures. Leaves are bipinnate, they are 3-6 in pairs, opposite, leaflets are many and alternate, they are ovate-oblong, obtuse, glabrous, unequal sided, dark green above. Flowers are pale yellow in colour. They are in short peduncled racemes. Fruits are falcately curved pointed pods. Valves are spirally twisted after dehiscence. Seeds are shining brilliant scarlet, lenticular globose.

Origin and Distribution

This is wild tree which is found throughout India. The native of tree is Southeastern Asia and Malaysia, but widely cultivated and introduced. It is

deciduous tree which belongs to arid and semi arid areas of the Asia. It grows on a variety of soils, more permanently moist climates throughout the tropics. (Swarbrick, 1997; p. 15).

Cultivation

This tree is propagated by seeds only. Seeds are sown in ploughed soil. Rows are prepared for its cultivation. When seedling gets proper height it is transplanted to the main field. Generally this tree is cultivated for its bark. Although leaves and seeds are also obtained for medicinal values.

Chemical Constitution

Seed contain non protein amino acids, via.dakua-methylene glutamic acid,dakua-methylene glutamine, and traces of dakua-ethylidene glutamic acid. The karnels contain a pale yellow fat. The fatty acid composition of the fat is: palmatic, stearic, arachidic, lignoceri, octacosanoic, oleic, linoleic, elcosenoic, and uncharacterized, the kernels also contain stigasterol and its glucoside, dulcitol and a polysaccharide the leaves contain octacosanol, dulcitol, glucosides of β-sitosterol and stigmasterol. Bark contains stigmasterol glucoside. The heartwood contains srobinetin, chalcone, butein, ampelopsin and dihydrorobinetin.

Medicinal and Economic Uses

- ✰ A decoction of the seeds and wood is used in pulmonary affections and externally applied in chronic ophthalmia.

- Seeds show inhibitory activity against trypsin and α-chymotripsin, but the activity is lost on subjecting the seeds to heat treatment.
- Kernels contain a pale yellow fat.
- The red heart wood is used as a substitute for true red sandal wood.
- Wood is also used for making building and cabinets.
- Bark and leaves are astringent, vulneralry and aphrodisiac and are useful in colonorrhea, haematuria, ulcers, pharyngopathy, vitiated conditions of Vata and pitta, burning sensation, hyperdipsia, vomiting, fever and giddiness.
- The seeds are bitter, astringent, sweet cooling, aphrodisiac, suppurative, antiemetic and febrifuge.
- Heart wood is useful in dysentery, haermorrhages.

12
Adhatoda vasica Nees (Ardusi)

Botanical Name: *Adhatoda vasica* Nees

Family: Acanthaceae

Local Names

- ☆ **Gujarati:** Ardusi
- ☆ **Hindi:** Basak, Arusa, Adulasa
- ☆ **English:** Malabar nut
- ☆ **Telugu:** Adasaramu

Introduction

Adhatoda vasica is perennial shrub which is grown in moist deciduous regions and sub mountain regions of the country. Plant is diffuse, branched, evergreen type. Adhatoda has mainly two species, *Adhatoda vasica* Nees and *Adhatoda beddomei* Clarke.Nees is also known as *A.zeylanica* Medic. Morphological and histological variation between these two species was given by Aiyer and Kollammal (1962). Adhatoda vasica is a large size herb with dense branches. This shrub can grow to a height of 2.5m or more. Leaves are broad, elliptic, entire, tapering or acute at both ends. Leaf is generally green or dark green on upper surface while pale at lower surface. Inflorescence is found to be long pedunculate, short spike dense flowered, prominently bracteate, and some times it is found as thysiform. In our country Adhtoda is grown in Punjab, Rajasthan, Assam, Gujarat and most of other states. In Gujarat Ardusi is widely used as medicinal source in most of the districts like Junagadh, Dang, Porbandar, Jamanagar etc.

Adhatoda vasica Nees vegetative

Adhatoda vasica Nees flowering

Origin and Distribution

Adhatoda is derived from two words 'Adu' and 'Thoda' which means that the plant is not eaten by goat. It is distributed to Sri Lanka, Singa pore, Malaysia, Indo – china and other South East Asian countries. In India it is grown in most of the states.

Cultivation

Plant can be cultivated with seeds or by using cuttings. Plants are developed in to polythin bags in the beginning. May-June or September-October is suitable month. Seeds or cuttings are planted on ridges and furrows in the ploughed land. When plant gets good height in one or two months it is transplanted to the field. Plant is highly resistant in any kind of land. Plant develops fast in loamy soil with good drainage and high organic content but it comes up in all type of climates. It can withstand drought to a great extent.

Harvesting

Harvesting is done after 6 to 7 months of planting. Leaves, roots and stems are collected individually. Generally leaves are harvested after 6 months and roots are collected after one and one and half year of planting.

Storage

Storage is done in the polythin bags or in dry or cold storages. They can be marketed fresh as well as dry.

Medicinal and Economic Importance

- ☆ Leaves, flower and roots of the plant are extensively used as treating cold, cough, whooping cough, chronic bronchitis and asthma. Leaves have anti-viral activity.
- ☆ Leaves contain alkaloid vaccine and essential oil. Leaves are source of drug used mainly as an expectorant in the form of juice, decoction or syrup.
- ☆ Alcoholic leaf extracts act as hypotensive, bronchodilator, and respiratory stimulant, antiseptic and anathematic.
- ☆ In the cigarette, dried leaf is smoked. According to traditional belief, leaves are useful in anemia and hemorrhage. Leaf juice cures diarrhoe and dysentery; it is also useful in glandular tumours. Leaf powder is helpful in rheumatic joints as counter-irritant on inflammatory swellings on fresh wounds.
- ☆ 'Zeetuss' is prompt symptomatic multipurpose herbal cough syrup which is made from extracts of *A.vasica, S.xanthocarpum, G.glabra, A.lebbeck, P.longum, O.sanctum* and *Z. officinale*. This syrup gives prompt symptomatic relief without any side effects. This is also proved to be useful in early stages of pulmonary tuberculosis.
- ☆ Drug is used against senile pruritus and as an antifatigue. The drug is one of the constituents of dry 'Geriforte'.

- ☆ Shoots are used in cases of liver enlargement. Roots are used in case of mild bronchitis. They are antiseptic and are given in intermittent fever and pulmonary afflictions.
- ☆ Root extracts are found as antibacterial, aqueous leaf extracts of *A. vasica* inhibited *Penicillium* sp. affecting oil seeds like sesame, groundnut and castor.
- ☆ *A.vasica* is effective in reducing bacterial population of raw water at pH 6.8 *A.vasica* leaf powder was effective in reducing population of Mycobacterium incognita in mint. Leaves are source of essential oil which is utilized as bronchodilator.

13

Adiantum capillus veneris Linn. (Hansraj)

Botanical Name: *Adiantum veneris* Linn.

Family: Polypodiaceae

Local Names

- ✰ **Gujarati:** Hansraj, Hanspadi
- ✰ **Sanskrit:** Hanspadi
- ✰ **Hindi:** Hansraj, Mubaraka
- ✰ **English:** Maidenhair fern

Introduction

This plant belongs to pteridophyta group of plant. They are mainly grow near hilly areas where there is lot of moisture remain present. In its structure it is very simple structure. Stem is blackish in colour. On the stem there are leaves produced on panducle on both the sides. This species of fern gets height upto 30 cm. There are numerous varieties in it. These plants occur in Mount Abu, Alwar, Gwaparnath of Rajasthan.

Origin and Distribution

This plant is distributed in areas where there is lot of moisture present. This plant is native to Europe and North America. It can be seen in crevices, cliffs by the sea on basic rocks in damp positions, particularly in the hilly areas. Plant is produced naturally from the crevices of the stones, from the corner of the plants *etc.*

Cultivation

The compost mixture consisting of two parts of organic manure, one part each of loamy soil and silver sand, mixed with charcoal is used for this species. Propagated by spores sown in fine sandy soil with leaf mould, and kept in moist and shades under glass or polythene cover. These plants are grown under glasshouses in humid situation. It likes a position with plenty of light but dislikes full sun. It prefers a sheltered shady position, If the plant dries out temporarily it will lose most of its fronds, though it will usually resprout from the base. Spores - best sown as soon as ripe on the surface of a humus-rich sterilized soil. Germination should take place within 6 weeks. It should be potted on small clumps of plantlets as soon as they are large enough to handle and keep humid until they are well established.

Harvesting

This plant is uprooted from the soil. Generally whole plant is used for medicinal value.We should not plant outside until the ferns are at least two years old and then only in a very well sheltered position. Division is found in spring or autumn. Best carried out in early spring.

Chemical Composition

Maidenhair fern contains flavonoides terpenoides, tannis and mucilage. It contains volatile oil, bitter principle, tanning material mucin, gallic acid, sugars and heterosides of kaempferol, quercetol and juteolol.

Medicinal and Economic Importance

- Maidenhair fern is still used by Western herbalists to treat coughs, bronchitis, catarrh, sore throat and chronic nasal catarrh.
- The plant also has a longstanding reputation as a remedy for condition of the hair and scalp
- This plant is also used in diabetes and skin diseases.
- It can be also applied in leprosy and hair falling.
- The decoction of leaves is taken for acute bronchitis and fever. Tea made from the leaves is widely used in coughs resulting from colds, nosal congestion or catarrh. It is also taken along with tea in abnormal stoppage of the menses.

14

Aegle marmelos correa ex Roxb. (Bilv)

Botanical Name: *Aegle marmelos* correa ex Roxb.

Family: Rutaceae

Local Names

- ☆ **Gujarati:** Bili, Bil, Bilv
- ☆ **Hindi:** Bel, Sirphal
- ☆ **English:** Bengal Quince, Bael tree
- ☆ **Telugu:** Maredu

Introduction

It is medium sized tree which has long life and gives fruits for long time. Some tree lives for 100 years and still they are able to give fruits. The trunk is stout and strong. It is light grey color with soft cork present. It is deciduous type of tree which shed its leaves during summer generally in the month of April – May. Although May-June is main flowering season. Leaves are green and trifoliate with strate branches. Spines are present which are generally sharp. Leaves usually grouped into 3 and rarely 5 are also found. People in India believe that leaves are another form of trinetra which means three eyed god. Tree is highly scared. Leaves are similar to the litchi and mango leaves. Flowers are fragrant and sweet honey, small inflorescence is small, lateral panicle and greenish –white flower.

Origin and Distribution

Plant has been originated from Eastern Ghats and Central India. Aegle marmelous is indigenous to Indian subcontinent and spread in tropical and

***Aegle marmelos* correa ex Roxb. with fruits and seeds**

subtropical regions. This plant is also found at 1200 meters height in Himalayas. In India it is grown in Uttar Pradesh, Bihar, Chathisgarh, Uttaranchal, Jarkhand, Madhya Pradesh, Deccan plateau and along the East Coast. Hindu community grows this tree as religious tree. This tree is also grown in the Srilanka, Pakistan, Bangladesh, Thailand, Vietnam and Japan. etc countries.

Cultivation

It suits sub tropical climate, particularly hot season in summer and mild cool winter. Although it can survive in any field but it grows quickly in dry land. Cultivation is generally done in rainy season. Seedlings are raised in polythin bags. One year old nursery raised seedlings are transplanted in the field after the oneset of monsoon at 7m x 5m spacing. 5meter space should be between plant to plant and 7 meter space should be there between lines. In the first year three weeding are done and in second and third year fertilizers are utilized in two dozes. Irrigation is useful but water logging is not too useful. Vegetative method and air layering method are also useful in cultivation.

Harvesting

In the 12 to 13 years tree start giving fruits. Generally fruits ripe in the October –November. In the initial stage a single tree can give 50 fruits but letter it can give up to 300 to 400 fruits. Ripe fruits are collected and inner pulp is taken out and it is dried. Dried pieces of the pulp are taken to the market.

Yield

Each tree gives 10 kg drugs and it is sold at the rate of 5 to 10 Rs. per kg. The net return of plantation is Rs.7, 000 per hactre which would go up to Rs. 20,000 when it reaches to age of 25 years.

Medicinal and Economic Importance

- Ripe and unripe fruits give drug which is known as Bel. Bel has been known to be useful in dysentery and diarrhea. It is proved useful in appetite and digestion.
- Drink is prepared from the fruit which is useful to the patient who has just recovered from bacillary dysentery.
- Root is known as Dashmul of Ayurveda and it used to cure palpitation of heart.
- Fruits provide substance which is gummy in nature-known as gluten which is highly useful as adhesive; it is also useful in making dye and varnishes. Outer rind of fruit gives yellow dye.
- Marmelosin is obtained from pulp of the fruit which acts as laxative diuretic. It is cardiac depressant when used in high dose.
- Infusion of root is used by tribal people to cure fever.
- Leaves are useful for fodder. Wood is utilized to make house hold articles and cattle sheds.
- Roots, stems and leaves contain tennis. Alkaloids, sterols, coumarin and aromatic components are also present in the plant.
- Aegelin, marmelosine, marmelin, o-methyl hayordinol, allomperatorin methyl ester, o-isopentanyl hayordinol and linolic acid are also found in the beal tree.
- Seed shows antimicrobecterial activity against bacteria and fungi. They show also antiprotozol activity (Henry and Brown).

15

Alangium salvifolium L.f. (Ankola)

Botanical Name: *Alangium salvifolium* L.f.

Family: Alangiaceae

Local Names

- ☆ **Gujarati:** Ankola
- ☆ **English:** Ankol
- ☆ **Hindi:** Ankol

Introduction

This plant is genus of shrubs or trees, comprising about 22 paleotropical species. Two species are found in India.

Description

This plant is small desiduous tree, shrub or strannngler, found in the drier parets of India and Cylon. It grows vigorously in the forest of South India and Burma. Flowers are white, fragrant and they are arranged in axillary fascicles. Fruits are 1-2 seeded. Berries crowned by calyx lobes, yellowish or red when ripe.

Origin and Distribution

This plant is believed to be originated from India. The plant is distributed in arid and semi arid forests of India and world.

Cultivation

This plant is cultivated by seed propagation method.

Medicinal and Economic Importance

☆ Root bark is anthelmintic and purgative. It is used in fevers and skin diseases. It is administered in the form of powder.

☆ Wood is olive brown, hard and close grained and weighs 42-56 ib. per c ft. In south India, it is used for pestles oil mills, cattle bells etc.It is suitable for inlaying and carving.

☆ Wood is used as fuel. Fruit is edible; seeds contain alkaloids, sterol and very little amount of fatty oil.

16

Allium sativum Linn. (Lasan)

Botanical Name: *Allium sativum* Linn.

Family: Lillyaceae

Local Names

- **Gujarati:** Lasan
- **Hindi:** Lashun
- **English:** Garlic
- **Tamil:** Vellai puntu
- **Telugu:** Tella- gadda, Velluli

Introduction

Garlic is perennial plant. It has a compound bulb, grayish white in colour and sub-globular in shape which contain about 4 to 6 cm diameter with 8 to 20 cloves. A thin, white or pinkish sheath covers group of small segments or cloves. These cloves are attached to wood axis with numerous roots. The flower of this clove is stronger than other bulb crops. Leaf is hollow which is similar that in onion. There are two varieties of garlic.

Desi and Acclimatized Varieties

1. **Deshi:** They are white colored with fairly big bulbs. They contain good keeping quality and higher yield. Red varieties are more pungent.
2. **Acclimatized varieties:** it can be divided in two varieties 1. Early white varieties and 2. Ooty-1 varieties.

(a) **Early White varieties:** Fawari, Rajalle Gaddi, Madrasi, Tabity, Creole, Eknalia, T-56-4, Jamnagar etc are very famous in Gujarat. Cultiver Jamnagar is best variety because it gives highest recovery of dehydrated peeled garlic and garlic powder of good pungency and antibacterial activity.

(b) **Ooty-1:** Ooty-1 is made from the germplasm assembled at Vijaynagargam, Horticulture Research Station, and Ooty (TNAU). It gives good yield than other varieties. Its bulb is big and weighs 30 to 40 gm. Clump has 20 to 25 cloves. Clones have flatterning surface inside. It has good resistant power. Chaubethia is a local selected clone which concludes 18 clones.

Different between Onion and Garlic

Although onion and garlic look similar in morphology but in the onion single large bulb is produced and in the garlic group of small bulbs which is called cloves are produced. This is basic difference between onion and garlic.

Origin and Distribution

South Europe and Central Asia is believed to be native place of Garlic. It is bulb crop which is useful as spice or condiment in India. In India garlic is grown in most of the states like Orrisa, Gujarat, Madhya Pradesh, Bihar, Rajasthan, Jammu and Kashmir, Karnataka, Maharastra, Orissa, Punjab etc. Orrisa is state which provides highest yield of garlic in India. Gujarat is at second place in the production of garlic.

Cultivation

For cultivation well drained loamy soil with presence of humus is good. Good amount of potash is useful to the garlic. It is generally grown in the late autumn. Short days are favorable. Not too cold or not too warm condition is suitable. Garlic is seldom grown in light sandy soils as they lack sufficient moisture holding capacity. Wet soils tend to produce large discolored bulbs, which causes injury to the plant. Garlic can also be cultivated from segments of large bulbs in which each bulb contains about ten small cloves which are planted. Cloves are sown in September to October. Sowing is done by dibbing, furrow planting, broad casting like methods. Irrigation is useful; soil should be irrigated regularly at the interval of 10 to 20 days.

Harvesting

Harvesting is done in January to February, generally 4 to 5 moths letter after sowing when green leaves become yellow or brownish and show signs of drying up. Plants are pulled out or uprooted with a country plough and are tied into small bundles then they are kept in the field or in the shade for 3 to4 days. Thus bulbs become hard.

Yield

India exports 3,776 tones of garlic annually, and gives earning of Rs. 4.90 crores of foreign exchange. In the United States annual growth of garlic is 750 ha annually.

Medicinal and Economic Importance

- Garlic is highly nutritious and used in making curries in India. It is useful for masking flavor in salted meal and fish.
- Against all kind of poisons, garlic is used as powerful antidote.
- Garlic has antibacterial activity as it contains allicin which is useful in intestinal disorders and for number of infectious disease. It inhibits growth of microorganisms.
- Bulbs of garlic are used for severe cough and fever; they also become useful in rheumatism, in improving lungs. It also gives benefit to the skin diseases.
- Garlic is antiseptic and destroys worms and externally it will rid the skin of parasites.
- Allistain I, Allistain II and garlicin have isolated from garlic plant which are widely used in medicine for animals and human beings both.

- ☆ Garlic is also useful in treatment of hypertension, diabetes, helminthiasis, chronic colitis and gastritis, amoebiasis, rheumatoid arthritis angina pectoris, fungal infection and bacterial infection and also it is used in cancer.
- ☆ Garlic powder is used as condiment, and serves as carminative and gastric stimulant.
- ☆ Garlic is also utilized in stomach disease.

17

Aloe vera Town ex. Linn (Kuvarpathu)

Botanical Name: *Aloe Vera* Town ex.Linn, *Aloe barbadensis* ex Linn.

Family: Liliaceae

Local Name

- ✰ **English:** Aloe, Barbados,Curacao Aloe
- ✰ **Gujarati:** Kuvar, Kuvar pathu
- ✰ **Hindi:** Gheekanvar
- ✰ **Tamil:** Thazhai, Kathalai

Introduction

Aloe is known as nature's sun screen. It is commonly grown in deserts and other dry situations. Although 275 spices of aloe are known but aloe barbadensis Mill is common in India. Variety is found in saurashtra coast is the source of Jafarabad aloe. Three varieties of aloes are official Indian Pharmacopoeia.

1. **Curaco aloe**- it is known as barbodes and obtained from *A. barbadensis*
2. **Socotrine or Zansibar aloe**- it is obtained from A. perryi
3. **Cape aloe**- it is obtained from A.ferox and its hybrids

Aloe is perennial plant, having shallow root with short stem. It gets height of 30-60 cm. Roots are tuberous type and supporting roots are also present. True stem is absent but it produces bloom stalks. Leaves are fleshy and are 60cm long, 10 cm broad and 3cm thick. Arrangement of fleshy leaves is dense. Plant is grown in typical rosette shape. Flowers are actinomorphic and are arranged in auxiliary

spikes. Flowers have yellow to rich orange color. Perianth is present in two whorls. In which outer whorl have long filaments than the inner whorl. Most of the species are male, sterile with scarcely any fertile pollen.

Origin and Distribution

Plants of Aloe genus are believed to be of old world whose native is Eastern and Southern Africa, the canary island and Spain. Plant was distributed from Mediterranean basin to West Indies, India, China and other countries. It is grown throughout in India. Andhra Pradesh, Gujarat, and Rajasthan are major states of India which grow aloe spices.

Cultivation

Sandy coastal to loamy soils of the plains with a pH of upto 8.5 are most useful. Water logging is harmful to the growth of aloe. Land is ploughed before cultivation. Cow dung is useful to fertile land. In March to June plants are cultivated. It can be grown in warm condition where yearly rainfall is almost 35-40 cm per year, although proper irrigation is also essential. Plants can be propagated by root suckers or rhizome –cuttings. Plants are planted at spacing of 60 x 30 cm and 60 x 45 cm.

Harvesting

Harvesting is done in December to February, Generally after 8 or 9 months of plantation. In the developed areas plants ate removed with help of tractor-drawn disc narrow or cultivator. In non developed are plants can be removed manually.

Yield

From second to five years yield is obtained, after which it needs replanting. On the fresh weight basis yield of the crop on a fresh-weight basis will be around 11,000 to 14,000.

Medicinal and Economic Importance

- ☆ Gel is obtained from leaves which contain polysaccharides while acetylated mannose sugar is the major bioactive component. Aloein is isolated from aloe which is highly useful in medicines.
- ☆ Aloe has bitter test which is useful against host of diseases related with digestive system.
- ☆ Aloe is used for wounds, burns and skin troubles.
- ☆ Gel is useful in eye troubles, spleen and liver ailments. it is used as endogenous purgative in indigogenous medicine so it is known as Musabbar
- ☆ Aloe has moisturizing and emollient properties as it is used as skin care. Extracts of aloe or aloin are used in sunscreens, X-ray burns, and other cosmetics preparation
- ☆ Aloe extracts are used as a flavoring ingredient primarily in alcoholic and non alcoholic beverages.

18

Alstonia scholaris (L.) R.Br. (Saptparni)

Botanical Name: *Alstonia scholaris* (L.) R.Br.

Family: Apocynaceae

Local Names

- ☆ **Gujarati:** Saptparni
- ☆ **Hindi:** Chatiun.
- ☆ **English:** Devil's tree, Dita bark tree
- ☆ **Tamil:** Elilupala, Pala, Palai
- ☆ **Telugu:** Edakulaphata

Introduction

Saptparni is tall evergreen tree and found in most part of the India. The name of genera alstonia is given in the memory of well known Botanist Professor C.Alton (1685-1760) Spices name is scholaris because in old time's wood of the tree were used to make slates of students. There is also one legend for this tree that it is the home of ghosts so it is also known as Devil tree too.Generally tree contains grey bark and verticillate branches. Leaves are arranged in 4 to 7 and some times 10 to 24 whorls of 2.5 to 4 cm. They are obovate, sub-sessile with short petiole (0.5 to 2 cm long) and glabrous. Leaf is pale green in color at lower side and dark shining at upper part. Greenish white flowers are present. They are arranged in compact umbellately- corymbose cymes, calyx is small which has 5 lobes. Fruits are long slender follicles, occurring in clusters. They are densely ciliate with long hairs all around. Bark of the tree is gray and rough which excused a bitter milky sap.

Origin and Distribution

Different views are there for origin of alstonia scholaris but no certain evidences present in the current time. It is distributed in whole peninsular India mainly in the moisture areas like Eastern and Western Ghats. In India it is found to be grown in the Rajasthan, Gujarat, Delhi, Maharastra and in most of the other states in India. In the east it is distributed to Java and Australia.

Cultivation

It can be grown in all types of soil but it grows well on lateritic soil. Usually it is propagated by seed. Seedlings are grown for one to two years and when plant gets proper height they are transplanted to the field where it gives good result. Rainy season is the best season for transplanting plant into field. Seeds are planted at 2 x 2 m distance, although seeds can be grown directly in the field but it takes more time to establish directly in the field. If good care is taken by removing weeding and providing proper fertilizers then within 5 to 6 years plant can be felled.

Harvesting

After 6 to 8 years harvesting can be done. Whole tree is cut to the ground level and bark of the tree is removed which is then dried under sun. This bark when becomes dried, small pieces of convenient lengths are prepared and they are taken to the market.

Yield

From one hactre, production of bark is around 2,500 kg. Drug is sold at the rate of 25 to 35 Rs.per kilograms. So return per hactre would be almost 40,000 to 50, 000 per hactre within 1 to 6 years.

Medicinal and Econonomic Importance

- ☆ Bark is used as tonic and drug which maintains normal function of the digestive system. Alcoholic bark has also showed anticancer activity.
- ☆ Drug is used in the treatment of intermittent and remittent fevers. Infusion of the bark is useful in the malaria, bowel complaints and skin diseases.
- ☆ Powder is prepared from bark of the tree which is used to cure diarrhea, catarrhal dyspepsia and dysentery.
- ☆ It is rich source of alkaloids. Echitennine, echitemine and diatamine have been isolated from the tree. Alcohol extract of the bark contains α-amyrin acetate and lupeol.
- ☆ Latex and coagulum both bear resin.
- ☆ Wood of the tree is soft so it is utilized in making furniture of schools, in constructing houses, in the boats *etc.*

19

Amaranthus spinosus L. (Kantanu Kanth)

Botanical Name: *Amaranthus spinosus* L.

Family: Amaranthaceae

Local Names

- ☆ **Hindi:** Cauleyi kateli
- ☆ **English**: Prickly amaranth
- ☆ **Sanskrit:** Tenduliyah
- ☆ **Tamil:** Mullukkirai
- ☆ **Telugu**: Mullutotakura

Introduction

It is an erect, glabrous, spinous herb, varying in colour from green to red or purple, 30-60 cm in height with grooved branches and sharp diversed spines in the leaf axils. Leaves are simple, alternate, ovate, lanceolate or oblong, entire, glabrous above, main nerves numerous, conspicuous below. Flowers are small, sessile, yellowish white or pale green in colour, they are numerous in numbers. They are arranged in dense axillary clusters and in terminal or interrupted spikes. Fruits ovoid capsules, membranous, circumscissile about the middle.

Origin and Distribution

The plant is native to tropical America and also found throughout in India Perticularly in weste lands. Generally it is arid or semi arid herb which grows in Madhya Pradesh, Gujarat, Maharastra and most of the other states of India.

Cultivation

The plant prefers light (sandy), medium (loamy) and heavy (clay) soils and requires well-drained soil. The plant prefers acid, neutral and basic (alkaline) soils. It cannot grow in the shade. It requires moist soil. This plant can be cultivated on any kind of land. There is no extra need of irrigation facility for the growth of this plant. Generally seed propagation is better method.

Seed are sown during late spring *in situ*. An earlier sowing can be made in a green house and the plants put out after the last expected frosts. Germination is usually rapid and good if the soil is warm. A drop in temperature overnight aids germination of growing plants root easily.

Chemical Composition

Higher alkanes and their methyl derivatives, higher aliphatic alcohols, acids and esters, amino acids, β-sitosterol, stigmasterol, campesterol, cholesterol, α-spinasterol, α-spinasterol octacosanoate, glycosides of α-spinasterol and oleanic acid have been reported in the plant.

Harvesting

Harvesting is done by uprooting, by cutting at ground level or by ratooning. If the crop is directly sown, a once-over harvest by uprooting or by cutting at ground level may be done 3–4 weeks after sowing. Second harvest is obtained 3 weeks later from the regrowth of the smallest plants. When ratooning is practised, the first cutting takes place about one month after transplanting, and then every 2–3 weeks for a period of 1–2 months. Cutting should be done at such a height that at least 2 leaves and buds are left behind for regrowth.The best cutting height is 10–15 cm

Yield

Commercial growers harvest 1.5–2.5 kg/m^2 of an uprooted crop (dry matter content 16 per cent, edible portion 35–50 per cent); the first cutting of a ratooned crop may yield 1–2 kg/m^2 (edible portion 70–80 per cent), subsequent cuttings about 1 kg/m^2.

Handling after Harvest

The harvested plants or shoots are bundled, the roots are washed, and the produce is packed for transport and then these are taken to the market. In markets and shops, it is sprinkled with water to keep a fresh appearance. If uprooted, the vegetable can be kept fresh for some days by putting it in a basin with the roots in the water. It is sold in bunches or by weight.

Medicinal and Economic Importance

- Whole plant is medicinally used. The plant is sweet, cooling, alexeteric, laxative, diuretic, stomachic, antipyretic, febrifuge, sudorific, galactagogue, haematinic, appetizer and tonic.
- It is useful in vitiated conditions of *pitta*, hyperdipsia, burning sensation, hallucination, leprosy, eczema etc.
- It is also used in bronchitis, leucorrhoea, menorrhagia, haemorrhoids, abscesses, boils, burns, strangury, nausea, flatulence, colic anorexia, fever, intermittent fever, agalactia, anaemia and general debility.
- Roots are thermogenic and haemostatic. They are useful in vitiated conditions of *kapha*, menorrhagia, haemoptysis, haematemesis and leucorrhoea.

20

Andrographis paniculata Nees. (Kal Megh)

Botanical Name: *Andrographis paniculata* Nees.

Family: Acanthaceae

Local Names

- ✰ **Gujarati:** Kal megh, Indrajav
- ✰ **Hindi:** Kiryat, Kalpnath
- ✰ **English:** King of bitters
- ✰ **Tamil:** Nilavempu
- ✰ **Telugu:** Nela vemu

Introduction

It is small annual herb which is found generally in wild condition but in most part of country it is also grown as medicinal plant. Stem is erect and bears nodes and antinodes. It is found in the scrubby and in deciduous forest and also in the ground under shade of trees and bushes. It grows up to 30 to 75 cms and contains four sided spreading branches. Leaves are short petioled, opposite lanceolate, glabrous, long and being narrow at both the ends. Leaf varies in size from 3 to 8 cms long and 12 to 14 mm wide. Axillary horizontal simple inflorescence is found. Flowers are white in color, dotted irregular bilabiate. Fruits are glabrous containing 22mm length and 3mm width. They are glabrous. Roots are of tap roots type.

Origin and Distribution

The plant is distributed in tropical Asia about 21 species occure in India, grown well in tropical and sub tropical region all over the India. It is distributed from Himachal Pradesh to Assam and Mizoram and all over in South India.

Cultivation

It can be grown in any type of soil because it is wild plant. Clay soil is more suitable to the plant. Although seed sowing is ideal but vegetative methods are also used for propagation. Sowing is done in the rainy season. Seeds are directly sown in the soil for propagation. Soil is prepared by repeated ploughing first which is followed by planting. Row method and Broad Casting method are used. Seeds are sown at distance of 15 x 15 cm. Proper weeding, irrigation and fertilizers are provided, plant gives yield in 3 to 4 months.

Harvesting

Harvesting is done after 120 days or 3 to 4 months after sowing when plant is in full boom. Lower leaves are harvested just after 2 months. Plant should be uprooted. Healthy plants are left in the field for seed production. When capsule matures seeds are removed. Then seeds are dried into open sun for drying. Stem are the important commercial product. The plants are uprooted and dried. The dried stems are them marketed.

Storage

Storage is done in the air tight container for further sowing.

Yield

Average yield per plant is about 2.5 tones per hactres. In the market it is sold at the rate of Rs. 15 to 20 per kg. It is estimated that a production of 4000 kg of Kalmegh from one ha. of plantation. Kalmegh is sold in the market at the rate of Rs. 5 per kg. Per hectare expenditure reaches upt Rs. 5000 and in the return net return goes upto Rs. 15,000 per hectare.

Medicinal and Economic Importance

- ☆ Entire plant is used for various diseases. The juice of leaves with certain species is dried and is used in infants to relieve irregular stool and loss of appetite.
- ☆ Decotion of whole plant is used by tribal peoples to cure malaria and dyspepsia and gaseous distention.
- ☆ Leaves contain glucoside like andrographolide, neo-andrographolide, panaculoside, flavanoides, andrographin, panicalin *etc.*
- ☆ It is used specially for children in digestive complains. 'Aluii' is made from the leaves and is given to children for stomach complaints in Bengal.
- ☆ Plant is bitter, acrid, cooling, laxative, antipyretic, antiperiodic, anti-inflammatory, expectorant, depurative, sudorific, antihelminthic, digestive and stomachic.
- ☆ It is useful in hyperdipsia, burning sensation, wounds, ulcers, chronic fever, inflammation, cough, bronchitis, skin diseases, leprosy, Purities, intestinal worm, diarrhea *etc.*

21

Annona reticulata L. (Ram Phala)

Botanical Name: *Annona reticulata* L.

Family: Annonaceae

Local Names

- ✰ **Gujarati:** Ram phala
- ✰ **Tamil:** Ramasita
- ✰ **Hindi:** Ram phala
- ✰ **Telugu:** Raamaaphalamu
- ✰ **English:** Bullocks heart

Introduction

It is small and deciduous tree or semi desi- duous tree. It gets height of 6.1-7.6 m. Leaves are acuminate, 12.7-20.3 cm. long and 3.8-5.1 cm wide. Flowers are cordate or subglobose, brownish or yellowish, with flat areoles, 7.6-12.7 cm diam. Flowers are produced in summer and fruting occurs in the end of rainy season. Ripe fruits are sweet and edible but inferior to those of *A. squamosa*.

Origin and Distribution

The custard apple is believed to be a native of the West Indies but it was carried in early times through Central America to southern Mexico. It has long been cultivated and naturalized as far south as Peru and Brazil. It is also naturalized in eastern and south India. It is distributed in Arid and Semi arid countries like India, Burma, Japan, Sri-Lanka *etc.*

Cultivation

The custard apple tree needs a tropical climate. It flourishes well in the coastal lowlands. In India, it does well from the plains up to an elevation of 4,000 ft (1,220 m). This species is less drought-tolerant than the sugar apple and prefers a more humid atmosphere. The custard apple does best in low-lying, deep, rich soil with ample moisture and good drainage. Seed propagation is good method. Nevertheless, the tree can be multiplied by inarching, or by budding or grafting onto its own seedlings or onto soursop, sugar apple or pond apple rootstocks.

Harvesting and Yield

The custard apple has the advantage of cropping in late winter and spring when the preferred members of the genus are not in season. It is picked when it has lost all green color and ripens without splitting so that it is readily sold in local markets. If picked green, it will not color well and will be of inferior quality. The tree is naturally a fairly heavy bearer. With adequate care, a mature tree will produce 34-45 kg of fruits per year. The short twigs are shed after they have borne flowers and fruits.

Chemical Constitutions

Anonaine is obtained from the tree. Nourshinsunine, tannic acid, oxoushinsunine and reticuline etc. have been also reported.

Medicinal and Economic Importance

- Mainly fruits are used for medicinal uses.
- Ripe and unripe fruits are used in diarrhea and blood dysentery.
- Leaves are considered as insecticide, anthelmintic and externally useful as suppurant.
- The leaf decoction is given as a vermifuge.
- Crushed leaves or a paste of the flesh may be poulticed on boils, abscesses and ulcers.
- The bark is very astringent and the decoction is taken as a tonic and also as a remedy for diarrhea and dysentery.
- Fragments of the root bark are packed around the gums to relieve toothache. The root decoction is taken as a febrifuge.

22

Annona squamosa Linn. (Sitafal)

Botanical Name: *Annona squqamosa* Linn.

Family: Annonaceae

Local Names

- ☆ **Gujarati:** Sitafal
- ☆ **Hindi:** Sitafal
- ☆ **English:** Custard apple, Sita phal
- ☆ **Sanskrit:** Agrimakhya
- ☆ **Tamil:** Sitil
- ☆ **Telugu:** Sitafalam, Sitapandu

Introduction

It is small and deciduous tree which is found in the dry and degraded forest areas. Leaves are simple and alternate, 5-15 x 2-5 cm, lanceolate or elliptical, tip of the leaf is blunt or slightly pointed, base is also acute and gland is dotted. Flowers are 2.5 cm long, solitary or several together which arise from the axils of the leaves, three fleshy petals hanging downwards; they are greenish, pubescent and feebly scented. Fruits are rounded with a deeply furrowed and wrinkled skin, 5-15 cm long, tubercled, green, fleshy, with dark brown or black seeds embedded in the edible pulp which is white in colour.

Origin and Distribution

Annona is native to tropical America and Africa, widely distributed throughout tropics and frost free sub tropics. About 8 species and a hybrid, *A.atemoya* is grown in India for fruit and root stock. *Anonna reticulata* and Anona sqamosa have completely naturalized there. *Annona squamosa* is the native of India from its occurrence in the early literature. In India most of the states are related with the marketing of the *Annona.*

Cultivation

It can grow on variety of soils including eroded, degraded areas. It grows in semi wild conditions of forest areas. It prefers warm and dry climate with moderate winter. Areas having rain fall of 500- 800 mm are considered best for this species. Normally seed propagation is good method for cultivation. Because Sitaphal is cross pollinated plant so there will be variation in size shape and quality of fruits. To avoid this vegetative propagation is used. Shield budding or Veneer budding on its own fruits give useful results. The budded plants will come to bearing early with uniform quality of fruits Plants are transplanted when they obtain optimum height. They are planted in the field at 2m x 2m spacing, at the beginning of the monsoon. Farm yard manure is added to each pit. Pits are prepared of 45x 45 x 45 cm before planting. Timely weeding and soil workings help the plants to grow better. During the rainy season, the plants may be given NPK mixture, at the rate of 50 grams/plants in two split doses, preferably along with soil workings. Watering the plants gives good yield but irrigation is not required.

Harvesting

Plant begins to give fruits in the 3 to 4 years. When fruits are ripen segments on the skin become flat and the interspace between segments turn whites. This is perfect time for plucking the fruits. Fully mature fruits ripe after 2-3 days of harvesting. The fruits are marketed in baskets. Good yield giving tree provide 60 to 75 fruits.

Yield

From one hactre of plantation, 40,000 fruits are obtained. They can be sold at Rs. 2 to 5 per each. Generally expenditure in five years reaches upto Rs. 5,000 and net return will be 15,000 Rs. per hactre.

Storage

Fruits develop black patches in the segments or on damaged spots on the skin during normal ripening at ordinary temperature, through these are not considered blemishes, cold storage is not promising either. It was found that hard and immature fruits chilled at 17°c or below become discoloured. Ripe fruits could be stored for 42 days at 5.5 °c, pulp of the fruit is good but has an unaatractive appearance.

Medicinal and Economic Importance

- ☆ Fruits bark roots and leaves are used in local areas for medicines.
- ☆ Sweet pulp is edible and it is very sweet in teste. Fruits is valuable in recovering after illness.
- ☆ Refreshing drink is made from the sweet pulp of the fruit.
- ☆ Roots are used as drastic purgative.
- ☆ Leaves are good insecticides. Leaves pounded with tobacco leaves and little quick lime is frequently used locally in ill-conditioned ulcers to destroy maggots.
- ☆ Leaves are also used in extraction of guinea worms.
- ☆ Seeds yield an oil and resin, latter is used as insecticide. Oil obtained from the seed is used in the manufacture of soaps, ointments, cosmetics, emulsifiers, water proofing agents, shoe and metal polishes, paints, etc.

23

Apium graveolens sub sp dulce (Mill) (Ajmo)

Botanical Name: *Apium graveolens* L.

Family: Apiaceae

Local Names

- ✰ **Gujarati:** Ajmo
- ✰ **English:** Celery
- ✰ **Hindi:** Ajmod
- ✰ **Sanskrit:** Ajmoda

Introduction

This is an annual or biennial herb. Stem reaches upto 2-4 m high, erect, branching. Radical leaves pinnate with deeply lobed segments, cauline 3-partite. Segments once or twice trifid coarsely toothed at the apex. Peduncle 6 mm or less. Umbell rays are 5-10, pedicles 6-16. Flowers are white, very small. Fruits are cremocarp. Each fruit with 2 mericarps. It is 1-1.5 mm long with narrow ridges, broad vittage.

Origin and Distribution

This plant is distributed throughout in India. It is mainly distributed in Punjab and Uttar Pradesh. It is also found in Europe and U.S.A.

Cultivation

This plant is cultivated by seeds only. It grows best on sandy and silt loams and is not suited to clayey soils. It requires well prepared and heavily manured

soil. It requires good drainage, but the soil with high moisture retentive soil. When grown as a garden crop on the hills, seeds are sown in March-April, seedlings are transplanted in May and the crop is ready in November to December. In the plains seedlings, preferably brought from the hills, are transplanted in September-October and the crop is ready in three months.

Seeds are sown at the rate of one kg per ha. Sowing is carried out in the middle of July to August in a months period. The land is ploughed thoroughly and a fine tilth is obtained. The nursery seedlings are transplanted in the field in rows of 60-70 cm apart an at 50 cm distance in the lines during November. The field is watered after transplanting. Soil is kept loose by hand hoeing. Fertilizer of 100 kg of Nitrogen per hactare is given in two split doses along with irrigation. The first dose is given during 30th and 35th day and the second during 60th and 75th day after transplanting. The crop becomes ready for harvesting by middle of March. In Punjab and U.P, nursery is raised in September-October, transplanting is done in January and harvesting is done in May.

Harvesting and Yield

The yield of seed (fruits) is about1000-1200 kg per hactare. The seeds yield volatile oil (2-3 per cent). Thus 25-30 kg of volatile oil (celery seed oil) is obtained from one ha. The oil is sold in the market at the rate of Rs. 600/kg. Total expenditure reaches upto Rs. 10,000 per hactare and Gross return per ha. goes upto Rs.18,000 while net returns per ha. is Rs.8,000.

Chemical Composition

Choline, Myristic acid, Myristitic acid, Myristoleic acid, Stearic acid, Apiin, Isoquercitrin, Apigenin; Apigravin, Apiumetin, Apiumoside, Bergapten, Celereoside, Rutaretin, α-eudesmol, β-eudesmol santalol, 3 n butyl phthalide, sedanenolide. Seeds contain d-limonene, Selinene, Oleoresin.

Medicinal and Economic Importance

- Whole plant is medicinally useful. Plant is carminative, stomachic, Anodyne, Nerve tonic, emmenogogue, Stomach trouble.
- For complain of depression and sleeplessness it is very useful.
- Obstinate retention of urine, throbbing headache, heartburn, rheumatic pain in muscles of neck and sacrum.
- It is used in Dysmenorrhoea with short pains in both ovarian regions better by flexing legs, left ovarian pain better by lying on left side.
- It is also used in itching blotches with burning.
- It is also used in intense constriction over sternum, Skin urticaria with shuddering.

24

Aquilaria agallocha Roxb. (Agar Agar)

Botanical Name: *Aquilaria agaliocha* Roxb.

Family: Thymeleacae

Local Names

- ☆ **Gujarati:** Agar
- ☆ **Sanskrit:** Aguruh, krsnaguruh
- ☆ **Hindi:** Agar
- ☆ **Tamil:** Agar Agalicandanam, Krsnaguru
- ☆ **English:** Aloe wood, Eagle wood
- ☆ **Telugu:** Krsnaguru

Introduction

A large evergreen tree about 21 m in height, leaves linear-lanceolate to ovate-oblong, 5-9cm long, silky glossy and faintly parallel nerved. Flowers are small, greenish on very slender pilose petioles in shortly peduncled umbles, on younder branchlets, perianth about 5 mm long, slightly hairy outside, stamens alternate the perianth, filaments red at the apex, ovary tawny-tomentose; fruits are slightly compressed, yellowish, tomentose and they are capsule type. The fungus infected trees furnish agarwood or eagle wood of commerce which occures as dark coloured resinous fragrant masses in the centre of the bole. The essential oil from agar wood is valued in high class perfumery as a fixative.

Origin and Distribution

This plant is found mainly in hilly areas where there is moisture present. In the forest of Burma, Bhutan, India it is mostly found. In the Assam this plant species is widely distributed. The genus of trees, distributed in South China and the Indo Malaysian region also. Two species are recorded in India, of which *Aquilaria agallocha* is commercial source of agar wood.

Cultivation

Seed propagation is main method. The seedlings raised and maintained in the nursery for one year period are good for raising plantation. At the beginning of monsoon, seeds are sown in the nursery beds. If they are looked after carefully, seedlings will be ready for the transplanting in the field at the commencement of next year's monsoon. Seedlings are transplanted in the field at 5m X 5m spacing in the pits of 45 X 45 X45 cm. The soil is kept loose and weeding is done. After 20 years period, the trees are injured aftificially at several spots and the fungi of *Fusarium,* Aspergillus and Pencillium species are introduced on the trees. The pegs from trees containing agar wood are driven into the trees, to infect those trees. This may result into inoculation of favourable fungi on the injured portion and the starting time of agar wood. All the plants are exploited at the end of 50 years, when they have the highest concentration of agar.

Harvesting

After 50 years trees are feeled and harvesting can be done of agar. Agar wood oil is obtained by steam distillation method. Agar is found in the form of chips and splinters of fine quality and also as blocks weighing about 0.4 kg. It is estimated that every tree on an average produce 250 kg. of agar wood. A production of 1,00,000 kg. of wood per ha would be possible which means 1,000 kg. of oil (at the rate of 1 per cent yield). The market rate of agarwood oil is revolving around Rs. 1,000/-per kg.

Yield

In 50 years total expenditure reaches upto Rs.1,00,000/-, in which gross return is around Rs.10,00,000/-. Ultimetly net return goes upto Rs.9,00,000. That means Rs.18,000/- per hectare is net return.

Chemical Constituents

The normal wood yeidls and essential oil wich contains selinene, dihydroselinene, two unidentified sesquiterpene-hydrocarbons, agarol, a sesquiterpene alcohol, a hydroxyl ketone, five isomeric decenes and rhombic as 5-isopropyl-7-methyl-5,5,5a,6,7,8-hexahydro-3H-naphtho. Agarwood, on distillation, yields an essential oil, known as a Agar oil.

Medicinal and Economic Importance

- ☆ The wood is acrid, bitter, thermogenic, digestive, carminative, deodorant, sudorific, anodyne, anti inflammatory, anti-leprotic, depurative, cardiotonic, rejuvenating and tonic.
- ☆ It is useful in vitiated conditions of vata and kapha, halitosis, dyspepsia, anorexia, cardiac debility, skin disease, leprosy and foul ulcers.
- ☆ It is also used in hypothermia, inflammations, rheumatoid arthritis, cough, asthma, highcough, albuminuria and general debility.
- ☆ An external application of agaru is very useful in vomiting in children, pectoralgia due to pneumonia and cephalalgia.
- ☆ The oil is astringent, acrid, bitter, thermogenic, depurative, alexeteric and antileprotic. It is useful in vitiated condition of vata and kapha.
- ☆ In rhemunatoid arthritis, cough, asthma, bronchitis, leprosy, skin diseases and foul ulcers this drug is used.
- ☆ It stimulates the nervous system.
- ☆ It helps in digestion and also reduces foul odor and tastelessness condition in the mouth.
- ☆ It stimulates heart and also purifies blood.
- ☆ It is helpful in expelling out the extra mucus in the respiratory tract thus enabling it to work properly. It is also an aphrodisiac in nature.
- ☆ It also works as body tonic
- ☆ Wood is used as third class timber because wood is not durable and it is used for making bows walking sticks and occasionally for dug outs.

25

Areca catechu Linn. (Shopari)

Botanical Name: *Areca catechu* Linn.

Family: Arecaceae

Local Names

- ✰ **Gujarati:** Sopari
- ✰ **Hindi:** Supari
- ✰ **English:** Betel nut palm, areca nut
- ✰ **Tamil:** Kamugu
- ✰ **Telugu:** Poka, Vakka
- ✰ **Sanskrit:** Gubak, Pulah

Introduction

Areca means cluster of nuts. As it is widely used with betelvine leaf, it is known as 'Betel palm'. The palm is adapted to the coastal belt up to 300 km from the sea. The tree requires heavy rain fall for its successive growth. Betel palm is erect, a slender- stemmed. Stem contains height of 30 meter and length of 15 cm. Leaves are similar to that of coconut. Length of the leaf is generally 1 to 1.5 meter. Fruit of areca nut is mono-locular or one seeded berry. It is orange red to scarlet when ripe and consists of a thick fibrous outer layer. This palm is grown for its fruits. Mature fruit contains three layers; exocarp, mesocarp and endocarp which are more or less in structure.

Origin and Distribution

Its native is believed to be of Malaysia but no certainty is there.It is distributed mainly in most of the countries like Srilanka, India, Pakistan and also in the West Indies and on other islands of the world. In India Sea coast bearing states like Karnataka, Kerala, Tamil Nadu, West Bengal, Assam and Maharashtra, Gujarat grow areca nut palm. In Gujarat districts like Jamanagar, Juagadh, Rajkot, Porbandar are known to grow areca nut palm.

Cultivation

In India areca nut is grown mostly in high rain fall regions. Laterite soil of red clay, acidic type is quite useful for the growth of this palm. Temperature required for this palm is between 4 to 15 c. Extreme temperature is not so useful for the growth of this palm. It is propagated by seed method only. Seed nuts are sown as whole. Germination is completed within 53 to 94 days. Direct sowing is useful rather than mudda or baskets. Sprouts are retained in the sand beds or primary nursery for about six months. Young seedling at this stage with two or three leaves is transplanted to secondary nursery beds of convenient width and length. Seedlings are transplanted up to 60 cm and 45 cm depth respectively. Seeds are sown at the spacing of 1.25 meter to 3.6 meter. Plant establishes in the month of September to October. Fertilizer and proper irrigation is essential for good growth.

Harvesting

There are two types of produce immature green nuts and ripe nuts. There are various methods for the harvesting of areca nut branches. In regularly spaced garden, the climber climbs a tree at one end of the garden, harvests the bunch and sends it down by a rope or gunny bag or drops down the bunch to the ground. Climber pulls the nearest palm with the help of a hook and swings to it. In some area of Karnataka and North Kerala, long bamboos with a sharp sickle or hook attached to the end is also being used for harvesting.

Yield

Each tree on an average yields two to three bunches per year, each containing 150 to 250 fruits. In certain large types number of fruits in a bunch may be less, varying from 50 to 100. Small size of nut is compensated by their large number yield is directly dependent upon several factors *viz.* nature of soil, cultural and manorial treatments, density of planting, age of trees, portion o bearing to non bearing trees in the garden, incidence of pest and diseases etc. Number of trees is largest variable factor governing yield per hectare. Trees yield 200 nuts annually.

Storage System

Areca nut is used either raw or cured. The following types are generally recognized: (1).raw nuts (2).Dried ripe nuts and (3) Processed nuts and (4) ripe nuts preserved in water or chemicals.

1. **Raw nuts:** – Ripe nuts are consumed during harvesting season. For off season use nuts with husk are stored in underground pits (1.2 to 1.5 m deep), nuts are moistened with water and placed in layers, covered at the top by a mat and a layer of earth.These nuts may also be kept immersed in water in large earthen wares or in concreate tanks. Water is changed occasionally to avoid bed smell.
2. **Dried ripe nuts:** – The whole ripe fruits are dried in the sun for six to seven weeks so as to get a moisture level of 10 per cent, thereafter husk is also removed. Drying ripe nuts on cement floor reduces fungas inferior in colour to the processed nuts and their test is also affected. A mechanical driver is used to produce challi supari of good quality.
3. **Processed nuts:** Curing is done to prevent amount of tannin within nut for curing nuts are gathered while they are three quarters ripe and the husk is still green. At the proper stage they are moderately firm to touch and the thumbnail can be pressed through husk up to the kernel. If the fruits are under ripe, the cured product is poor and shrunken in appearance and if over ripe, hard and light coloured.

4. **Preservation of ripe nuts in water and other chemicals:** Fruits washed first in chlorinated fungicidal spray, then blanched for three minutes in 0.1 per cent sodium benzoate and 0.2 per cent potassium metabisulphite, acidified to a pH of 3.5 to 4.0 by dilute hydrochloric or phosphoric acid.

Medicinal and Economic Importance

- ☆ The areca nut is common masticatory nut, popularly known as supari. It is commonly used along with betel vine leaf. Some times it is used with tobacco for chewing.
- ☆ Nut comes from tall trees which has greater demand because it yields good amount of tannin. These tannins are employed for manufacturing black and red inks.
- ☆ Nuts are used as tooth powder.
- ☆ Leaves are used for thatching the houses and the wood are used as poles for small and big houses.
- ☆ In the treatment of leucoderma, leprosy, cough, fits, worms, obesity and anemia it is used as purgative, stimulant and an appetizer.

26

Argyreia nervosa (Burm.f.) Boj. (Samudra Sos)

Botanical Name: *Argyreia nervosa* Burm.f. Boj

Family: Convolulaceae

Local Names

- ☆ **Gujarati:** Samudra-sos
- ☆ **English:** Elephant creeper
- ☆ **Hindi:** Samandar-ka-pat, Samudra sos, Tameshar
- ☆ **Tamil:** Samuttirappaccai, Samuttirappalai
- ☆ **Sanskrit:** Vrddhadarukah
- ☆ **Telugu:** Candrapada

Introduction

It is very large, woody, climbing herb. It has white and tomentose type stems. Leaves are simple type, large and ovate. They are acute, base is cordate type, glabrous above and white tomentose beneath. Flowers are large and purple in colour. They are silky pubescent without in long-peduncled cymes inflorescence. Fruits are generally dry and globose type. They are apiculate.

Origin and Distribution

Origin of this plant is Asia. It is believed that plant was originated in Asia and then naturalized in Hawaii. This is very common climber throughout India, grows well in sub tropical and semi arid, also found at 900 meter elevation. It is found in

moist localities of dry deciduous and moist deciduous forest areas. It grows well on all types of soil, but prefers well drained sandy loam soil.

Cultivation

This plant is wild plant plant so it can be easily found any where but its cultivation is also done for its medicinal uses. It can be propagated by seeds. The seeds are collected from the previous years crop and kept ready. The field is ploughed thoroughly and pits are dug at 3 m x 3m. Since this is a vigorous climber, pandal of wooden poles or trellies is erected. At the beginning of monsoon the seedlings attain a height of 30 cm, one weeding around the plants is carried out. The plants are tranined to climb up the pandal/trellies. The plant is fertilized with 50 grams of urea/super phosphate.

Harvesting and Yield

The first harvesting of the roots can be done at the end of the 2nd year and every year thereafter. At the end of the second year, alternate plants are uprooted in the month of May. The roots are separated from the stem and dried under sun. The dried roots are then marketed. In the place of the plants exploited, fresh seeds are sown at the onset of monsoon. Thus, every year 50 percent of the plants are exploited and that vacant place is replenished with new plants. It is expected that each plant will produce one kilogram of root. Thus, a production of 500 kilogram of roots from one hactare is estimated.

Chemcical Composition

Tannin, and amber-coloured acid-resin soluble in ether, benzole and partly soluble in alkalies.

Medicinal and Economic Importance

- Roots are acrid, bitter, astringent, sweet, emollient, thermogenic, roborant, appetizer, digestive, carminative, aperient, cardiotonic, anti-inflammatory, expectorant, diuretic, aphrodisiac.
- They are also rejuvenating, intellect promoting, brain tonic, nervine tonic and tonic. They are used in vitiated conditions of Kapha and vata, emaciation,, colic, constipation, cardiac debility, inflammations, cough, bronchitis, strangury, seminal weakness, nervous weakness, cerebral disorders, synovitis, heamorrhoids, obesity, hoarseness, syphilis, anaemia, diabetes, tuberculosis, arthritis, ascites, leucorrhoea and general debility.

27

Artocarpus heterophyllus Lam. (Fanash)

Botanical Name: *Artocarpus heterophyllus* Lam.

Family: Moraceae

Local Names

- ✩ **Gujarati:** Fanash
- ✩ **Hindi:** Kathal
- ✩ **English:** Jack fruit tree
- ✩ **Tamil:** Pilapalam
- ✩ **Telugu:** Panasa
- ✩ **Sanskrit:** Panasah

Introduction

It is large evergreen tree with dense crown containing more than 10 meter height with light black motted green or dark brown and rough bark. Leaves are 5 to 7 long and 2 to 10 cm wide, leaves are elliptic and oblong or obovate and glabrous. Flowers are variable in size, dark green in color and always produced on trunks. Fruits are 30 to 60 cm long, 15 to 30 cm in diameter. When fruits are immature they are pale green but becomes yellow to brownish when it ripes. Fruits contain many seeds. It is drought resistant and frost tender species.

Distribution

It grows well in semi humid regions. It is grown into India Pakistan Sri Lanka and other countries. In India this tree is grown in the Himalayas, jack fruit is also

grown in the Rajasthan, Gujarat, Hariyana, Maharastra and some other states as well. In Gujarat it is cultivated in nurseries and gardens. It can be seen in Junagadh, jamanagar, Panchmahal and some other districts of the states. Origin of jack fruit is still not clear.

Cultivation

There are two varieties of jack fruit Kapa and Barka. Kapa variety has fleshy and crisp peri carp while Barka has thin and sour pericarp. Seed are sown after their collection. Seed germination rate goes onto 100 per cent. Seedlings are transplanted in the field after one year of sowing seeds. Seedlings are planted at spacing of 10 x10 meter. Pits are prepared and seedlings are sown into these pits. Farm yard manner and urea are quite useful.

Harvesting

When tree becomes 10 years old it starts bearing fruits. May – August is the main fruiting season. Harvesting of fruiting is maximum at 20 to 25 years.

Yield

Fruits give yield of 5000 per hactre and they are sold at a rate of Rs. 10 to 15 each.Net return after 11 th year is around 45,000.

Storage

Mature jackfruit may be stored for three to six weeks at 11-12°C. Above this temperature there is increase in acidity decrease in ascorbic acid content and increase in total soluble solids. Slices of fruits in 50 per cent sugar syrup with 0.5 per cent citric acid frozen at –29^{oC} and stored at –18°C can be kept for even upto one year.

Medicinal and Economic Importance

- ☆ Jack fruit is used to obtain Vitamin B_1 and Vitamin B_2.
- ☆ From the bark and fruit milky latex is obtained which is used as a varnishing material.
- ☆ The plantation of this species can be profitably combined with cultivation of other crops.
- ☆ Fruits are eaten as vegetable or they are used in pickles.

28

Artocarpus lakoocha Roxb. (Barhal)

Botanical Name: *Artocarpus lakoocha* Roxb.

Family: Moraceae

Local Names

- ☆ **Gujarati:** Barhal
- ☆ **Tamil:** Irapala
- ☆ **Hindi:** Barhaal
- ☆ **Telugu:** Kammaregu
- ☆ **Sanskrit:** Lakucha

Introduction

This is large deciduous tree which gets height of 20-25 meter. Colour of bark is dark brown, exfoliating in small round woody peels, reddish inside, softy fibrous with faint streaks of white lates. Leaves are elliptic, oblong or ovate, broadly cuneate, pinnatified, serrate, thick like a skin, rough, having veins in 8-12 pairs. Inflorescence is axillary and solitary. Male heads are with styles exerted to 1-1.5 cm via low papilla everging between peltate bracts. Fruits are syncarp type, sub-globose, shallowly lobed, 6-12 cm long. They are yellow and drying to brown in colour. Surface of fruit is pubescent, irregularly papillate with numerous persistent bracts.

Origin and Distribution

A native of the humid sub-Himalayan regions of India, it grows up to 1,200 m altitude.This species of plant is found in evergreen, semi evergreen and moist deciduous forests. Although this plant is distributed throughout in India.

Cultivation

This plant is cultivated by seeds and root suckers. Generally seed propagation is good method. But tree population of lakoocha is gradually decreasing due to poor seed viability and extensive exploitation for food, timber, and other uses. Seeds, once extracted from the fruit, quickly loose viability within a week. Vegetative propagation methods such as rooting of hardwood or softwood stem cuttings have not been successful. Although micro-propagation method is successful for its rooting and shooting.

Chemical Composition

Wood contain galangin, artocarpin, cyclortocarpin, norcycloartocarpin, norartocarpin and resoucinol. Bark gives β-sitosterol, cycloartenol, cycloartenone, α-amyrin and lupeol acetate. The heartwood contains artocarpin, norartocarpin, norcycloartocarpin, cycloartocarpin, resorcinol and oxyresveratol.

Medicinal and Economic Importance

- ☆ Externally powdered bark is applied to sores to draw out purulent matter.
- ☆ A paste of the bark found eneficial by local application in boils and small pimples.
- ☆ Lakoocha seeds and milky latex are purgative. Seeds contain artocarpins, the isolectins which exhibit high haemagglutination activity
- ☆ The edible fruit pulp is believed to act as a tonic for the liver
- ☆ Raw fruits and male flower spike (acidic and astringent) are utilized in pickles and chutney (sauce).
- ☆ The lakoocha tree is also valued for feed and timber.
- ☆ The hardwood sold as lakuch, is comparable to famous teak wood. Lakuch which is durable outdoors as well as under water is used for construction,

furniture, boat making, and cabinet work.

☆ Tree bark containing 8.5 per cent tannin is chewed like betel nut, and is also used to treat skin ailments.

☆ It yields a durable fiber good for cordage.

☆ The wood and roots yield a lavish color dye.

29

Asparagus racemosus Willd. (Shatavari)

Botanical Name: *Asparagus racemosus* Willd.

Family: Liliaceae

Local Names

- ☆ **Gujarati:** Shatavari
- ☆ **Hindi:** Sat muli, Satawar, Phusar.
- ☆ **English:** Shatavari, Wild asparagus
- ☆ **Tamil:** Ammai Kodi
- ☆ **Telugu:** Calla Gadda
- ☆ **Sanskrit:** Abhiru, Shatamuli

Introduction

It is straggling or scandent, much branched and climbing shrub which grows to 1 to 2 meter in length. The leaves are like pine needles. They are small and uniform, reduced to minute spinescent structures subtending leaf like cladodes which are falcate, slightly compressed, channeled beneath. They are born in axillary clusters of 2 to 8. Flowers are small and they are arranged in spike type of inflorescence. Roots are tuberous and succulent; they are 30 cm to 100 cm in length. Generally roots are finger type and clustered. Flowering time of plant is late November. Fruits are of barrises type and contain one, two or three seeds. Seeds are spherical or hemispherical. Barries are globose type, purplish black in color when it ripes.

Distribution

Asparagus is found throughout India, in all the states like Gujarat, Maharastra, Rajasthan etc. It is also cultivated up to 1500 m elevation in Himalayas also. It is also distributed in Africa, through South Asia to China, South Malaysia and Northern Australia. Almost 150 species of asparagus are distributed all over the country. In the Gujarat Shatavari is grown in all the districts. It is mainly grown in Junagadh, Jamanagar, Bhavanagar, and Porbandar etc. like districts.

Cultivation

Fertile, moist, sandy- loam soils are ideal for its cultivation though it grows in a wide range of soils. Asparagus is best grown from its stem disc but commercially propagated through seeds. Seeds usually start germination after 40 to 45 days. Seeds are sown at the spacing of 30 x 30 cm. Best time of planting is June-July. If proper fertilizer and regular irrigation is provided then we can obtain good yield in two to three years.

Harvesting

Harvesting is done after two years of sowing. In which higher root yield is obtained. If proper irrigation is provided then harvesting of root tubers is very easy. Roots are cut into peacies of 5 to 15 cm. Then roots are taken to the market. Roots are used either fresh or dried, for marketing.

Yield

Yield of root tubers per hector is generally 10 to 15 tones. Although over 60 t/h yield is also reported.

Diseases Control

During May to November stem blight potato is caused by *Alternaria tenuissima*. Lesions start appearing on to stem and branches becomes necrotic that ultimately results into drying of stem. Dithane M-45 at 3 g/liter of water controls the disease.

Medicinal and Economic Importance

- ☆ Roots and leaves are used as medicine in the Indian system of medicine. Roots contain protein, vitamin, fat, carbohydrate, and several alkaloids
- ☆ Asparagin and asparagamine are isolated from shatavari which are useful in cancer. Asparagamine contains four antioxytoxic saponins shatavarin I to IV.
- ☆ Roots are useful as aphrodisiac, diuretic, anti dysenteric and as demulcent in veterinary medicine.
- ☆ Mucilage is also present in roots. Diosgenin has been reported in its leaves.
- ☆ It is used as tonic and because of presence of saponin it is also used as antioxygotic drug.
- ☆ It is used to prevent abortion and pre-term labour on place of progesterone preparations. Generally boiled powder with milk is used to prevent abortion.
- ☆ Its preparations in milk help in increasing breast milk in lactating women and it is also useful to avoid excessive blood loss during periods.
- ☆ Fresh roots fed to cattle increase milk yield.
- ☆ Young shoots are eaten raw, made into preserve and candied. Leaves contain rutin, diosgenin and flavonoid glycoside which is identified as quercetin-3-glucuronide.

30

Azadirachta indica A. Juss (Neem Tree)

Botanical Name: *Azadirachta indica* A.Juss

Family: Meliaceae

Local Names

- ✰ **Gujarati:** Limbado
- ✰ **Hindi:** Neem, Nimb
- ✰ **English:** Neem tree
- ✰ **Tamil:** Arulundi
- ✰ **Telugu:** Nimbau, Nimbi
- ✰ **Sanskrit:** Nimbah, Neta

Introduction

Neem is evergreen tree which is 13 to 20 meter in height and 1.5 to 2.5 m in girth, it has straight stem. Branches form broad crown, Tap root is seen in this tree. Roots are long and deep, thick grey bark reddish brown inside, with numerous longitudinal furrows and scattered tubercles. Flowers are bisexual, greenish white, pentamerous, bracteate and small. Aestivation is imbricate or valvate. Inflorescence is a long, slender, axillary or terminal panicle. Leaves are alternate, imparipinnate, swollen at the base, 7 to 13 foliolate containing short leaves which are glabrous green above and pale at the lower side. Young leaves are reddish green which turns into lush green. The genus Azadirachta consists of two species, *A. indica* and *A. excelsa*. Fruit of the neem tree is ellipsoid drupe which is fleshy which becomes yellow on ripening with single seed. Seed is exalbuminous.

Origin and Distribution

It is believed to be native of Upper Myanmar which is found apparently wild on Siwalic Hills in dry forests of, Tamil Nadu Karnataka and Andhra Pradesh. It is found in most of the states of India. In Gujarat it is grown in the most of the districts like Junagadh, Jamanagar, Rajkot, Porbandar, and Ahmedabad, Val sad, Dang etc. This tree is spread to Pakistan, Bangladesh, Srilanka, Malaysia, Indonesia, Thiland, Sudan and also into Middle East. Presently neem trees are also grown into Mauritius, Central and South America, Caribbean Island and also in Australia.

Cultivation

Neem tree grows on to dry, slurry, shallow, moderately saline. It suits alkali soil including black cotton soils, cotton soils, clays and lateritic crusts. Neem tree is sown by seed method. During June to August, seeds are cultivated. In the nursery seeds are sown 2.5 cm apart in rows and covered with rotten straw. Germination takes place in 15 to 30 days. Farm yard manure is applied rather than chemical fertilizers.

Harvesting

Harvesting is done in the February to March. Generally flowering occurs after 5 years during January to May. And leaves are shed during February to March. Fruits are hand picked to avoid dirt. Fruits are dried under shade. Fully dried fruits, both whole and decorticated, store well without deterioration for a year.

Storage

Seeds are stored in to air tight gunny bags or tin containers. It has to be ensured that seeds don't contain excess moisture.

Yield

A full grown tree produces about 350 kg leaves. Fruits get matured in June August and produce about 50 kg berries per tree. Fresh fruits yield 60 per cent dry fruits which yield 10 per cent kernel.

Medicinal and Economic Uses

- ☆ Plant is very useful as it has great medicinal value. Plant is regarded as the Village Dispensary. Whole plant is used.
- ☆ Fruits are antiperiodic, anthelminitic, purgative, emollient, tonic, astringent, diuretic and they are used against piles. Dry fruits are used to treat cutaneous infections.
- ☆ Seeds are used against snake bite.
- ☆ In the chronic skin diseases and ulcers, karnel oil is used. Oil is also used in the cosmetic preparations like cream, soaps *etc.*
- ☆ Flowers are stomachic with tonic effect. Flowers are eaten in the fever.
- ☆ Bark contains nimbidin which is a penta-nor-triterpenoid. Bitter ness of bark is due to the nimbidin. Bark of the tree is useful in malaria and typhoid fevers in local areas. Bark is antiemetic, astringent and tonic and also used in liver disorders.
- ☆ Twigs are used as tooth brushes for its antipyrrhoeal property. They are effective carminatives and digestives.
- ☆ Seeds contain Azadirectin which is major active ingredient and have anti feedant property. It is effective in acute and chronic inflammation including psorasis.
- ☆ Gum is obtained from the tree which is known as East India gum, this gum is highly useful in spleen enlargement.
- ☆ Neem is very useful insecticide. Seed extract is repellent against rice weevil and flour beetle of wheat. Crushed seeds protect pulses from the bruchid.
- ☆ The shell from seeds can be utilized to produce activated carbon and tooth powder. It is also used as a fuel and in thermosetting moulding compositions.

31

Bacopa monnieri Linn. (Barami)

Botanical Name: *Bacopa monneiri* Linn. Pennell

Family: Scrophulariaceae

Local Names

- ☆ **Gujarati:** Barami, Sarasvti
- ☆ **Hindi:** Jalnim, Barami
- ☆ **Tamil:** Nirpirami, Piramiyapundu
- ☆ **Telugu:** Sambranicettu
- ☆ **Sanskrit:** Brahmi, Saraswati

Introduction

Brahmi is annual herb. It grows throughout in India particularly in wet places up to 1,200m elevation. It is prostrate or creeping, juicy, succulent. Roots are produced at the nodes with numerous ascending branches; herb is simple, opposite, decussate, sessile. Flowers are pale blue or whitish in color, axillary and solitary; they are on the long slender pedicles. Fruits are ovoid, acute; capsules are 2-valved and tipped with style base. Seeds are many. There are three species of bacopa in India.

Origin and Distribution

There is no certainty about the origin of Bacopa monnerie. This herb is distributed in warmer parts of the world like Africa to America, West Indies and in India it is grown into moist and marshy places in all plain districts throughout India. It is abundant on borders of water channels and particularly where irrigation

facilities are available. It is grown in Himalayan regions, Rajasthan, Gujarat, Punjab and most of the other states.

Cultivation

Generally propagation is done in the months of June-July. Cold and humid climates are suitable for its cultivation. Generally sandy, light black soils are useful. It is propagated through vegetative method generally with the help of branches. Developed branches are planted in row at 60 x 60 cm distance.

Harvesting

Harvesting is done from January to March season which comes after 6 or 7 months letter of planting. Plants are pulled by hands and some plants are kept into field for its seeds. Pulled plants are dried under sunlight. It takes 8 to 10 days for them to dry.

Storage

Storage is done in the bags in the store houses.

Yield

Per acre we can get 60 to 70 quintals. Its present market value is Rs. 30.00 to 35.00.

Medicinal and Economic Uses

- ✰ Whole plant contains drug. Plant is astringent, bitter, sweet, cooling, laxative, intellect, promoting and digestive.

- ☆ Herb contains the alkaloids brahmine and herpestine. Brahmine which is a cardiotonic is toxic at high doses producing a fall in blood pressure.
- ☆ Brahmi is reputed as brain tonic, sharpening dull memory. It is very effective in anxiety neurosis. It is also safe cardiac stimulant.
- ☆ It is also useful in inflammations, epilepsy, insanity
- ☆ Juice of the herb is prescribed to children against bronchitis and diarrhea.
- ☆ In children stomouch troubles, brahmi is used along with ginger juice. A poultice made of boiled plant is applied to chest of children suffering from chronic cough.
- ☆ It dispels poisonous affections, splenic disorders and impurity of blood. Brahmi is used in Ayurvedic preparations like Brahmighrtam, Saraswatarishtam, Brahmitailam and Mirakasneham.

32

Balanities aegyptica (L.) Delite (Ingoriyo)

Botanical Name: *Balanities aegyptica* (L.) Delite

Family: Simarubiaceae

Local Names

- ☆ **Gujarati:** Ingoriyo
- ☆ **Hindi:** Hingor, Hingan, Hingot
- ☆ **English:** Balanities plant
- ☆ **Tamil:** Nanjunda
- ☆ **Telugu:** Gari

Introduction

Ingoriyo is spiny shrub or a small tree, with young tomentose or pubescent part. Yellowish green branches are seen in the ingoriyo plant. Spines are present. These spines are very sharp and strong. Leaves are bifoliate, petioles are short, leaflets are lanceolate and base is acute. Small white flowers are seen some times greenish white flowers are also seen. Inflorescence is short peduncle cymes or fascicles pubescent externally and they are silk inside. Fruit is 5 to 6 cm long, in a large drupe, 5- grooved, 1-celled and they are covered with a light grey to dry rind.

Distribution

In India it is cultivated in dry and arid areas. Particularly in Rajasthan, Punjab and Gujarat like states are known to grow ingoriyo. Although in most of the states Balanites plant is grown. In Gujarat ingoriyo is grown into Junagadh, Rajkot, Porbandar, Kutchh *etc.* like districts.

Cultivation

This plant suits black soils. It does not thrive in rocky areas. It withstands water logging for long periods. Generally this plant belongs to semi arid and arid region. This plant can also grow on alkaline and saline tracts in which no crop or plant grows. Generally ingoriyo is propagated by seeds. The plant produces tap root which is long. The plantation can be done by direct dibbing method or in the nursery they can also be raised. When seedlings are grown in nurseries, they can be transplanted in field in the next monsoon. Generally seeds are sown at a spacing of 4m x 2m. Pits are prepared for sowing the seeds. Farm yard manure and urea is also used for better growth of plants.

Harvesting

Harvesting can be done after 6 to 7 years of planting when fruits are ripen. Generally in summer, they are collected. Seeds are attached with fruits so heavily that it is very difficult to remove the seeds from the fruits. The fruits may be buried in the soil for one or two months or some times submerged under water for 15 days for easy removal of the seed.

Yield

From one hector of plantation 3000 kg of seeds can be expected. Seeds yield good oil which is used in soap making, Thus around 1200 kg of oil can be sold at a rate of Rs. 10/kg. After 9 to 10 year later net return will be around 8000 per hector.

Medicinal and Economic Importance

- ☆ The pulp of the fruits is useful in cleaning silk, cotton and hair.
- ☆ The seeds, fruits, bark and leaves are said to be anthelmintic and purgative.
- ☆ The woody fruits with seeds extracted are used as a kind of bomb shell in native fire works.
- ☆ The seeds are expectorant and colic.
- ☆ Juice from the bark is used as fish poison.

33

Baliospermum montanum (Wild.) (Danti)

Botanical Name: *Baliospermum montanum* (Wild.) Muell-Arg.

Family: Euphorbiaceae

Local Names

- ☆ **Gujarati:** Danti
- ☆ **Sanskrit:** Danti
- ☆ **Hindi:** Danti
- ☆ **Tamil:** Nakatanti
- ☆ **English:** Danti
- ☆ **Telugu:** Adaviamudamu, Kondaamudamu

Introduction

This plant is looked stout undershrub which obtains height of 0.9-1.8 meter with herbaceous branches from the roots. Leaves are simple, sinuate-toothed, upper ones are small, lower ones are large, sometimes palmately and 3-5 lobes. Flowers are numerous and are arranged in axillary racemes with male flowers above and a few females below. Fruits are capsules type and 8-13 mm long, obovoid, seeds are ellipsoid, smooth and mottled.

Distribution

This plant is distributed throughout in India. It is found throughout in India near water streams and riverside in its natural habitat. It occurs abundance in North

Bengal, Chhatishgadh, Bihar, Jharkhand, Nagpur, Gujarat, outer Himalayas from Kashmir to Bhutan.

Cultivation

Generally cultivation is done with the help of seeds and vegetative method. It thrives best in porous and well drainage soils (pH 5.5-7.0) however also tolerant to alkaline conditions. Sandy loam and black cotton soils are well suited for its successful growth. Field is ploughed 2 to 3 times and brought to a fine tilth. Farm yard manure at 25 tones/ha is applied and mixed well. Ridges and furrows are opened at 1.5 m apart and planted 1.2 m within the row. It can be propagated by seeds and terminal cuttings.

Seeds are sown in well prepared, raised nursery beds of 1 x 6 m in June-july. They are then covered with a light soil and leaf mould mixture and watered to keep the bed moist. About 5-6 kg of seeds is enough to raise seedlings to cover a one hectare area. About 75 percent germination is obtained. Nursery raised seedlings are ready for transplanting after 15-20 days. Seedlings are planted before onset of monsoon or they are directly sown in the main field. If the upper portion of the shoot is taken, inflorescence should be removed and only terminal bud is to be retained for satisfactory results. Crop needs irrigation once in three days during summer and weekly interval during winter season.

Harvesting

Crop becomes ready for harvesting in 6-7 months after planting. The crop is harvested by digging and uprooting the individual plants. Then roots are cleaned and dried.

Yield

On an average a yield of 15 tonnes/ha of dried roots can be obtained yielding 60-65 per cent of dry matter under proper management conditions.

Chemical Composition

Roots yield 5 new phorbol esters belonging to diterpene gydrocarbo, tigliane skeleton *viz.* mountain, balliospermin etc. Leaves contain β-D-glucoside and hexacosanol.

Medicinal and Economic Importance

- ✰ The roots are acrid, thermogenic, purgative, anti-inflammatory, anodyne, digestive, diuretic, diaphoretic, rubefacient, febrifuge and tonic.
- ✰ They are useful in anasarca, dropsy, flatulence, constipation, jaundice, haemorrhoids, leprosy, skin diseases, strangury, vesical cacculi, wounds, splenomegaly, anaemia, leucoderma, fever and vitiated conditions of vata.
- ✰ The leaves are good source of asthma and bronchitis.
- ✰ Seeds are drastic purgative, rubefacient, hydragogue and stimulant, and are useful in vitiated condition of vata.
- ✰ It is also applied in inflammation and flatulence.

34

Basella rubra Linn. (Palak)

Botanical Name: *Bassela rubra* Linn.

Synonym: . *Basella alba* L. Stewart

Family: Basellaceae

Local Names

- ☆ **Gujarati:** Palak ni Bhaji, Palak, Poi
- ☆ **Sanskrit:** Upodika
- ☆ **Hindi:** Poi, Lalbaclu
- ☆ **Tamil:** Vasalakkiral,
- ☆ **English:** Indian spinach
- ☆ **Telugu:** Baccali

Introduction

This is perennial herb which is succulent glabrous twining with white or red branches. Leaves are simple, alternate, broadly ovate, acute or acuminate, cordate at base, thick lamina which becomes narrowed into petiole. Flowers are white or red in spikes inflorescence. Bracteoles are longer than perianth. Fruits are red in colour, sometimes white or black, globose, utricle enclosed in the perianth.

Origin and Distribution

This plant is believed to be originated from East indies. It is distributed in India, Malasia and Phillipines, it is also found in tropical Africa, the Caribbean, and tropical

South America. Although it is cultivated plant but it is grown throughout in India. It grows well on all types of soil. Plant does not prefer high elevation.

Cultivation

This plant is cultivated by seeds or by stem cuttings or by root. They are planted at 50 x 20 cm in strips of 1.5 m apart. The plants are made to grow onb pergolas erected over the planted strips. The pergolas will be of 1.5 meter wide, with a distance of 1.5m (unplanted width) in between the two adjoin ing pergolas. They should be of permanent nature and erected preferably in east west direction. During non rainy season, crop is irrigated freely. Fertilizers are also applied liberally at monthly intervals along with irrigation. The leaves will be ready for first picking in 80-90 days. Later on, they can be picked up at 15 days interval in rainy and cold seasons and at 30 days interval in summer months.

Harvesting and Yield

Every year, around 12-15 pickings are carried out. A minimum production of Rs.5000 kg. of leaves can be expected per ha/year.

Chemical Constituents

It contains iodine, fluorine, carotenoids, organic acid, and vitamin-K.

Medicinal and Economic Importance

☆ Stems and leaves are sweet, cooling, emollient, aphrodisiac, laxative, haemostatic, appetizer, sedative, diuretic, demulcent, maturate and tonic.

- ☆ They are useful in vitiated conditions of pitta, burning sensation, constipation, flatulence, anorexia, haemorrhages, haemoptysis.
- ☆ It is also used in sleeplessness, pruritus, leprosy, urticaria, ulcers, dysentery, gonorrhea, balanitis, strangury, fatigue and general debility.
- ☆ They are especially useful as a laxative in children and pregnant woman.

35

Bauhinia racemosa Lamk. (Ashitro)

Botanical Name: *Bauhinina racemosa* Lamk.

Family: Cesalpinaceae

Local Names

- ☆ **Gujarati:** Ashitro, Ashatro
- ☆ **Hindi:** Kasundro
- ☆ **English:** Apta, Sona
- ☆ **Tamil:** Archi, Arka
- ☆ **Telugu:** Ari
- ☆ **Sanskrit:** Ashmantaka, Anupuspaka

Introduction

It is plant of dry and arid areas, this is tree has bluish brown color, rough bark with deep vertical cracks. Leaves are simple, broader than long, generally 2-6 cm long. They are leathery grey, pubescent beneath. Leaves are pubescent beneath, cordate and petiole 1 to 2 cm long. Flowers 8 to 13 cm long and arranged in long racemes inflorescence. They are arranged opposite to the leaves, small, white or yellow. Fruits are pod type, 12 to 25 cm long and glabrous; they are turgid and irregularly reticulate outside. In each pods there are 12 to 20 seeds present.

Cultivation

It generally grows on the dry lands well but grows well on clay, loam and sandy soils. Plant can be propagated from the seeds. Stump planting is also good

method. Germination capacity of the seeds is 60 to 90 percent. Seeds are pretreated by soaking the seeds in hot water taken off the fire, for 24 to 48 hours before sowing. In the pits dug one year old nursery raised seedlings are planted in the field with the onset of monsoon at 2 x 2 meter spacing.

In the first year 3 to 4 weedings are required, second year 2 to 3 weedings are required while after 4 to 5 years one to two weedings are required according to soil requirement. Proper fertilizers should also be given to the soil for obtaining good results. In the 10 to 12 years pollarding of trees is carried out to get good quality and quantity of leaves. Care is taken to see that sharp implements are used for pollarding and no heavy pollarding is done. Ploughing is done between two plants to conserve rain water.

Harvesting

Leaves are obtained twice, once in chaitra and other is Bhadarvi. Chaitri crop usually are heavy and superior, these leaves after plucking from the trees are tied into bundles and then they are dried.

Yield

The market rate of these leaves will be around 4 per kg. Therefore the gross revenue provides Rs.12, 800. Generally 10000 to 12000 Rs are needed for plantation and after 10 to 12 years they will be recovered.

Medicinal and Economic Importance

- ☆ The leaves are used for making for beedi wrappers.
- ☆ Bark is economically useful as it is used to make ropes of good quality, particularly inner bark is useful.
- ☆ The pods are eaten during famine.
- ☆ In South and Eastern Gujarat are this tree is grown specially to get good income.

36

Benincasa hispida (Thunb). Cogn. (Petha)

Botanical Name: *Benincasa hispida* (Thunb). Cogn.

Synonym: *Benincasa cerifera* Savi.

Family: Cucurbitaceae

Local Names

- ☆ **Gujarati:** Petha
- ☆ **Sanskrit:** Kusmandah
- ☆ **Hindi:** Petha, Raksa
- ☆ **Tamil:** Pusanikka
- ☆ **English:** Ash gourd, White gourd
- ☆ **Telugu:** Budidagummadi

Introduction

This is large trailing gourd climbing by means of tendrils. Leaves are large, hispid beneath. Flowers are yellow, unisexual, male peduncle 7.5-10 cm long, female peduncle is shorter. Fruits are broadly cylindrical and they are 30-45 cm long, hairy throughout, ultimately covered with a waxy bloom.

Origin and Distribution

This herb is cultivated throughout in India. Generally this plant is very common in hilly areas. In the world China, Iraq, Malaya, Singapore, Turkey are countries where plant is grown. Plant is believed to be originated in South Asia.

Cultivation

This plant is cultivated by seeds only. It needs tropical climate. It grows well from sea level to hill upto 1300 m. It comes up on all types of soil but sandy loams and loamy soils with good water holding capacity are the best. It is an annual and can be propagated by seed. The seeds are dibbled *in situ*..(2-3 seeds per pit) at the beginning of monsoon at a spacing of about 3m x 3m. The vines may be allowed to trail on the ground. There is no need to erect any trellies for this. Before sowing of seeds, pits are dug and farmyard manure is added to each while refilling the pit with soil. After the germination, 1 or 2 vigrous plants are retained per pit. The vine is supplimented with 10 gram of urea once or twice during July-August and August-September. The fruits will be ready for plucking in the months of October-November.

Harvesting and Yield

The fruits are collected without damaging the vine and marketed. An average yield of 10 fruits is obtained from each vine. Thus, a minimum yield of 5,000 fruits can be obtained fom one hectare.

Chemical Composition

It contains glucose, rhamnose, mannitol, n-triacontanol, lepeol and β-sitosterols are found in ash gourd. Seeds contain pure protein and arginine, histidine, lysine, tryptophan, phenylalanine, cystine.

Medicinal and Economic Importance

- The fruits are sweet, cooling, styptic, laxative, diuretic, tonic, aphrodisiac and antiperiodic.
- They are useful in asthma, cough, diabetes, haemoptysis, haemorrhages from internal organs, epilepsy, fever and vitiated conditions of pitta.
- The seeds are sweet, cooling and anthelmintic, and are useful in dry cough, fever, urethrorrhea syphilis, hyperdipsia and vitiated conditions of pitta.

37

Bixa orellana L. (Sinduri)

Botanical Name: *Bixa orellana* L.

Family: Bixaceae

Local Names

- ☆ **Gujarati:** Sinduri
- ☆ **Sanskrit:** Sinduri
- ☆ **Hindi:** Sinduriya, Latkan
- ☆ **Tamil:** Sappira viral, Uragumanja
- ☆ **English:** Annato, Lipstic tree
- ☆ **Telugu:** Jafracettu

Introduction

This is small handsome evergreen tree. Leaves are large, cordate, acuminate, glabrous on both surfaces. Flowers are white or pink and they are arranged in terminal panicles. Fruits are reddish brown or green capsules, clothed with soft bristles. Seeds are trigonous covered with a red pulp. The term "annatto" referes series of food quality colouring carotenoids produced from mature seeds.

Origin and Distribution

The plant is originated from tropical America, is widely distributed throughout in India. This plant is of tropical climate, but it can be grown very easily everywhere. It can not withstand frost. It grows well on variety of soils.

Cultivation

Generally this plant is cultivated by seeds only. But vegetative propagation has also been noted. The field to be planted is ploughed once or twice and later on staking is done at 2m x 2m spacing well in advance. Just before the onset of monsoon, seeds are collected from good trees are dibbed at every stake. The seeds start germinating within a period of 10 to 15 days. When seedlings are still small, the gaps due to failure of germination of seeds are filled up by taking out the excess seedlings elsewhere and transplanting carefully. Three weedings and soil workings are essential. The plants are fertilized at the rate of 40 grams per plant in two split doeses of 20 grams each, immediately after the 1st and 2nd weeding.

In the second year, casualties are replaced by planting branch cuttings. Two weedings and soil workings are carried out. Application of fertiliazers is carried out at the rate of 30 grams per plant in one dose, along with 1st soil working. In the third year, one weeding and soil working is carried out and one dose of fertilizer at the rate of 30 gram/per plant is given.

Harvesting

Though plants start flowering in the first and second year, it is advisable to pluck off the flowers to encourage vegetative growth for the better performance of the plantation in subsequent years. The spiny pod contains from 30-50 small seeds, which are surrounded by a scarlet aril. The aril yields the bright yellow Annato dye. The average annual yield of seeds per plant is one half kilogram in the third year and one kilogram in the subsequent years. The plant flowers at the end of August, and continues upto middle of October. The time taken by flowers to convert into fruits is about 90 days. The fruits should be harvested when a crack found in the stigma and of the capsule, otherwise the pods burst on the tree and result in the loss of seeds. After harvesting, seeds and pods are packed in gunny bags and kept closed for some days and then exposed to the sun. The seeds are taken out by beating with sticks. The seeds are sun dried and packed in gunny bags for marketing.

For extracting the dye, the seeds are bruised and the pulp macerated with hot water in wooden vats, and soaked in it for several days, till the colouring matter forms a fine suspension. The seeds are then removed, and the brei, which contains the pigment, is allowed to ferment for a week. The dye, annatto, that settles at the bottom is separated and dried into cakes.

Yield

The yield is 4.8-6 percent by weight of the seeds. Another method consists in boiling the seeds with sodium carbonate solution, is filter pressed, washed and dried. Bixin, when crystallied from glacial acetic acetic acid, is obtained in the form of rhombic needles.

Chemical Constitution

This plant contains important chemical bixin.

Diseases and Pests

For the most part no insect past or pathogens cause serious damage to Annato. One reason that insects generally do not cause extensive damage to this plant is that extra floral nectars attract ants, which keeps intruders at bay. The other reason is due to the sporadic and highly dispersed pattern of annatto plantings in backyards and gardens.

Medicinal and Economic Important

- ☆ The roots, bark and seeds are antiperiodic, antipyretic and astringent.
- ☆ They are useful in intermittent fevers and gonorrhoea.
- ☆ The pulp surrounding the seed is a mosquito repellent, and is useful to treat dysentery.
- ☆ A non-toxic dye, Annatto dye is obtained from the pulp is used for colouring edible materials.

38

Boerhavia diffusa Linn. (Shatodi)

Botanical Name: *Boerhevia diffusa* Linn.

Family: Nyctaginaceae

Local Names

- **Gujarati**: Shatodi
- **Hindi:** Gadahpurna, Sant, Santhi
- **English:** Horse-Purslane, Hog wood
- **Tamil:** Mukaratte, Mukkurattai
- **Telugu:** Atik, Atikamamidi
- **Sanskrit:** Punarnava, Sophagni

Introduction

This is annual herb which is grown through out in India. It also grows wild in arid and semi arid region. Stem is branched diffusely. Not so much thick stem but small branches some times long branches are produced. Leaves are oblong and ovate with thick arrangement present. Leaves are 1.5 to 3.5 cm long and 2 to 2.5 cm wide. Leaves are generally fleshy. Flowers are minute and small and subcapitate. Inflorescence is axillary and terminal penicles. Roots are tap roots.

Distribution

Genus *Boerhavia* consisting of 40 species of weeds, distriubuted in tropical and sub tropical regions. Six species are found in the India. The plant is known as rakta punarnava on account of its pink flowers.

Cultivation

Plant suits slight dumping places and also sown into sandy soil. In the Soil which is fertile or with clay or loam, this plant can grow quickly. Propagation is done by cutting or grafting method. Although seed propagation is also useful. Generally plant is propagated in the monsoon.

Harvesting

Harvesting is done when roots are dried. It is particularly done by hand pulling. Generally early summer is best time for harvesting. Roots are dried under sun and they are stored and taken for market.

Medicinal and Economic Importance

- Whole plant is used for medicinal purpose because Boerhevia contains punarnava which is useful drug. It has active principal of punarnavine.
- Drug possesses diuretic properties it is used in asthma and dropsy.
- Roots are very useful; hypoxanthine-9-L-arabinofuranoside has been isolated from the roots. Roots are known to cause anti inflammatory action.
- Sterols and ecdysterone has isolated from this plant, ecdysterone is insect moulting hormone.

- Roots have diuretic action. Roots boiled with milk are used in a single daily dose for maintaing good health.
- In dropsy and Chronic renal failure, Juice of leaves is used. Decoction of the plant is used to cure fluid retention, general weakness, anemia and viral hepatitis.

39

Bombax ceiba Linn. (Semlo)

Botanical Name: *Bombax ceiba* Linn.

Family: Bombacaceae

Local Names

- ✰ **Gujarati:** Semlo, Palas
- ✰ **Hindi:** Simul, Semur
- ✰ **English:** Red silk cotton, Flame of the forest
- ✰ **Tamil:** Kongu, Agigi
- ✰ **Telugu:** Adavi burgu, Burgu
- ✰ **Sanskrit:** Chirjivi, Apurani

Introduction

This tree is known as flame of the forest. Plant is annual large. Mainly it is found in the forest region but is also cultivated in the garden and other places as well. This tree contains beautiful flowers. Height of tree can be up to 25 meters some times but generally it gets height of 15 to 18 meters. Specific spines are present on to stem which is hollow. Leaves are large and its petiole is 2.5 to 3 cm long. Flowers are large red in color, white flowers are also found and generally they are produced before the leaves. Fruits are long capsule type which has 15 to 20 cm length. Fruits are silky. Flowers and fruits are eaten in India.

Distribution

Tree is distributed in the arid and semi arid regions of the world. In India, tree

is widely distributed in Andamans, ascending the hills upto 1500 m or even more. Tree is grown in Sri Lanka, Pakistan, Burma, Africa and other countries. In India Assam, West Bengal and Adaman are major areas to grow *Bombax ceiba*.In Gujarat plant can be seen in Girnar forest, Dang forest and some other areas.

Cultivation

For cultivation, three methods are there. (1). Direct sowing (2) Entire sowing and (3) Stump planting.

1. **Direct sowing:** Seeds are collected from middle of the March to middle of the May. There are 25, 300 to 38,500 seeds to a kilogramme. Sowing is done at the commencement of the rains in prepared mouns of loose, mineral soil at a spacing of 3.7m x 3.7m using 3-6 seeds per mound.
2. **Entire sowing:** Sapling for entire transplanting as well as stump planting are raised in the nursery, they are sown during May to June. Generally without any pretreatment at spacing of 5 cm in lines which are 23 cm

apart. Irrigation is given till monsoon break. Germination is completed within two months after sowing.

3. **Stump planting:** For stump planting, like entire planting stumps are prepared from one or two year old seedlings. Stump normally consists of 3.8 cm of short and c-30 cm. of root. Extra root as well as shoots is cut off and side roots are pruned. Bigger stumps like 60 cm give good results. Planting is done in crowbar holes or in pits of 30 cm x 30 cm x 30cm after the break of monsoon.

Harvesting and Yield

Plantation can be established in 3-5 years with good care and in moist regions are mature for final fellings on a rotation of 20-40 years depending upon the growth within 50 years total production including intermediate yield from thining reaches upto 407 cu.m/ha.

Storage

Storage of timber is done in open stacks for 4-5 months. Timber may be air seasoned, but kiln seasoning is preferable. Air seasoning process gives best results on green conversion followed by immersion of the sawn material in running water for six weeks.

Medicinal and Economic Uses

- ☆ Almost all the parts of the tree are used for different medicinal purposes like gum, seeds, leaves, fruits or its capsule, bark, cotton and flower.
- ☆ Plant contains mucilage, tannin cells and cells and resette crystals of calcium oxalate in the ground tissue of the root and stem. Gum is used for Vat diseases in ayurveda.
- ☆ Fruits are aphrodisiac and ground spike-paste is recommended for acne.
- ☆ Seeds provide non drying oil. Which is quite useful
- ☆ Soft cotton is obtained from the tree is used for filling pillows and mattresses.
- ☆ Gum is known as Supari ka phul and contains tannic and Gallic acid.

40

Bosewellia serrata L. (Shaledi)

Botanical Name: *Bosewellia serrata* L.

Family: Burseraceae

Local Names

- ☆ **Gujarati:** Shaledi
- ☆ **Hindi:** Salai
- ☆ **English:** Indian olibanum
- ☆ **Tamil:** Parangi sambrani
- ☆ **Telugu:** Anduga

Introduction

This tree is known for their fragrant resin which has many pharmacological usesparticularly as anti inflammatories. There are four main species of Boswellia serrata.The gardens depend on the time of harvesting and the resin is hand sorted for quality. Leaves are deciduous, alternate towards the tops of branches, one opposite, oblong, obtuse, serrated, pubescent, sometimes alternate. Very short petiole is present.

Flowers are pale rose or white in colour and arranged in single axillary racemes shorter corolla with five obovate-oblong, very patent petals, acute acstivation slightly imbricative. Stamens are ten in numbers. They are inserted under the disk, alternately shorter; Filaments are subulate and persistent. Anthers are caduceus, oblong and ovary is oblong and sessile. Style is one and caduceus, the length of the stamens. Stigma is capitate and three lobed. Fruit is capsular, three angled,

three celled, three valved, septicidal, valves are hard. Seeds are solitary in each cell surrounded by a broad membranaceous wing. Cotyledons intricately folded multifid.

Cultivation

This tree is of dry, rocky and plateau areas, often in gregarious form. It grows easily on arid and semi arid regions. It also grows on all types of soil, but in good drainage soil plant can grow quickly. It is propagated by vegetative means. The branch cuttings of 15-20 cm length and thickness of a finger size are used for preparing nursery. The cuttings are planted in polypots (20 X 30 X 300) filled with good soil mixed with farm yard manure in June. The cuttings sprout soon and develop root system. The plants are looked after properly by timely weeding and regular watering and addition of fertilizers etc. When plants attain height of 80-90 cm. in next June, when they are ready for transplanting in the field. They are planted in pits of 45 X45 X 45 cm at a spacing of 5m X 5m at the beginning of monsoon. They are kept free of weeds and the soil around the plants is kept loose. Generally they are not required any watering. In the beginning of rainy season, when they attain 90cm girth at breast height, which is achieved at the ege of 15-20 years, the trees will be ready for tapping.

Harvesting

The light tapping method is suggested which involves making a moderate number of blazes in the trees. In this method is trees with more than 90 cm girth are only selected for tapping. The initial blaze of 20cm wide and 30cm in length/height may be made at 15cm above ground level. The blaze may be made horizontally leaving approximately equal space between the blazes. The blaze should not have any loose fibre. In the length/height, about 1.6cm may be freshened every time towards the upper edge. Thus, at the end of season, the final size of the blaze in length/height will be 50cm. surface of the blaze may be slightly scraped each time taking care that depth is not increased.Each blaze yield 1 kg of oleoresin

For continuous tapping, the bole is divided into three zones, each zone being tapped for one year. Thus the tree zones are covered in 3 years. For making another horizontal row of blazes in the subsequent years, 7.5 cm space is left above the blazed portions. The blazes of the subsequent year should be alternating with the previous year's blazes. *i.e.* The previous and fresh blazes will not be in the same longitudinal row. The size, shape etc of the blaze before and after freshening, may be same as the first year. Similarly in the third year the same method will be followed. In the fourth year, blazing may be continued in the zone, where it was done in the first year. But, this time, the blazes may be between the previous ones. In 25th year, 400 trees will be available for tapping from one ha. The oleoresin is sold lat a rate of Rs.50/-per kg. The oleoresin, when freshly collected is stiky and on exposure to air, it slowly dries under shade only. When the drying process is over, the oleoresin assumes crystal form of different colours like green, blue, brown etc.

Yield

For the first 24 years total expenditure per year reaches upto Rs.30,000 and after that every year 5,000 Rs. are needed. Gross return reaches upto Rs.20,000 and net return per hectare comes about Rs.15,000 per hectare.

Medicinal and Economic Uses

- Tree exudes oleoresin on tapping and it is known as Salai guggal, Indian olibanum of commerce. Good quality light coloureed gum oleoresin is used to cure rheumatism.
- It is stimulant, but seldom used now internally, though formerly was in great repute. According to Pliny it is an antidote to hemlock.
- According to Avicenna it is good for tumours, ulcers, vomiting, dysentery and fevers.
- It is used for leprosy in china. It is also used in bronchitis and laryngitis.
- The ceremonial incense of the jews was compounded of four sweet scents of which pure Frankincense was one, pounded together in equal proportion.
- It is frequently mentioned in the Pentateuch. Pure Frankincense formed part of the meet offering and was also presented with the shew bread every Sabbath day.
- Boswellia has long been used in Ayurvedic medicine. Recently, the boswellic acids that are a component of the resin it produces have shown some promise as a treatment for asthma and various inflammatory conditions.
- In West Africa, the bark of *Boswellia dalzielli* is used to treat fever, rheumatism and gastro intestinal problems.

41

Borassus flabellifer Roxb. (Tad)

Botanical Name: *Borassus flabellifer* Roxb.

Family: Palmae

Local Names

- ✰ **Gujarati:** Tad, Tad mad, Div- Tad
- ✰ **Hindi:** Tad, Tal
- ✰ **English:** Palmyra palm, Brab Palm
- ✰ **Tamil:** Panai, Panam
- ✰ **Telugu:** Tadi, Tali
- ✰ **Sanskrit:** Taladrumah, Tade

Introduction

This is very tall palm with a crown of fan like leaves at the top, attaing a height of 25 to 30 meters with a trunk of 50 to 60 cm in diagram. The bark is black in color. Male and female flowers are found on to both male and female trees. Flowers are originated between leaves. Male flowers are pink in color and they are arranged in spike type inflorescence while female flowers are greenish in color and they are arranged in paniculate spikes, some times they are looked like small fruit. Fruits are large drupe type; they are 10 to 15 cm in diameter with an outer leathery covering. Fruit is covered with covering of calyx at the top.

Origin and Distribution

This plant is indigenous to India and it is found in most of the other areas of

world which are warmer. In India this palm tree can be seen in Gujarat, Maharastra, Tamil Nadu, West Bengal and southern part of India. In Gujarat Tad is seen in Junagdh, Jamanagar, Dang and some other parts. It is also grown in gardens, Hotels and some important places as attractive tree.

Cultivation

It can grow from sea to altitudes of 500 meters. It grows well in arid areas with good rain fall. The best method of propagation is direct dibbing of seeds. Seeds are dibbed at 2 to 2.5 meter distance in lines, which are spaced at 3 meter apart. Only fresh seeds are used. Normaly in 5 to 6 years tree gets its shape. Between two crops of this palm space is kept for the cultivation of other crops like vegetables, annuals *etc.*

Harvesting

Tree starts giving sap at the age of 12 to 13 years which remain continuous for next 40 to45 years. Male and female tree both the trees provide sap. In the male trees flowering shoots are bruised. They are scrapped with a tapping knife after two days and their tips are freshened everyday. After one week of bruising sap start oozing out. In female plants tapping done when nuts are small. From the male tree sap can be collected twice in a day and it generally gives 60 to 70 percent yield than that of female tree. Pots are cleaned, heated and smoked inside and then it is used.

Yield

After 12 to 13 years tree start giving yield. Tree gives 2 liters of average sap everyday which means that total yield per season is 200 liters. Excess sap is converted into jaggery. Tad jaggery gives 15 to 18 percent of the sap. Fibers and leaves are also obtained from the tree. Per annuum average 8 to 10 leaves are available. For almost 100 leaves we can earn 30 to 40 Rs. Sap is sold at the rate of Rs. 5 to 10 per litter in the market. After 16 to 17 years letter net return will be around 2, 50,000 Rs.

Diseases and Control

Pests Rhinoceros beetle- Beetles bore into unopened tender parts and spathes and throw the chew fibrous mass. This fibrous mass can be seen in the holes which are made by beetle at crown region. To control this disease beetles are hooked and are killed out. Spray of BHC (10 per cent) and sand in equal proportion is useful. Bud rot is another disease in which fungus attacks on to bud eats the growing point and ultimately kills the tree. For the control of the diseases we should cut and remove all the dead roots. These cut ends are drenched with 1.5 kg of aureofungisol and 1 g Copper sulphate in 5 lit. of water.

Medicinal and Economic Importance

- ☆ It is multy purpose tree in rural areas because whole plant is useful.
- ☆ The trunk is often used as pipe after removing the central core, by splitting into two halves it is used as a drain pipe for conveying water.
- ☆ Neera or toddy is obtained from the tree which is sweet sap, generally neera should be taken before sunrise other wise it becomes fermented or frothy.
- ☆ Tad sugar or Jaggery is obtained from the sap. Arrack is another product which is prepared.Jaggery made out of the juice is very useful for health.
- ☆ This is good product which is utilized in bronchitis.
- ☆ Fruits are useful because they have a laxative effect and the sap is good tonic. Pulp of the ripe fruits is used for pastry items and also eaten raw and also mixed with rice flour, baked and eaten.
- ☆ Radicle is full of starch and is a delicacy when baked or roasted.
- ☆ Dried male flowers, leaf stalks, dried unripe fruits and all other parts are a good source of fuel.
- ☆ Spathes are used for making brushes. The fibrous web at the base of young leaves is used for straining toddy and to make torches. This is also utilized on to cut wounds.

42

Butea monosperma (Khakhro)

Botanical Name: *Butea monosperma* Lam.

Family: Fabaceae

Local Names

- **Gujarati:** Khakhro
- **Hindi:** Dhak, Palas
- **English:** Flame of the forest
- **Tamil:** Parasu
- **Telugu:** Moduga, Tellamoduga

Introduction

Butea monosperma is medium sized tree. This tree is grown in all over India in Warmer regions. Flowers are orange- red in color and they are 4.5 cm long on leafless branches. Best flowering time is February to April. Leaves are trifoliate and leaflets are 7.5 to 40 cm wide and 3.5 to 60 cm long, leaves are glabrous, obovate, ovate rhomboid or elliptic oblong. Fruits are 7 to 20 cm long and oblong, sandy brown in color, fruits are one seeded, densely hairy which contain dark brown color. Seeds are rounded on edges. This tree is attached with so religious matters. This tree was described by Valmiki in Ramayana to express Ravana's body during war against Rama. Ved Vyas also took example of palas to describe war of Mahabharata.

Origin and Distribution

This medium sized tree is believed to be originated in India. Although it is distributed in most of the Asian countries like Burma, Pakistan, Srilanka, Nepal,

Bangladesh etc. In India this tree is distributed in most of the states like Gujarat, Maharastra, Jammu and Kashmir, Punjab, Himachal Pradesh, Tamil Nadu etc. most of the districts in Gujarat are known to grow Khakhro in garden, farm and other places. It can be seen in Rajkot, Jamanagar, Porbandar, Junagadh and in most of the area of Saurastra and Gujarat regions.

Cultivation

This tree can be grown in all types of the soil, but it excels in moist black and alluvial soils. It is drought resistant and frost hardy species, it is valuable for reclaiming saline soils and waste lands. Seed propagation is good method. Plantation can be raised both on irrigated and dry lands. Pods are collected and raised in lines which are 2 meter apart from one to other. Germination of the seeds takes place within one or two weeks. Germination capacity of the seed is 75 to 80 percent. Root suckers are also useful in vegetative propagation; there is need for fertilizer because this is very slow growing spices. The plant start to produce leaves on 5 to 6 years on large scale. Weeding and pruning are very useful for grooving plants. Stump planting has been found to be useful in *Butea*.

Harvesting

Harvesting done in February to March. Generally plants are cut to the ground level. The fully grown leaves are plucked with petiole and they are tied into 25 leaves each. They are then sun dried. Storage is done in dry places because dry leaves are to be used, they are washed with water to get flexible and soft leaves.

Yield

From the one hactre around 2, 00,000 leaves are obtained in a year; they can be sold at Rs.10 to 15 per 500 leaves that mean we can earn Rs. 3,000 per hactre as net return.

Medicinal and Economic Uses

- ☆ Whole plant is medicinally useful, the flowers and leaves of *Buotia monosperma* are astringent, depurative, diuretic and aphrodisiac which are used against pimples and boils.
- ☆ Gum is obtained from the bark of the tree which is useful in diarrhea and dysentery especially in children and weak women. Gum contains tannins.
- ☆ Decoction of the gum as a rectal enema would yield instant relief in pain.
- ☆ .Flowers and seeds are mixed in a decoction and used as wormicide against tapeworms and ringworms. Flowers give yellow dye.
- ☆ Leaves are useful in making dining plates; they are also used in beedi factories, making parcels *etc.* Decoction of leaves is used to cure Leucorrhoea.
- ☆ Bark and roots give specific fibers which are used to prepare ropes; bark can be utilized for tanning.
- ☆ Seeds are useful in the treatment of skin diseases like dhobi's itch, ringworm etc. Decoction of leaves obtained by boiling them in water, may be used as a mouth wash in the treatment of congested and septic throat.
- ☆ Tree is used to make the commercial cultivation of lac which stays as host on this tree. This tree is also useful as source of fuel; leaves are also useful in coppice management at wide spacing on heavily grazed grassland.

43

Caesalpinia bonduc Linn (Kant Karanj)

Botanical Name: *Caesalpinia bonduc* Linn.

Synonym: *Caesalpinia cristata*

Family: Caesalpinaceae

Local Names

- ✫ **Gujarati:** Kant karanj, Lata karanj, Kakasiyo
- ✫ **Hindi:** Kant karenj, Kantikaranja
- ✫ **English:** Fever nut
- ✫ **Tamil:** Kaliccikkai
- ✫ **Telugu:** Gaccakaya
- ✫ **Sanskrit:** Kantkikkarajah, Kuberakshi

Introduction

It is large straggling, throny shrub, branches are armed with hooks and straight hard yellow prickles. Leaves are bipinnate, large, stipules, foliaceous, pinnate 7 pairs, leaflets 3-8 pairs with 1-2 small recurved prickles between them on the underside. Flowers are yellow in color and in dense long peduncled supra-axillary racemes at the top. Fruits are inflated pods, covered with wiry prickles, seeds 1-2 per pod, oblong or globular, hard, gray with a smooth shiny surface.

Distribution

It grows through out in India, in the plains on waste lands and coastal areas. It is distributed in most of the states like Maharstra, Gujarat, Madhya Pradesh, Uttar

Pradesh, Punjab etc. In Gujarat *Caesalpinia bonduc* is grown in Junagadh, Porbandar, like distraction of Saurastra regions and whole state.

Cultivation

This is cultivated in soft, red black and with adequate water escaping drainage conditions. Generally every kind of land is suitable. Only the land where water remains stagnated is not suitable. Generally plant is propagated through seeds. These are sown around the field hedge in 4 to 5 rows. This will protect the field from stray animals in addition to giving extra income. In June-July after first rain seeds are sown at 0.5 x 0.5 meter distance in the trenches made along the hedge. In this way these can be sown in as many farmers desire. There are 400 to 500 seeds in per kg weight. No manner or chemical fertilizers are necessary for its cultivation. The plant belongs to caesalpiniaceae therefore, it improves the fertility of soil and enable in the nitrogen fixation from the environment into soil. Land is prepared by preparing drains and ridge in the field. It is not irrigated in the monsoon months. In summer, irrigation is required at 15 days intervals. Initially when plant is small

1 or 2, weeding and hoeing is needed to ensure that the weeds do not grow wild enough to hamper its growth. These after no need to do weeding or hoeing.

Harvesting

After two year when plants start giving some fruits, harvesting can be done. After 3 years each plant can give 4 -5 kg of seeds. If care is not properly taken then fruits come up in 40 to 50 years.

Yield

Good quality seeds can be sold at Rs. 35.00 to Rs. 50.00 per kg.

Medicinal and Economic Importance

- ✰ Root bark, leaves and seeds are used. The root bark is emmenagogue, febrifuge, expectorant, anthelmintic and stomachic.
- ✰ It is useful in amenorrhoea, dysmenorrhoea, fevers, cough, asthma, intestinal worms, colic, flatulence and dyspepsia.
- ✰ Seeds are bitter, astringent, acrid, thermogenic, anodyne, anti-inflammatory, anthelmintic, digestive, stomachic, liver tonic, depurative, expectorant, contraceptive, antipyretic, aphrodisiac.
- ✰ Leaves are anthelmintic, emmenagogue and febrifuge, and are useful in elephantiasis, intestinal worms, splenomegaly, hepatomegaly, amenorrhoea, dysmenorrhoea, fevers and pharyngodynia.
- ✰ This plant is also useful in vitiated conditions of tridosa, arthralgia, inflammations, hydrocele, cough, asthma, leucoderma, leprosy.
- ✰ Lata karanjh can also be used in skin diseases, dyspepsia, dysentery, colic, haemorrhoids, intestinal worms, hepatopathy, splenopathy, diabetes.
- ✰ The de-fated kernels contain α-, β-, dakua-, δ- and ε-caesalpins, caesalpin F and a homoisoflavone, bonducellin.

44

Caesalpinia sappan Linn. (Pataranjka)

Botanical Name: *Caesalpinia sappan* Linn.

Family: Caesalpiniaceae

Local Names

- ✰ **Gujarati:** Patang, Patarangika
- ✰ **Tamil:** Patungam
- ✰ **Hindi:** Patang, Bakam
- ✰ **Telugu:** Bakamu
- ✰ **English:** Sappan wood, Brazil-wood, Bukkum wood

Introduction

This is small tree which gets height of 5-7 m with hard wood, brownish-red. Stems are prickly. Young shoots are tomentose. Branches are glabrous covered with short spines. Leaves are alternate, pinnate leaflets trapeziform, glabrous above, tomentose beneath. Inflorescence is terminal raceme type. Flowers are yellow in colour and it is on the penducle with a ferruginous tomentum. Pod is compressed, with hard shell and sharp horn. Seeds are yellowish-brown in colour.

Origin and Distribution

This plant grows wild in forest areas. Perticularly in mountains or hilly areas this tree is very common. It is also cultivated in many places as a hedge plant.

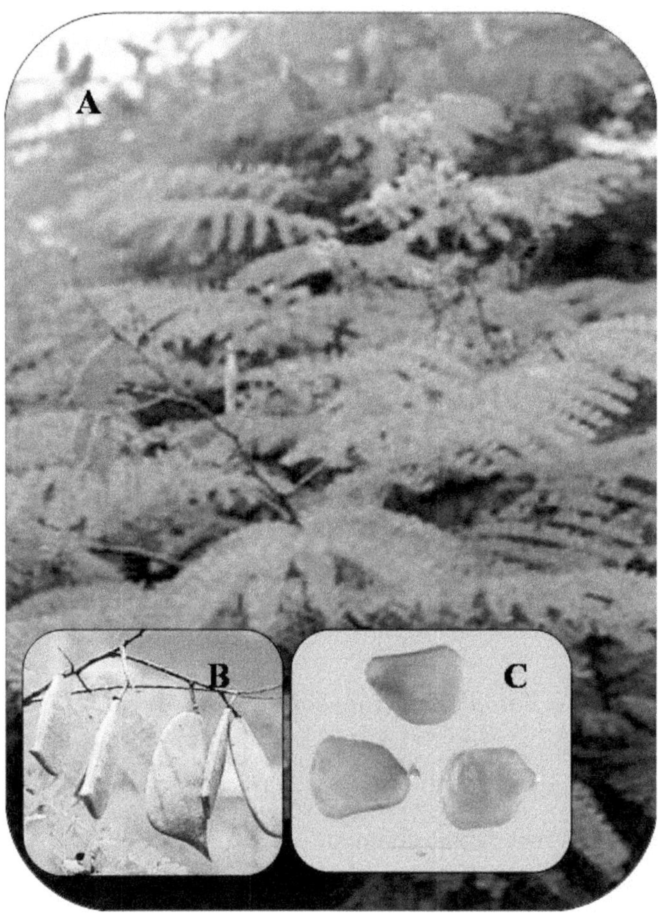

Cultivation

This tree is cultivated by seed propagation method only.

Chemical Composition

The woody part contains brazilin and brasilein and an essential oil consisting of D-a-phellandrene, ocimene, tannin gallic acid and saponin.

Medicinal and Economic Importance

- The trunk wood possesses antibacterial, demulcent and haemostatic properties.
- It is indicated for the treatment of dysentery, diarrhea, intestinal and uterine haemorrhages.
- It is also used in post-partum haematometra, contusions, wounds, dysmenorrhoea, colic furnuculosis, impetigo, leucorrhoea and anaemia.

45

Calophyllum inophyllum Linn. (Sultan Champa)

Botanical Name: *Calophyllum inophyllum* Linn.

Family: Clusiaceae

Local Names

- ✰ **Gujarati:** Undi, Sultan champa
- ✰ **Hindi:** Sultan chamapa
- ✰ **English:** Alexandrian Laurel
- ✰ **Sanskrit:** Surangi, Punaga
- ✰ **Tamil:** Punai
- ✰ **Telugu:** Namera

Introduction

It is moderate sized sub maritime, evergreen tree. It has smooth bark with grey, dark-brown or blackish colour. Leaves are simple and green which are shining on upper side. They are broadly elliptic, rounded and often notched at the apex with wavy margins and very close elliptic and lateral nerves, base is acute and petioles are 1-1.6 cm long, stout, flat. They are 10-18 x 7-10 cm is size.

Origin and Distribution

It grows wild in the coastal regions of south India, Andaman islands, Burma and Ceylon. It is distributed in most of the states in India like Karnataka, Maharastra, Gujarat, Punjab and others. In Gujarat this plant is grown in the Junagadh, Jamanagar, and Porbandar like areas of Saurastra and other regions.

Cultivation

It grows well in deep soil but often slightly blackish soil is suitable. It is littoral tree of tropics. It can be grown in all types of soils and climates. It is sensitive to frost and fire. Too much water logging is not useful.

One year old potted plants give results artificial regeneration. The fruits and seeds weigh 137 and 212 a kg respectively. The germination percentage is 60 per cent. (With seed coat) and 90 per cent (when seed coat is removed). Because of seed coat germination is poor. But when seed coat is ruptured germination occur in 30 days, generally in month of June. Seedlings attain 30 to 40 cm height in one year. After one year of germination plant is transplanted in to field. Pits are prepared of 30 x 30 x 30 cm. they are dug up in the field by the end of April, at a spacing of and m x 5m. Between two lines 7 meter distance is kept and 5 meter distance is kept between two plants. Soil workings and weeding is done in the first year and 2 weeding and soil working is done in the second year and soil working in the third year and subsequent years are carried out. Fertilizers are applied to the plant at the rate of 20 grams per plant in two split doses in first year and 30 grams per plant in two split doses in 2nd year while 30 grams per plant in single dose in 3rd years and later.

Harvesting

Harvesting in Sultan Champa can be started from 10 years. Plant start to give fruits. Harvesting is done by hand. Fruits are collected and stored in the storage. Seeds are obtained from the fruits.

Yield

Yield of seeds per tree per year is around 10 kg; 20 kg of fruits can be obtained from the 15 year and onwards. Expenditure is Rs. 3000/- per ha. Up to 10 th year and later on Rs. 1,000/- per annum, the gross return will be Rs. 7000/- per hectare in 10 th year and 20kg of fruits in 15th year and onwards. Karnels (43-52 per cent of the fruits) yield 50 to 73 per cent of a dark green viscous oil known by various names – Domba, Laurel Nut, Dillo, Pinnay or Poon seed oil.

Medicinal and Economic Importance

- Oil is obtained from the fruits which are known as poon seed oil. Both the extracted and expressed oils possess a disagreeable colour and teste.
- Oil is used as bio dieseal.
- Resinous substances are found in the oil which varies in quantity from 10 -13 per cent.
- Oil is useful in making soap. It is used as illuminant. Expressed oil contains 10-13 per cent resins it may be useful as varnish.
- Bark is also useful economically because it contains 11.9 per cent tannin. Pounded bark is applied in orchitis, and its juice is used as purgative.
- A decoction of it is applied as a lotion for indolent ulcers. A yellowish green aromatic resin possessing emetic and purgative properties, is obtained as an exudation from the bark.
- Leaves are useful as they contain saponin and hydrocyanic acid, and are poisonous to fish.
- The Cack would be probably suitable for manulel use.

46

Calotropis procera (L.) R.Br. (Akdo)

Botanical Name: *Calotropis procera* (L.) R.Br.

Family: Asclepiadaceae

Local Names

- ☆ **Gujarati:** Akdo
- ☆ **Sanskrit:** Arkah, Aditya, Mandare
- ☆ **Hindi:** Ark, Madar
- ☆ **Tamil:** Erukku
- ☆ **English:** Gigantic swallow wort, Madar milk weed
- ☆ **Telugu:** Jiledu

Introduction

This plant is large hard much-branched milky shrub, very pale in colour, the branches, leaves and inflorescence covered with loose soft white wool. Leaves are opposite, subsessile, ovate, cordate at base. Flowers are beautiful lilac, rosy or purple tinted in umbellate lateral cymes. Fruits are fleshy follicles, green. Seeds are with abundant with coma. Another species of the Calotropis genus is *Calotropis gigantia.*

Origin and Distribution

This is semi arid and arid area plant. This is found throughout in India. Particularly in dry places and weste areas this plant is very common.

Chemical Constituents

The dried whole plant is good tonic, expectorant, depurative and anthelmintic. The dried root bark is a substitute for ipecacuantha. The root bark is febrifuge, anthelmintic, depurative, expectorant and laxative. It is also useful in cutaneous diseases, intestinal worms, cough, ascites and anasarca. The powdered root promotes gastric secretions and is useful in asthma, bronchitis and intermittent fevers. The flowers are bitter, digestive, astringent, stomachic, anthelmintic and tonic. They are useful in asthma, catarrh, anorexia, inflammations and tumours. In large doses it is purgative and emetic.

Medicinal and Economic Importance

- ☆ The dried whole plant is good tonic. It is expectorant, depurative and anthelmatic.
- ☆ Dried root bark is a substitute for ipecacuanha, the root bark is febrifuge, anathelmatic, depurative, expectorant and laxative, and is useful in cutaneous diseases, intestinal worms, cough, ascites and anasarca.
- ☆ The powdered root promotes gastric secretions and is useful in asthma, bronchitis and dyspepsia.
- ☆ Leaves are useful in the treatment of paralysis, arthralgia, swellings and intermittent fevers.
- ☆ Flowers are bitter, digestive, astringent, stomachic, anthelmintic and tonic. They are also useful in asthma, catarrh, anorexia, inflammations and tumors.
- ☆ In large doses it is purgative and emetic.

47

Capparis decidua (Forsk.) Edgew. (Kerdo)

Botanical Name: *Capparis decidua* (Forsk.) Edgew.

Family: Capparidaceae

Local Names

- ☆ **Gujarati:** Kerdo
- ☆ **Tamil:** Senkam, Sirakkali
- ☆ **Hindi:** Karil, Kurel, Ker
- ☆ **Telugu:** Enugadanta
- ☆ **English:** Caper, Berry

Introduction

This is densely branching strangling glabrous shrub with smooth terete reen branches. Leaves are simple, caudaceus, found only on young shoots, linear-oblong, acute, spinous pointed, stipular thorns long, sharp, straight, orange yellow. Flowers are red in many flowered corymbs on old branches or short lateral shoots, gynophore about 12 mm long. Fruits are globular, glabrous, beaked. Seeds are numerous embedded in the pulp.

Distribution

This plant is distributed throughout India. This is dry and semi dry plant which is wildly grown in nature.

Chemical Constituents

It contains capparin, capparilin, and capparinin, l-stachydrine, capparidisine, capparisine, capparisinine, n-pentacosane, n-triacontanol and β-sitosterol.

Medicinal and Economic Importance

- The roots are acrid, bitter, thermogenic and anodyne, sudorific, expectorant, digestive, carminative, anthelmintic, purgative, antibacterial, stimulant, emmenagogue, aphrodisiac and tonic.
- They are useful in boils, eruptions, swelling, chronic and foul ulcers, cough, hiccough, asthma, vomiting, haemorrhoids, intermittent fevers.
- It is useful in arthritis, odontalgia, lumbago, dysmenorrhoea and general debility.
- The fruits are bitter, sweet, astringent, acrid, thermogenic and constipating.
- They are useful in halitosis, cardiac dis orders, urethrorrhea and vitiated conditions of pitta.

48

Cardiospermum halicacabum L. (Kag Doliyo)

Botanical Name: *Cardiospermum halicacabum* L.

Family: Sapindaceae

Local Names

- ☆ **Gujarati:** Kag dolio
- ☆ **English:** Ballon vine, Heart pea
- ☆ **Hindi:** Kan phuti, Kapalphoti
- ☆ **Tamil:** Mudukkottan, Modikkottan
- ☆ **Sanskrit:** Sakralata, Indravalli, Kaphamadinka
- ☆ **Telugu:** Vekkudutiga.

Introduction

A pubescent or nearly glabrous annual or perennial with slender branches climbing by means of tendrillar hooks. Leaves are ternately bicompound, leaflets acuminte at the angles. Seeds are black with a large white heart-shaped aril.

Distribution

This plant is wildly grown in the whole in arid and semi arid regions of India. This is distributed throughout in India.

Chemical Constituents

It contains alkaloid, 1-triacontanol, and n-pentacosane, n-triacontane, n-triacontanol, n-triacontanol, n-non-acosane, pelargonidin-3-galactoside, glucocappasalin, β-sitosterol and phthalic acid.

Medicinal and Economic Importance

- The roots are diuretic, diaphoretic, emetic, mucilaginous, laxative and emmenagogue.
- They are useful in strangury, fever, arthritis, amenorrhoea, lumbago and neuropathy.
- The leaves are rubefacient and are good for arthritis, otalgia and ophthalmodynia.
- The seeds are tonic and diaphoretic and are good for arthritis and fever.
- The plant has sedative action on the central nervous system.

49

Cassia alata Linn. (Kruminasak Senna)

Botanical Name: *Cassia alata* Linn.

Family: Caesalpiniaceae

Local Names

- ✰ **Gujarati:** Kruminasak senna
- ✰ **Hindi:** Senna
- ✰ **English:** Ring worm senna, winged senna, candelabra bush, craw-craw plant.

Introduction

This is small shrub, about 1.5m high, with horizontal branches. Leaves are paripinnate, alternate. Leaflets are 8-12 pairs and broadly rounded, oblique at the base. Twings and petioles usually reddish brown in colour. Inflorescence is axillary and terminal erect spike. Flowers are yellow in colour. Pods are long, slightly compressed, with winged margin. Seeds are numerous and black in colour.

Distribution

This plant is grown naturally in wet places and is also cultivated for its ornamental foliage and showy flowers. This plant is grown throughout in India.

Chemical Constituents

The leaves contain anthraglucosides, chrysophanic acid and rhein.

Medicinal and Economic Importance

- The leaves and the stems have antiseptic and laxative properties.
- They are prescived for constipation, oedema, hepatitis and icterus in a tea like infusion.
- Dermatomycosis, tinea imbricate, ringworm, scabies and impetigo are treated externally by rubbing with pounded fresh leaves or by applying fresh leaf juice on the diseases parts.

50

Cassia angustifolia Vahl. (Sona-mukhi)

Botanical Name: *Cassia angustifolia* Vahl.

Family: Caesalpiniaceae

Local Names

- ☆ **Gujarati:** Sonamukhi, Aval
- ☆ **Sanskrit:** Swarnpatri, Bhumiari, Bhupadma
- ☆ **Hindi:** Sonamukhi
- ☆ **Tamil:** Nila virai
- ☆ **English:** Senna

Introduction

It is a small shrub which gets height of 61-91 cm. Leaves are paripinnate. Leaflets are in 7-8 pairs, glabrous. They are yellowish green and 2.5-5.1 cm X 0.4 – 1.3 cm. Flowers are yellow in colour. Pods are greenish-brown to dark-brown in colour. They are 3.4 to 6.8 cm long and 1.9 cm broad. Seeds are 5-7, obovate and smooth, dark brown in colour.

Distribution

This is grown in most of the states in India. Particularly it is cultivated in South India, Rajasthan and Gujarat. West Germany is the largest importer of senna from India. Other major importers in order of importance are Hungary, Japan, U.K., U.S.A., Netherlands, France and Czechoslovakia. In addition to senna, there is a demand for calcium sennoside of high purity in these countries. In India, there are

very few units manufacturing calcium sennoside of 60 percent purity, while the rest are producing material of 16-20 percent purity only.

Cultivation

It can be cultivated as both rain fed as well as irrigated crops, however, very heavy rainfall, cool climate and dew formation during the growing season are not suitable for normal growth and development of plants. It can be grown as monsoon crop successfully on soils with good drainage. In Tamil Nadu, it is cultivated in heavy soils after paddy. Water stagnation should not take place in the plantation, especially during the early stages of crop. This plant is propagated by seed. The direct sowing of seeds in the soil gives poor germination. The best way to overcome this problem is to soak the seeds in water for about 10-12 hours and then used only swollen seed for germination. This procedure will ensure not only nearly 100 percent plant population. About 25 kg seeds are required for 1 hactare under rainfed conditions and 15 kg/ha. under irrigated conditions. After onset of monsoon, the land is ploughed and the field is cleaned from the weeds. The ridges are formed at a distance of 45 cm after applying fertilizers at the rate of 25 kg of nitrogen per hactare (Urea) and 25 kg of phosphorous (superphosphate). The beds of 3.15m x 6.0 m with one meter wide channels are suitable for irrigation and also for drainage if required.

The crop grows slowly during the early stages. It is important to ensure that water logging does not occur, especially during early growth. Two weedings, one after about 20 days and another 30 days after the first, are essential. After the first weeding 25 kg Nitrogen per hectare in the form of Urea is applied. The crop is hardy and requires at the most 3-4 irrigations, if the rain is normal. In absence of sufficient moisture at the time of addition of fertilizers, the field is irrigated.

Harvesting

Harvesting of leaflets is carried out by stripping the leaves and that of pods by hand picking. Three strippings of the leaves are possible at an interval of minimum 20 days. Maximum yield of leaflets could be obtained by initiating stripping on day 90th and subsequently on 110th and 130th day. Similarly, when the strippings are delayed, the yield is more, but the quality is poor. An average yield of 1500 kg per hectare of dry leaves and about 750 kg per hactare of pods can be obtained.

Good quality final produce is green in colour. The method of drying affects the quality of leaflets. The leaflets dried in an oven at 40°C is found to be superior to air dried leaflets. Sennoside content also decreases when leaflets are dried in direct sunlight. Quick drying method is most suitable. For the bulk drying, the harvested leaflests are immediately spread in thin layers in well ventilated, shaded area and the stuff is stirred atleast 3 times per day for quick and uniform drying. It takes about 4-5 days for the drying of leaflets. The leaflets when baled, as is done for cotton, will reduce transportation cost.

Chemical Constituents

The pods contain sennosides A and B, sennidin-8.8-diglucoside, glycosides of rhein and chrysophanic acid, aloe-emodin, its dianthrone diglucoside and emodin glucoside. Occurrence of oxymethyl-anthraquinone has been reported from the fruits. Leaves contain flavanols, isorhamnetin, kaempreroI, rhein, emodin and sennosides A, B, C and D.

Medicinal and Economic Importance

- ☆ Plant is laxative and purgative.
- ☆ Plant is used in constipation, loss of appetite, hepatomegaly, spleenomegaly,
- ☆ It is also applied in digestion, malaria, skin diseases, jaundice and anaemia.
- ☆ Because of its beautiful flowers it is cultivated especially for flowers also.

51

Cassia auriculata L. (Tanner's Cassia)

Botanical Name: *Cassia auriculata* L.

Family: Caesalpiniaceae

Local Names

- ☆ **Gujarati:** Avarttak, Tarval
- ☆ **Sanskrit:** Avarttaki, Hemapushpam
- ☆ **Hindi:** Tarval, Anwal
- ☆ **Tamil:** Avaral, Avirai
- ☆ **English:** Tanner's cassia, Tanner's senna
- ☆ **Telugu:** Tangedu

Introduction

This is small branched shrub with reddish brown branches. Leaves are with subculate glands between all the 8-12 pairs of leaflets and a pair of large obliquely cordate stipules at their bases. Flowers are bright yellow in colour and they are arranged in sub-terminal axillary corymbs. Fruits are pods type, flat and thin; papery they are pale brown in colour. Fruits are deeply depressed between the seeds, transversely veined. Seeds are 10-20 per pod, obovate, dark brown with hard shiny seed coat.

Distribution

Senna is native of India, Myanmar and Srilanka. Although it has successfully introduced into African countries. The plant is distributed throughout central,

Western and south India. This is arid and semi arid plant which requires loamy soil.

Cultivation

Cassia auriculata can be propagated by seed and stem cuttings. For quick germination seeds are scarified and held in running water. The seedlings are fairly resistant to desiccation. Stem cuttings are planted 5–12.5 cm apart in rows. Thinning is necessary one year after sowing. Weeding and cultivation stimulate growth but are not absolutely necessary. Limed soil is reported to increase the amount of tannin. Coppiced plants regrow well.

Harvesting

Starting the third year after establishment, the twig bark of *Cassia auriculata* can be stripped and used. Twigs that have not developed a corky bark are best. Coppiced plants can be harvested annually.

Handling after Harvest

The bark is sun dried in small pieces and stored or marketed. Unstripped twigs can be directly used by the tanners to make a tanning extract which is as effective as when made from dried stripped bark. To prepare a yellow dye, flowers (about twice the weight of the textile to be dyed) are boiled in water. Then the cloth, previously mordanted with alum, is immersed into the bath, which is kept boiling until the desired shade is obtained. In Andhra Pradesh (India), 2.5 kg of broken and cooked *Senna auriculata* seeds are added to a vat of indigo of approximately 227 l with 750 g of indigo.

Yield

The yield of green bark of *Senna auriculata* averages 1500 kg per ha in a 4- year-old plantation of about 9000 plants/ha.

Chemical Constituents

Emodin, chrysophanol, rubiadin are found in the Tanner's senna.

Medicinal and Economic Importance

- ✰ Roots are astringent, cooling, alterative, depurative and alexeteric and are useful in skin diseases, leprosy, tumours, asthma and urethrorrhoea.
- ✰ Bark is astringent and alternative and a decoction of this used as enemas and gargles.
- ✰ The leaves are depurative and anthelminticm and are recommended for leprosy, skin diseases and ulcers.
- ✰ The flowers are used in diabetes, urethrorrhoea, nocturnal emissions and pharyngopathy.
- ✰ The seeds are astringent. Sour, cooling, constipating, depurative, aphrodisiac, anthelmintic, stomachic and alexeteric and are useful in diabetes, chyluria, ophthalmia, dysentyery, diarrhea, swellings, abdominal disorders, leprosy, skin diseases, worm infestations and chronic purulent conjunctivitis.

52

Cassia fistula Linn. (Garmalo)

Botanical Name: *Cassia fistula* Linn.

Family: Cesalpinaceae

Local Names

- ☆ **Gujarati:** Garmalo
- ☆ **Hindi:** Amaltus, cirimalah
- ☆ **English:** Amaltus, Indium labernum
- ☆ **Tamil:** Aragvadamu, Kopagona
- ☆ **Telugu:** Rela
- ☆ **Sanskrit:** Rajataru, Kritamalah

Introduction

Medium sized tree *Cassia fistula* is deciduous tree. This tree flowers in April and May. At the time of flowering tree is completely nude by shedding its leaves Process of flowering remains continuous till fresh leaves are born. Bark is greenish white which remain smooth up to middle age but turn into rough and brown in old age. Leaves are large, paripinnate type, they are 0.25 to 0.5 meter long and terete rachis are seen. On the upper side copper colour is found while the inner side is soft browny. Flowers are bright yellow to golden yellow in color and inflorescence is simple raceme type. Bud opens towards tip of inflorescence. Generally flowers open in acropetal manner. Fruit is dark, cylindrical, woody and 30 to 50 cm long which contains 40 to 100 seeds. Seeds are brown colored and separated from one another by septa.

Origin and Distribution

Garmalo is common tree in India and its native is believed to be India. Although it is found in Myanmar, Thai land, Vietnam, Java and Philippines like other countries. In India tree is grown into deciduous forest of hilly and plain regions like in Siwalik region of Himalayas and other parts of India. In Gujarat most of the districts are known to grow this tree like Junagadh, Kutch, Jamanagar, Baroda, Dang *etc.*

Cultivation

Plant refers well drained soils. It can survive drought condition. Natural regeneration of plant is occurred by root suckers. Before sowing seeds are immersed into hot water for 5 minutes for breaking dormancy. Seeds are first sown into beds. The seedlings are transplanted to polybags when they are 3 to 4 months old. After one year tap roots grows very quickly so pulking of seed should be done very quickly. When plant gets height of 30 to 50 cm, plant is sown into field. Pits are prepared in the fields at the distance of 3 x 3 m. Seedlings should be regularly weeded till it gets height of weeds.

Medicinal and Economic Importance

- ☆ Most parts of the trees are useful medicinally like leaves, stem, fruits *etc.*
- ☆ Pulp of Cassia fistula is very useful as laxative.
- ☆ In West Bangal it is used to flavour tobacco.
- ☆ Bark of the tree is utilized for tannin extraction which is used in leprosy, jaundice, syphilis and cardiac ailments
- ☆ Bark is also used in hides and dyeing of leather and jute fibers.
- ☆ Its heart wood is brick red or yellowish in color which is used for making carts, agricultural implements, posts and beams of buildings.
- ☆ Pulp of the ripe fruits is a strong purgative.
- ☆ The gum is also extracted from the stem; this gum is used as astringent.
- ☆ Twings are useful to cattles for fodder.
- ☆ Flowers are used as vegetables and food by *Santal tribals* of West Bangal.
- ☆ Cassia pulp is obtained by extraction of fruit pulp by percolation with water and evaporation of the strained percolate under reduced pressure to the consistence of a soft paste, is official pharmacopoeia. It is used as an ingredient of confection of senna.

53

Cassia occidentalis L. (Kasundro)

Botanical Name: *Cassia occidentalis* L.

Family: Caesalpiniaceae

Local Names

- ☆ **Gujarati:** Kasaumdi, kashundro
- ☆ **Sanskrit:** Kasamardah
- ☆ **Hindi:** Kasaumdi, Barikasaumdi
- ☆ **Tamil:** Ponnavirai, Peravirai
- ☆ **English:** Negro coffee, Stinking weed, Coffee-senna
- ☆ **Telugu:** Kasinda

Introduction

This is diffuse offensively odorous undershrub with furrowed subglabrous branches. Leaflets are 3-5, flowers are yellow in colour. They are arranged on short peduncled few flowered racemes. Fruits are cylindrical type or compossed, hard, smooth and shiny dark olive-green or pale brown.

Distribution

The plant is originated from tropical America. Plant grows abundantly on wastelands immediately after the rains and found throughout in India particularly in the areas of arid and semi arid regions. The plant is of tropical and subtropical climates. It grows well on variety of soils including poor and eroded soils. It comes

up in areas with a rainfall ranging from 400mm to 2000mm. It thrives well even in areas with heavy biotic pressures.

Cultivation

This plant is propagated through seeds. The area is ploughed and after obtaining fine tilth, furrows are made at one metre apart. This work is completed by the end of May. The seeds are sown in the furrows continuously in the month of June on the onset of monsoon. Fifteen to twenty kg of seeds is sufficient to cover 1 hectare area. One weeding and hoeing are carried out in July. Extra weedings from the congested patches may be uprooted, so that the remaining seedlings are spaced at 45-60 cm apart. The plants start flowering in September and the pods may be collected from December-January.

Harvesting and Yield

Pods are collected when they become matured. They are then dried, threshed to separate the seeds. Yield of 2000 kg, seed can be expected from one hectare.

Chemical Constituents

It contains chrysophanol, emodin, glycosides, physcion, metteucinol-7-rhamnoside, jaceidin-7-rhamnoside, and 4,4,5-tetrahydroxy -2,2-dimethyl1-1,1-bianthraquinone, crude protein, ether extra, crude fibre, ash, calcium phosphorus, iron, niacin, and ascorbic acid, β-sitosterol, emodin, physcion.

Medicinal and Economic Importance

- The plant is bitter, sweet, thermogenic, purgative, expectorant and febrifuge, and is useful in vitiated condition of vata and kapha, cough, bronchitis, constipation, fever, epilepsy and convulsions.
- The roots are bitter, acrid, thermogenic, diuretic, anti inflammatory, digestive, stomachic, and tonic.
- They are useful in vitiated condition of vata, inflammation, diabetes, strangury, elephantiasis, ring worm, colic flatulence, dyspepsia, epilepsy, convulsions and scorption sting.
- The leaves are bitter, acrid, sweet, thermogenic, depurative, vulnerary, anodyne, expectorant, alexeteric and aphrodisiac, and are useful in vitiated condition of *vata* and *kapha*, leprosy, erysipelas, pruritus, wounds and ulcers.
- This plant is also used in bronchitis, high cough, asthma, pharyngodynia, fever and hydrophobia.
- The seeds are bitter, sweet, acrid, depurative, diuretic, expectorant, vulnerary, purgative, stomachic and febrifuge and are useful in leprosy, erysipelas, ulcers, strangury, cough, bronchitis, highcough, constipation, flatulence, dyspepsia and fever.

54

Cassia tora Linn. (Kuvadiyo)

Botanical Name: *Cassia tora* Linn.

Family: Caesalpinaceae

Local Names

- ✰ **Gujarati:** Kuvadiyo
- ✰ **Hindi:** Chakunda, Panevar
- ✰ **English:** Sickle senna, Sickle pod, coffee weed, Tavara
- ✰ **Tamil:** Tagarai
- ✰ **Telugu:** Tantemu
- ✰ **Sanskrit:** Ayudham, Dadamari

Introduction

Kuvadiyo is very common in the midlands and mountains. It is annual shrubby weed which gets height of 50 to 90 cm. Stem is erect and thin. Leaves are pinnate and alternate. There are 3 pairs of obovate leaflets in *Cassia tora*. Flowers are yellow in color; they are 1 to 3 in numbers, some times four flowers are also seen. An arrangement of flower is axillary raceme which is generally shorter than leaf. Pod is slender and very long, curved pod are also seen. Seeds are many in numbers; they are shining dark brown in color.

Origin and Distribution

There is no certain clue regarding the origin of Cassia tora but it is distributed in most of the parts. It can be seen in India, Pakistan, Myanmar, Sri Lanka, Japan,

Korea and some other parts of the world. In most of the states of India Cassia tora is wildly seen on both the sides of road. It is found in Gujarat, Maharastra, Tamil Nadu, Karnataka and most of the other states. In Gujarat it is wildly seen in Saurastra region particularly in mountainous area like Girnar Hills and Barda hills. It is also seen in Jamnagar, porbandar, Rajkot and other areas of Gujarat.

Cultivation

Plants are propagated by seed propagation method. Seeds are sown either in pod or directly into soil, it can grow on any kind of soil but semi arid land with good rain is suitable to the *Cassia tora* plant. Seeds are sown at 0.50 meter distance in lines. Although there is no specific farming method for cultivation of Kuvadiya.

Harvesting

Hand pulling is the best method for harvesting the plant. After two to three months when plant starts to give pods at that time harvesting is done. Generally harvesting is done in September to October. Plants are dried in sun light and then they are tied together. Then they are taken for storage.

Yield

It is estimated that 1500 kg. of seeds can be obtained from 1 ha. of area. The seeds can be sold in the market at the rate of Rs.6/kg. Expenditure per hectare reaches upto Rs.3,000 while gross return goes upto Rs. 9,000 and net return will be Rs.6,000 per ha.

Medicinal and Economic Importance

- Whole plant is medicinally useful. Whole plant contains anthraglucosides which on hydrolysis give emodin and glucose, chrysophanos and rhein.
- Seeds are used in treating insomnia, headache, constipation, oliguria, cough, ophthalmia, dacryoliths, ambylopia, ocular congestion and hypertension.
- Raw seeds are utilized as laxatives.
- Fresh leaves are used in eczema dermatomycosis.
- Seeds also give fatty oil consisting of oleic, linolic, palmitic and lignoceric acids and sitosterols.

55

Casuarina equisetifolia L. (Sharu)

Botanical Name: *Casuarina equisitifolia* L.

Family: Casuarinaceae

Local Names

- ✰ **Gujarati:** Saru
- ✰ **Tamil:** Savukku,
- ✰ **Hindi:** Jangali saru
- ✰ **Telugu:** Sarugudu, Chavuku
- ✰ **English:** Iron wood, she – oak, beefwood

Introduction

This plant is tree which gets height of 25 meters. Branching are drooping and they are needle like. Leaves are highly reduced and scale like giving the branchlets a pine needle like appearance. Flowers are anemophilous (wind pollinated), in which male flowers are borne in spike inflorescence. Where each flower is relatively inconspicuous. Female flowers are borne in globose heads. Fruit is globose woody aggregate type which resembles with a small pine cone. It encloses many small winged nuts.

Distribution

This plant is very common along the coast on beaches, rocky coasts, limestone outcroppings, dry hillsides and open forests in both wet and dry zones from sea level to mid montane.

Cultivation

Casurina is well adapted to sandy soils of semi arid and coastal humid climate.

It can grow from sea level up to 1300-1500 m elevation. Tree can tolerate frost, drought, water logging conditions. It grows well under rainfall varying 900 to 4000m. It can be grown by seeds, also it can be propagated by stem cutting and air layering. In the month of January-March seeds become ready for collection. They are very small in size. Seeds are sown in nursery beds. Adequate amount of sand should be added in nursery soil. By broadcast method seeds are sown, and then covered with ½ cm layer of soils. They are also sown in polythin bags. Pricking is done when seedlings obtain proper height. They are taken transffered to the main field. Seedlings are planted at the distance of 10 x 10 x 10 cm or 15 x 10 cm. Generally monsoon planting gives good result.

Harvesting

The final harvesting of crop depends on the purpose and the situation for which casuarinas was planted. In general it is grown in the rotation of 6 to 8 years. When planting of casuarinas is done on farm boundary or grown as windbreak and shelterbelts, trees are not felled up to 15 to 20 years of age. But time to time these are lopped and pruned so that shading effects on arable crops are avoided.

Chemical Composition

This plant contains ellagic acid, beta sitosterol, kaempferol and glaycosides, quercetin, cupressuflvone, isoquerctrin, several common triterpenoids, trifolin, catechin and epicatechin, cholesterol, stigmasterol, campesterol, cholest-5-en-3 beta-ol derivatives, tannin, proantho-cyanidins, juglanin, citrulline and amino acids, afzelin, casuarine, gallicin, catechol derivatives, gentisic acid, hydroquinone, nictoflorine, rutin, trifolin.

Medicinal and Economic Importance

- The plant is used to treat nervous disorders, diarrhea and gonorrhoea.
- This plant is used to treat coughs, ulcers, stomachaches and constipation.
- Bark of the tree is used in dysuria and menorrhagia.
- An infusion of the bark is used to treat throat infections, coughs and stomachaches. Bark infusion is good remedy for coughs, asthma and diabetes.
- Use of infusion of the grated bark is given in the mouth infections and urinary tract infection.

56

Catharanthus roseus (L.) G. Don (Bar Mashi)

Botanical Name: *Catharanthus roseus* (L.) G. Don

Family: Apocynaceae

Synonyms: *Lochnera rosea* (Linn.)Reichb; *Vinca rosea* Linn

Local Names

- ☆ **Gujarati:** Bar mashi
- ☆ **Hindi:** Sadabahar
- ☆ **English:** Madagascar perwincle, rosy flowered,
- ☆ **Tamil:** Sudukadu
- ☆ **Telugu:** Mallikai
- ☆ **Sanskrit:** Nitya Kalyani

Introduction

It is common plant in coastal areas and it is also cultivated. Plant is perennial herb type which gets height of 30-80 cm. High stems are pinkish red, much branched. Leaves are opposite, obovate, glabrous on sides, dark shining above flowers pink or white in the axil of the leaves. Follicle cylindrical, narrow, slightly arched-recurved in pairs; seeds are numerous, tiny, blackish-brown.

Distribution

This plant is native of Madagaskar and now it is introduced throughout the tropics of both hemispheres. This plant is distributed in most of the countries of Asia

like India, Pakistan, Japan, Sri Lanka, Myanmar etc. In India this plant is distributed in most of the states like Maharastra, Gujarat, Madhya Pradesh, Tamil Nadu etc.

Cultivation

Plant grows easily in marginal lands. The areas where annual rainfall is about 100 cm are suitable for its cultivation. Soil may be ploughed 2 to 3 times and cow dung or compost or 250 kg of bone powder or rock phosphate in a hectare of land. Seed propagation is the best method. During June July seeds are sown in nursery beds. Sprouting occurs in 8 to 10 days time. Such prepared plants are transplanted in the field in rows, keeping 45 cm distance between row and 30 cm between plant to plant. For one hectare of land 5 to 6 kg seeds are required. Irrigation should be provided after rainy season at 20 to 25 days intervals. From 60 days of sowing, weeds are removed by weeding and hoeing practices.

Harvesting

Leaves are harvested twice in a year. After 6 months of sowing day and second afte 9 months and finally dig out the roots after 12 months. These roots and leaves are dried under sun light. On maturity of crop the beans are plucked and dried under sun. There after seeds are taken out and stored.

Yield

From one hectare of land, yield of dry leaves, stem and roots can be 30, 10 and 8 quintals. The market values of dry roots, leaves and stem are Rs. 50 to 60, Rs. 25 to 100 per kg, respectively.

Diseases and Protection

Sometimes die backm patimodak, phugarium, murgan diseases affect the crop. These can be controlled by spray of dithian Z-78 at 10 to 15 intervals.

Medicinal and Economic Importance

- ☆ Mainly roots and leaves are used. The leaves are used in treating oliguria, haematuria, diabetics mellitus and menstrual disorders.
- ☆ The roots and the leaves are used in the form of decoction or extract are active on hypertension.
- ☆ The purified alkaloids extracted from leaves are effective in treating leukemia.
- ☆ Alkaloids obtained from roots are used to induce cerebrovascular dilatation and for hypertensions.
- ☆ Leaves contain alkaloids, serpentine, ajmaline, catharanthine, catharanthinole, vindoline, vincaleucoblastine, leurosidine and vincristine.

57

Celastrus paniculatus Willd. (Mal Kankadi)

Botanical Name: *Celastrus paniculatus* Willd.

Family: Celastraceae

Local Names

- ☆ **Gujarati:** Mal kankadi
- ☆ **Sanskrit:** Jyotismati, Pitataila
- ☆ **Hindi:** Malkangani, Malkunki
- ☆ **Tamil:** Valuluvai, Siruvaluluvai
- ☆ **English:** Climbing staff tree, balck oil plant
- ☆ **Telugu:** Danti cettu, Gundumida

Introduction

It is large climbing unarmed shrub with long slender elongating branches which are reddish brown and covered with elongate white lenticles. Leaves are simple, alternate, ovate or obovate, crenulate, coriaceous, glabrous, lateral nerves arching. Flowers are greenish-white in colour. They are arranged in terminal drooping panicles. Fruits are capsules types, depressed globose. When they become ripe they look yellow in colour.

Distribution

This plant is distributed throughout in India. It is also found on Hills upto altitude of 1,200 meter. Shrub is commonly found all over the hilly tracts of India.

Cultivation

Celastrus paniculatus is propagated through seed. It is found in the hilly tracts but not above the altitude of 1300 m. It prefers moist shady localities, shady localities. It grows on all types of soil but needs good drainage. It does not tolerate salinity/ alkalinity in the soil. The land is ploughed thoroughly till a fine tilth is obtained. Then continuous furrows and ridges are formed at four meter apart. The seeds are sown on the ridges at three meter apart at the beginning of monsoon. At every spot, 3-4 seeds are dibbled. 5-6 kg of seeds are required to cover one hectare. When the seedlings are 2-3 weeks old, they are thinned out, leaving one good vigorous plant at each spot. The seedlings are encouraged to grow on stakes or threads connected to the wires on the top. These wires rest on poles erected at 5 x 5 m distance. These G.I. wires are drawn criss-cross, to form one meter squares.

Weedings and soil workings are carried out twice *i.e.* in July and August. Thirty grams of NPK mixture is applied per plant in two split doses along with the soil workings. In the second and subsequent years also, two soil workings are carried out in June –July and August-September. Fertilizers are also added as in first year. As an alternative, the plants can also be grown in older tree plantations and they can be made to climb on those trees.

Harvesting and Yield

The plants flower during April-May and the fruits ripe during October.The ripe pods are collected and washed in bamboo baskets with water, so as to wash away the aril. The seeds thus separated are dried and marketed. It is estimated that an annual yield of 1600 kg/ha. of seeds is expected from one hectare.

Chemical Constituents

This plant contain Sesquiterpene alkaloids, celapagine, celapanigine and celapanine, hydrolysis gave polyalcohol A, polyalcohol C, Polyalcohol D.

Medicinal and Economic Importance

- The bark is abortifacient, depurative and a brain tonic.
- The leaves are emmenagogue and the leaf sap is a good antidote for opium poisoning.
- The seeds are acrid, bitter, termogenic, emollient, stimulant, intellect promoting, digestive, laxative, and emetic.
- This plant is useful in vitiated condition of pitta and kapha.
- This is useful in abdominal disorders, leprosy, pruritus, skin diseases, paralysis, cephalalgia, arthralgia, asthma, leucoderma, cardiac debility, inflammation, strnagury, nephropathy, amenorrhoea, dysmenorrhoea and fever and for stimulating the intellect and sharpening the memory.
- The seed oil is bitter, thermogenic and intellect promoting.
- It is also used in abdominal disorders, beri-beri and sores.

58

Chlorophytum borivillianum Sant and Fernand (Safed Musli)

Botanical Name: *Chlorophytum borivillionam* Sant and Fer.

Family: Liliaceae

Local Names

- ☆ **Gujarati:** Safed Musli
- ☆ **Hindi:** Safed Musli, Dholi Musli
- ☆ **English:** Safed Musli
- ☆ **Marathi:** Sufed Musli
- ☆ **Malayalam:** Swedeveli

Introduction

During rainy season small seedlings of musli are found in forest. Leaves are yellowish and white flower with 6 petals are arranged on the flowering stalk, which emerge from the center of the plant. About 20-25 flowers on the flowering stalk appear in July. The seed is small, black and enclosed in the holes. 10 to 20 seeds are there in one hole. Seed is very light in weight. Tubers emerge at the bottom of the plant, the thickness is around 0.9 cm and the length is 8 cm. Number of tubers vary from plant to plant, age of the plant and on an average 5-30 tubers per plant can be seen. Tubers are white in color that's why this plant is called safed musali.

Origin and Distribution

Safed Musli is found in sub tropical countries like India, China, Japan, Pakistan, Burma, Sri Lanka etc. In India it is seen in forest of Himalayas as well as in Gujarat, M.P, U.P, and Karnataka etc like parts. In Gujarat safed musli is seen in the forest of Girnar, Barada and Dang specially.

Cultivation

Soil which contains light to medium texture is useful for the growth of this plant. Well drained fertile soil is most suited. The land should be ploughed and harrowed so that there should be proper aeration. Well decomposed FYM about 15 – 20 tones/ ha should be mixed in the soil properly before planting. Bed furrows of 90 cm broad and 15 cm height are prepared, water canals are also prepared. Usually safed musli is propagated by tubers. However the propagation of musli can be done by seed also. But germination percentage by seed is very less which takes 18 months for harvesting tubers. Hence sprouted are economical for planting. It takes 6 months for harvesting in the rainy season. Previous year tubers are used for cultivation.

The bunch of the tubers should be removed from soil in May and tubers are separated from the bunch in such a way that some portion of crown should remain attached to each tuber. Tubers separated should be in gunny bags filled with moisture of fine sand and silt with moist condition, generally under shade. It can also be kept in saw dust under moist condition. In June tuber get sprouted naturally and these are used for planting in seed propagation, where germination is very poor, the treatment with calcium chloride was found to be good for seed germination. Planting should be done in June when there is lot of moist, planting is done at the spaces of 30 x 30 cm. Tubers are placed deeply in the soil. Irrigation should be given when there is required but water should be stagnant for long. Weeding should be done by hand.

Harvesting

After 6 months of planting, tubers are ready for harvesting. Generally harvesting is done in December-January, when the matured leaves drop down. This is the best indication for harvesting tubers. Tubers branches are removed by digging the soil. Branches when removed from soil, should be thoroughly washed with water. Such branches are dried in shade for 2 days. The skin on the tubers should be removed by pressing the tubers by holding between 2 fingers. Such peeled musli should be again washed with water and dried under shade. Hard and dried musli are taken to the market. Musli should be packed in polythene bags.

Yield

Fresh and dried tubers are yielded. Fresh tubers are obtained 1.8-20 tones per hectare and dried tubers 0.3 -0.4 tones per hectare dried fruits are obtained. Dried roots are sold about at Rs. 500-800 per kg. Market rate of musli can reach up to Rs.1000-1200 per kg. The cost of cultivation per hectare is about Rs. 1 lakh. One can get the minimum profit around Rs. 60,000-70,000 per hectare.

Medicinal and Economic Importance

- ☆ Tubers are greatly used as medicine as it contain steroid sapoginine (1-2 per cent), protein (10-20 per cent) and calcium to some extent with some extent with some water soluble minerals.
- ☆ Tubers are useful in *Pitta* and *vata.* as they are fat free. it is effective on fatigueness, daaha and in blood purification.
- ☆ It is also useful in certain diseases like renal calculus, dhupani, sangrahani, leucorrhoea and diabetes.
- ☆ It is basic ingredient of *chyanprash*. It is lacting, energetic to heart
- ☆ Powder is prepared which is used in providing healthiness. It contains steroids, resin, phonetics, tannins, carbohydrate, calcium, magnesium and potassium.

59

Cissampelos pariera Linn. (Pathavel)

Botanical Name: *Cissampelous pariera* Linn.

Family: Menispermaceae

Local Names

- ☆ **Gujarati:** Pathavel
- ☆ **Hindi:** Patha, Patha vel
- ☆ **English:** Abuta

Introduction

The plant is twining shrub. Stem is short, throwing out long herbaceous tomentose branches. Usaually peltate, orbicular to reniform, more or less cordate at the base, apex is obtusely mucronate. Male cymes are long peduncled, axilary or nearly so, many fid, hairy, bracts are minute. Raceme of female flowers is 1-2, axillary, with large reniform or orbicular bracts. Drupe type fruits are subglobose, hirsute, scarlet is red. Plant can be collected after rainy season.

Distribution

This plant is distriubuted in the warmer part of the world. It is found in tropical and sub tropical India and also in Sri Lanka and Singapore. This is very common climber of the deciduous forest of the region.

Cultivation

It is found naturally growing in tropical, arid and semi arid areas on hedges, fences, bushes and trees. It comes up on all types of soil. Water logging at the root

zone is to be avoided. It is very easily propagated by seed. The seeds can directly be sown in the field at 3 x 1 m spacing, at the beginning of monsoon. The soil should be made loose in about one square foot area with the help of a pick – axe and then the seeds are dibbled.

The germination will be complete within a week's period after the first good shower. Since this plant is a climber, some supporting material like Euphorbia plant or bamboo stake with side branches on, or trees/bushes or some other means, will have to be provided, Euphorbia plants may be erected, just like a fence in parallel strips at four meter. apart running in East West direction. The seeds may be sown in between the strips, and half a meter away from the Euphorbia on both sides of the strip at one meter distance. Thus, there will be about 5000 plants in one hectare. The plants are trained to go up the Euphorbia strips. At the end of one year, in the month of May, the plants may be uprooted and the roots are harvested.

Harvesting and Yield

In the first forthnight of May, the plants are uprooted and the roots are taken out and dried. Then, they are cut into pieces according to market requirement and dispatched for marketing. In the second fortnight, the seeds should be sown for the next year's crop. It is estimated that a production of 500 kg of air dried roots would be available from one hectare.

Medicinal and Economic Importance

- ☆ Plant is antiperiodic, diuretic, purgative, stomachic.
- ☆ It is used in diarrhea, dropsy, dyspepsia, in cough, and urinary troubles
- ☆ Plant is also used in fever, diarrhea and dysentery
- ☆ Paste of leaves and seeds are used on sores, itches *etc.*
- ☆ It is also fish poison.
- ☆ Plant yields a strong fibre.

60

Cissus quadrangularis L. (Had Shankal)

Botanical Name: *Cissus quadrangularis* L.

Family: Vitaceae

Local Names

- ☆ **Gujarati:** Had shankal
- ☆ **English:** Edible stem vine,Adamant creeper, Bone-setter
- ☆ **Hindi:** Asthi shrinkhala
- ☆ **Sanskrit:** Asthisrnkhala
- ☆ **Tamil:** Pirantai, Vajjiravalli
- ☆ **Telugu:** Nalleru, Nulleratiga

Introduction

It is tendril climber which has stout fleshy stem. Stem is quadrangularis. Tendrils are simple, long, slender and leaf opposed, in addition to the normal roots, some aerial roots arising from the jointed nodes grow downwards and strike the soil. Leaves are simple and broadly, they are reniform, entire or toothed, rounded, truncate or cuneate at the base. Flowers are green in colour and small in size, they are arranged on shortly peduncled cymes inflorescence. Petals are four, hooded at the apex. Fruits are ovoid or globose and they are berries type. Seeds are ellipsoid.

Origin and Distribution

The plant is distributed throughtout in India. Particularly hadshankal is found in the arid or hot and semi arid area.

Cultivation

This plant is cultivated by vegetative propagation. This plant is basically of tropical climate. It is seen growing in semi arid and arid zones also. It comes up on all types of soil. However, it does not flourish in highly saline areas and water logged conditions. It can tolerate partial overhead shade.

It is easily propagated by vegetative propagation. The stem cuttings, when planted give satisfactory results. The branch cuttings 30-45 cm in length are planted in the field at 1m x 0.5 m spacing. The plants need some support for climbing. The branch cuttings planted directly in the field at the beginning of monsoon establish soon and start growing. The plants should be given fertilizer doses (DAP/Urea) at the rate of 100 gms per plant in two split doses during monsoon.

Harvesting and Yield

The first crop can be harvested at the end of 2nd year. Area is divided into two parts and harvesting is done in one plot in a year. The aerial portion of the plants is cut and dried in sun. Since the root is in tact, the plants start growing again. Thus, 10,000 plants will be available in one ha. area each year. Each plant is estimated to yield about 100 grams, of dry matter (drug). A production of 1000 kg. of the drug would be available every year.

Chemical Constituents

The stem contains two unsymmetric tetracyclic triterpenoids, onocer-7-ene-3α-,21 β,21α-diol and two steroidal prindiples I and II. Prescene of β-sitosterol, δ-amyrin and δ-amyrone is also reported.

Medicinal and Economic Importance

- Whole plant is used medicinally.
- Plant is bitter, sweet, sour, thermogenic, alternate, laxative, anthelmintic, carminative, digestive, stomachic, depurative, haemostatic.
- It is aphrodisiac, anodyne, ophthalmic and union-promoting and it is useful in vitiated condition of vata, helminthiasis, anorexia and dyspepsia.
- It is also useful in colic, flatuelence, skin diseases, leprosy, haemorrhages, haemoptysis, ophthalmopatht, otorrhoea, chronic ulcers, tumors, haemorrhoids, epilepsy, convulsions, spanomenorrhea, fractures, swellings and vitiated conditions of kapha.
- The shoots are useful in colonopathy, scurvy, otorrhoea, asthma, burns and wounds.
- Powdered roots are as well as the stem paste for specific bone fracture.

61

Citrus limon Linn. (Motu Limbu)

Botanical Name: *Citrus limon* Linn.

Family: Rutaceae

Local Names

- **Gujarati:** Mota Limbu, Motu Limbu
- **Hindi:** Baranimbu, Jambira
- **English:** Lemon tree
- **Tamil:** Malaielumichai, Periya elumichal
- **Telugu:** Bijapuram

Introduction

The lemon tree is small tree or a straggling bush, 3-4.5 m (10-15 ft) in height, with thorny branches and scented white flowers tinged with pink. Leaves are light green, oblong to elliptical to ovate, lanceolate, sharp-pointed, subserrate, petiole narrowly winged, spines small and stout. Fruits are fleshy rind and abundant acid pulp, generally green but gets yellow when they ripe.

Origin and Distribution

It is believed to be Indian Origin of native of India. It is found growing wild along the foot of North West Himalayas. The lemon grows in many part of the world but till recently Sicily and some extent Calabria, Italy has been the home of lemon oil industry. It is also cultivated in California, USA. In India, the lemon is cultivated in household gardens and small sized orchards in the Punjab, UP, Maharastra, Gujarat,

Andhra Pradesh, Tamil Nadu, Kerala and Karnataka. In Gujarat lemon is grown in most of the districts like Ahmedabad, Surat, Dang, Junagadh, and Jamanagar.

Cultivation

This plant is suitable for shady or sheltered climate. It doesn't like water logging. Dry climate is preferable since the plant is susceptible to various roots and fungal diseases common in a hot and humid climate. Moderate rain fall is good for the crop. Naturally drained soil is best for growth of this plant but it can be grown on wide range of soils. It may be propagated through cuttings or seeds. Budded plants are planted out in field 6-18 months after bud-insertion. Lemon is best propagated by budding on a suitable stock.

Harvesting

If plant is grafted then fruits can be picked in 4-5 years and when it is planted by seeds fruits can be collected when they ripe in 6 to 7 years. Generally Harvesting is done throuout the year. Lemon fruits when become yellow they are picked with hands from the plant. Then they are stored. Oil is obtained from the fruits of the lemon.

Oil Extraction from Lemon Fruits

Oil is extracted in three processes. (1) Spong process (2) Rcuelle process and (3) the Machine process including the employment of centrifugal separator. Spong process is the most common process. Fruits in this method are cut into halves, the pulp removed with a spoon and oil pressed out with a cup shaped sponge. The fruit is cut into positions, the pulp imperfectly removed and then pressed out with large round sponge.

Yield

Oil is the main yield of lemon plant. Following are the statistics of export of lemon oil. Lemon fruits are sold in the market at Rs. 1-2 per fruit. Following are the statistics of export of lemon oil

Year	*Quantity (tones)*	*Value in Lakh Rs.*
1994-95	0.15	0.21
1995-96	0.06	0.29
1996-97	0.29	1.71

Medicinal and Economic Importance

- ☆ It has characteristic flavoring and odorous constituent. It is because of citral which is 4-5 per cent, the main constituents are terpens and sesquiterpenes particularly α-and β-lemonene, which are in largest amount (up to 90 per cent). The following have also been reported: α-and β pinepine, l-camphene, nonyl alhedyde, citronellal, bisabolene, cadinene, methylheptenone, terpineol, geranyl acetate, linalyl acetate, limettin (a sort waxy residue- 4:6 dimethoxycoumarin) and possibly traces of methyl anthranilate.
- ☆ The lemon fruits are used for various culinary purposes such as in pickles, chats, lemon pies, lemon cakes, lemon ices and as a flavoring for candies, jellies, jams.
- ☆ Ripe fruit is largely used in the preparation of cooling beverages and effervescent draughts.
- ☆ It is used as bleaching agent and stain remover. Citric acid, rectin and lemon oil are also obtained from the fruits.
- ☆ The lemon is used for flavorings of all kinds of beverages and soft drinks and food products like cakes, parastries, pies, candies, desserts, ice creams, confectioneries *etc.*
- ☆ It is also employed in perfumes, toilet waters, eav de cologne, soaps *etc.* to which it imparts a refreshing top note.

- ☆ Lemon extract which is made by dissolving 5 parts of oil in 95 parts of alcohol is well known flavoring material nest in importance to only vanilla.
- ☆ The oil is utilized for the preparation of terpeneless oil and concentrated oils which are more stable on storage and have a better solubility in dilute alcohol.

62

Cleome viscosa L. (Pili Talavani)

Botanical Name: *Cleome viscosa* L.

Family: Capparideaceae

Local Names

- ☆ **Gujarati:** Pili talavani
- ☆ **Sanskrit:** Pasugandha
- ☆ **Hindi:** Hulhul, Hurhur
- ☆ **Tamil:** Naivelai, Naikkaduku
- ☆ **English:** Wild mustard, Dog mustard, Sticky cleome
- ☆ **Telugu:** Kukkavaminta,Nallavaminta

Introduction

This is an annual herb which is very strong and hairs are there on its surface. Leaves are with 3-5 foliate, gradually becoming shorter upwards. Flowers are yellow in colour and are arranged in raceme inflorescence. Fruits are hairy and compressed, they are capsule type. Seeds are brownish type they become black when they ripe.

Distribution

This is very common in arid and semi arid areas of the India. It is found throughout India. In any waste land this plant is very common. It occurs in northern tropical Africa, from Cape Verde and Senegal to Egypt, Ethiopia and Zanzibar. It is also present in Madagascar and other Indian Ocean islands. Outside Africa

it is widespread in peninsular Arabia, tropical Asia, Australia and tropical and subtropical America.

Cultivation

This plant is cultivated for its medicinal values. Propagation is done by seeds and by vegetative method. It can be grown as rainfed crop. It is propagated easily by seeds. The area is ploughed throughoutly and a fine tilth of the soil is obtained. Furrows are made at 30 cm apart and the seeds are sown in the furrows continuously, at the beginning of monsoon. The germination is complete in ten days. One weeding and hoeing are carried out in July-August. Fifty kg of nitrogen, is added to the soil immediately after the hoeing. The plants start flowering and fruiting from October.

Harvesting

The fruits are collected in month of December. The whole plant is cut and dried. Then the seeds are separated by threshing the garbage.

Yield

The yield of 1200 kg of seeds is obtained from one hectare.

Chemical Constituents

The aerial parts contain a macrocyclic diterpene, 20-oxocembra-3, 7, 11, 15-tetra en-19-oic acid and a bicyclic diterpene, cleomeolide. Seeds contain coumarino-lignans, cleomiscosin, A, B, C and D.

Medicinal and Economic Importance

- ☆ Whole plant is medicinaly useful, Plant is acrid, thermogenic, anthelmintic and sudorific.
- ☆ The roots are stimulant, antiscorbutic and vermifuge.
- ☆ The leaf juice is digestive and is good for otalgia.
- ☆ The seeds are anthelmintic, carminative, constipating, febrifuge and cardiac stimulant.
- ☆ Seeds are useful in fever, diarrhea, worm infestations, worm infestations, cardiac disorders and dyspepsia.

63

Clerodendrum phlomidis L.F. (Linn.) (Arani)

Botanical Name: *Clerodendron phlomidis* L.F.(Linn.)

Synonym: *Clerodendrum multiflorum* (Burm.f.)

Family: Verbenaceae

Local Names

- ☆ **Gujarati:** Arani
- ☆ **Sanskrit:** Laghuagnimantha
- ☆ **Hindi:** Arani
- ☆ **Telugu:** Takkeda, Takkali
- ☆ **English:** Arni

Introduction

The plant is perennial shrub. It attains a height of 3 meters and has creamy white to grey bark with numerous lenticels. Leaves are broadly ovate, deltoid-ovate or rhomboid-ovate, rarely nearly suborbicular; they are 0.6-8.3 X 0.3-6.5cm in size. Thin hairs are found on the surface of the leaves. Flowers are creamy white containing pinkish tinge. They are arranged in axillary dichotomous cymes and terminal panicles inflorescence. Fruits are drupe type, 0.4 - 0.6 cm across, obovoid, glabrous or sparsely hairy.

Distribution

This plant is found in the arid, semi arid, and sub-tropical climatic areas. It

can be grown in any kind of soil. It can withstand with pH value of 0.8. It is a light demanding species.

Cultivation

Seed propagation is good method. Seeds are collected fresh and sown in the polypots of 15 X 25 X 200 cm. Polypots are filled with Farm Yard Manure. After one year when seedlings get proper height, are transplated in prepared pits of 30 X 30 X 30 cm at 2m X 2m distance. Thus, near about 5000 seedlings are needed for one hectare land. In the beginning of monsoon planting is done, seedslings require good care of weeding and soil workings. The plants may also be watered during non rainy season if possible. The plants may also be given fertilizers.

Harvesting

The roots and stem bark etc. can be harvested at four years duration. Plants are uprooted and separated, are cut into pieces of proper length and dried into sun light. Leaves are separately dried by removing stem bark under shade. Out of 5,000 plants in hectare, around 1250 plants are harvested every year and planted again in next season. Thus, in the period of 4 years, one hectare can be harvested.

Yield

An average production of roots from a plant is taken as one kg. Thus a production of 1000 kg of roots and 400 kg of stem bark and leaves can be expected from one hectare. The market rate of root is about Rs.15 per kilogram. Total expenditure per hectare per year is around Rs. 5,000 while gross return goes upto Rs.19,000. Thant means net return per hectare is around Rs. 14,000 per hectare.

Medicinal and Economic Importance

- ☆ The root is taken in Dashmula of Ayurveda system of medicine.
- ☆ The leaves, roots and stem bark are used in the treatment of venereal diseases.
- ☆ The root is aromatic, astringent. Its decoction is used as demulcent in gonorrhoea.

64

Clerodendrum serratum L. Moon (Bharangi)

Botanical Name: *Clerodendrum serratum* L. Moon

Family: Verbenaceae

Local Names

- ☆ **Gujarati:** Bharangi
- ☆ **Sanskrit:** Kharasakah, Bharangi
- ☆ **Hindi:** Bharangi
- ☆ **Tamil:** Sirutekku
- ☆ **English:** Bharangi
- ☆ **Telugu:** Gantubarangi

Introduction

This is slightly woody shrub, stem looks quadrangular in shape. At each node there are three leaves found, they are oblong or elliptic and some times they are opposite. They look sharply serrate. Flowers are blue in colour. They are many in numbers. They are in long cylindrical thyrsus with a pair of acute bracts at each branching and a flower in the fork. Fruits are generally four lobed and they are purple in colour. At each lobe, fruits look succulent with one pyrene.

Distribution

This plant is found throughout in India. It is found in arid and semi arid region of the world. It has also been distributed up to height of 1,500 meter elevation.

Medicinal and Economic Importance

- ✰ The roots are bitter, acrid, thermogenic, anti-inflammatory, digestive, carminative, stomachic, anthelmintic, depurative, expectorant, sudorific, antispasmodic, stimulant and febrifuge and are useful in vitiated condition of kapha and vata.
- ✰ It is also used in inflammation, dyspepsia, anorexia, colic, flatulence, helminthiasis, cough, asthma, bronchitis, highcough, tumours, tubercular glands, dropsy, consumption, chronic nasitis, skin diseases, leucoderma, leprosy and fevers.
- ✰ The leaves are useful as an external application for cephalalgia and ophthalmia. Seeds are aperient and are used in dropsy.

65

Cochlospermum religiosa Linn. (Kumbhi)

Botanical Name: *Coclospermum religiosum* Linn. Alston

Family: Cochlospermaceae

Local Names

- ☆ **Gujarati:** Kumbhi
- ☆ **Sanskrit:** Girisalmalika, Silakarpasika
- ☆ **Hindi:** Kumbhi
- ☆ **Tamil:** Kattupparutt
- ☆ **English:** Yellow flowered cotton tree
- ☆ **Telugu:** Konda gogu, Kongu

Introduction

It is a small desiduous tree which gets height of 2.4 to 5.4 meter. Bark is deeply furrowed and contains fibers. Fibers are smooth and ash coloured. Bark provides gum. Leaves are palmately containing 5-lobes.They are tomentose beneath. Flowers are large, golden yellow in colour and they are arranged in terminal panicles appearing after the leaf fall. Fruits are large, pear shaped and 3-5 valved capsules type. Numerous seeds are found in the fruits. They are covered with wooly hairs.

Distribution

This plant is distributed throughout in India. Perticularly in dry and semi dry forests of the Asia and India. It is distributed in Burma, China, Japan, Nepal *etc.*

Cultivation

This plant is cultivated by seed propagation method.

Medicinal and Economic Importance

- ✰ Dried leaves and flowers are used as stimulant and the young leaves are used as a cooling wash for the hair.
- ✰ The gum is known as Katira gum, it is sweet, thermogenic, anodyne and sedative, and it is useful in cough, diarrhea, dysentery, pharyngitis, gonorrhoea, syphilis and trachoma.

66

Cocculus hirsutus Linn. (Patal Garodi)

Botanical Name: *Coculus hirsutus* (L.) Diels.

Family: Menispermaceae

Local Names

- **Gujarati:** Patal garodi
- **Sanskrit:** Patalagarudah, Chilihindah
- **Hindi:** Patal garodi, Jaljamni
- **Tamil:** Kattukkoti
- **English:** Broom creeper, Ink berry
- **Telugu:** Dusaratiga

Introduction

This is scandent shrub with straggling soft villous young parts. Leaves are simple, alternate, ovate-oblong, obtuse, speculate, subcordate or truncate at the base. This plant is soft and villous on both surfaces. Petioles are also densely vilous. Male and Female flowers are present. Male flowers are small and arranged in axillary cymose panicles inflorescense. Female flowers are combined together and form axillary cluster of 2-3 flowers. Petals at the apex are devided into triangular lobes. Fruits are purple coloured and black drupes type. They are transversely rugose.

Distribution

This plant is distributed throughout India. It is mainly found in the areas of of arid and semi arid conditions. Desert forest, deciduous forests are some of the source of cocculus.

Cultivation

This plant is cultivated by two methods (1). By seed propagation and (2) By vegetative propagation.

Chemical Constituents

Coculous contain cohirsine, haiderine, hamtine N-oxide, trilobine, isotrilobine, syringaresinol and protoquercitol. The triterpenoid alcohols, hirsudiol and nonacosan-10-ol have been reported from the plant. Cohirsitinine, hirsutine, shaheenine, cohirsinine and jamtinine. The plant also contains cintaons, coclaurine, magnoflorine, sitosterol, ginnol and monomethyl ether of inositol.

Medicinal and Economic Importance

- The roots are bitter, acrid, themogenic, laxative, alternate, emollient, aloexeteric, depurative, demulcent, digestive, carminative, diuretic, aphrodisiac, expectorant, anodyne, antipyretic and tonic.
- They are also useful in vitiated condition of *kapha* and *vata*, poisonous bronchitis, gout, cephalalgia, intermittent fevers, spermatorrhoea,

urethrorrhea, burning sensation, tubercular glands, fractures, hypertension and general debility.

- The leaves are mucilaginous, cooling, demulcednt, anodyne and expectorant and are used in eczema, gonorrhoea, prurigo, impetigo cough, ophthalmia, cephalalgia and neuralgia.

67

Coleus forskohlii (Willd.) Briq (Pathar Choor)

Botanical Name: *Coleus forskohlii* (Willd.) Briq.

Family: Labiatae

Synonymous: *Plectranthus barbatus.* (Benth).

Local Names

- ✰ **Gujarati:** Patha-choor, Gar-mar
- ✰ **Hindi:** Pashan bhadi, Pathar-choor
- ✰ **English:** Coleus

Introduction

It is an aromatic perennial herb and gets height of 60 cm. It grows well on dry slopes of the Indian plains and in the foothills of the Himalayas. Roots are tuber like while stem is erect. Flowers are dark violet to bluish-black.

Origin and Distribution

This plant is originated from tropical Africa. It is distributed in Europe to Asia. Plant can be seen in Pakistan, Sri Lanka, Japan, Myanmar, India like countries. In India mainly this plant is cultivated in south India mainly. It is also cultivated in Gujarat, Madhya Pradesh, Uttar Pradesh and Maharastra states. In Gujarat plant is grown in Baroda and other districts.

Cultivation

Hot and humid climatic conditions are suitable for the growth of the plant. It can be easily being grown in less humid climatic conditions.

This can also be grown in the tropical as well as sub-tropical climates. For this crop the humidity in the air should be around 83 to 85 per cent.for the cultivation fertile soil is needed. The red sandy and sandy soils are suitable. Field is made soft by tilling deep with plough prior to monsoon season. Cow dung is added in to the field. Beds are prepared in the field; raised beds are made at 60 x 60 cm distance. Generally seed propagation is good method. But vegetative propagation is also useful. When nursery plants are of one month age and sufficient number of roots grow on them. Then these are to be transplanted in the bed at 20 x20 cm distances. If there is no rain at this time, irrigation is given properly. Irrigation should be given every third day and gradually changed to once a week, which will enable speedy growth of the plants. The sowing through stem cuttings is easier and better. The stem cutting of 10 to 12 cm long (Approx. 4 -5 inch) pieces with 5-6 live leaves are cut and planted in nursery under shade. This helps in early growth of new roots. Sufficient is provided to plants in the nursery. In one month of old plants stem cutting sufficient number of roots grows, they can be transplanted into the field. For one acre land around 33,600 stem cuttings are needed.

Harvesting

Plants get matured in about 8 month's period from the day of transplantation.

The plants are uprooted along with roots. Thereafter, the roots are separated from plants; they are cleaned, and cut into small pieces and dried under a shad. Dried roots are filled in bags and are taken to market for selling.

Yield

From one acre of land an average, 600 to 1000kg of dry roots can be obtained.

Medicinal and Economic Importance

- ☆ Roots are used as spice and condiment.
- ☆ They lowers blood pressure, it is antispasmodic.
- ☆ Roots dilate bronchioles, blood vessels. It also works as heart tonic.
- ☆ Roots contain volatile oil.
- ☆ Generally leaves are also of economic importance.

68

Coleus amboinicus Lour. (Pathar Chur)

Botanical Name: *Colius amboinicus* Lour.

Family: Lamiaceae

Local Names

- ☆ **Gujarati:** Pathar chur, Garmar
- ☆ **English:** Countryborage, Indian borage
- ☆ **Hindi:** Patta ajavayin, Patharcur
- ☆ **Tamil:** Karpuravalli
- ☆ **Sanskrit:** Karpuravalli, Sugandhavalakam
- ☆ **Telugu:** Sugandhavalkam

Introduction

This plant is aromatic herb which grows in arid and semi arid regions of the world. It is perennial fleshy stem. Leaves are simple and opposite, broadly ovate, crenate, fleshy, and very aromatic. Flowers are pale purplish indense whorls at distant intervals in a long slender raceme. Fruits are orbicular or ovoid nut lets.

Distribution

This plant is distributed throughout in India. It is also cultivated in the gardens for its medicinal uses. The plant is distributed all over dry and semi dry condition where even at 3000 feet at the Himalayan region.

Cultivation

This plant is cultivated throughout in India. Plant is grown by cutting method.

Chemical Constituents

This plant contains thymol, carvacrol and camphor.

Medicinal and Economic Importance

- ☆ The leaves are bitter, acrid, thermogenic, aromatic, anodyne, appetizing, digestive, carminative, stomachic, anthelmintic, constipating, deodorant, expectorant, lithontriptic, diuretic and liver tonic.
- ☆ They are used in cephalagia, otalgia, anorexia, dyspepsia, flatulence, colic, diarrhea and cholera especially in children, halitosis, convulsions, epilepsy, cough, chronic, asthma, hiccough, bronchitis, renal anc vesical, calculi, strangury hepatopathy and malarial fever.

69

Commiphora whightii (Arn.) Bhand. (Guggal)

Botanical Name: *Commiphora whightii* (Arn.)Bhand.

Family: Burseraceae

Local Names

- ☆ **Gujarati:** Guggal
- ☆ **Hindi:** Guggal,Gugal
- ☆ **English:** Indian Bedellium tree
- ☆ **Tamil:** Gukkulu, Kiluvai
- ☆ **Telugu:** Konda mamidis
- ☆ **Sanskrit:** Ilkatah

Introduction

Guggal is small tree or herb which is 3 to 4 meter in height, it has crooked, knotty and aromatic braches which results into sharp spines. Younger plants are pubescent and globular. Older bark is papery and peels are seen in strips from the older parts of the stem. Leaves are sessile, alternate or fascicied, 1 to 3 foliate, terminal is sessile or subsessile, obovate, serrate (sometimes serrate only towards apex), they are 1-5 cm long 0.5 to 2.5 cm broad; lateral when preset, sessile, serrate or entire, it is less than half size of terminal leaflet. Plants are dimorphic, having bisexual and male flowers, female flowers alone and male flowers alone. Bisexual and male flowers are sessile, 3-5 mm long, generally red color to pinkish white. Sessile flower appear singly or in groups of two or three. The calyx is fused with disc basally; it remains tubular or urceolate, 1-2 mm long. Petals are reflexed, acute,

3-5 mm long. There are 8 stamens some times 10 stamens are seen, they are 3-5 mm long. The disc is conspicuously toothed. Ovary is 2 locoed with a sessile 2 lobed stigma. Bracts are 2 in numbers, opposite, glandular and hairy. Fruit is ovoid, up to 1 cm- long drupe. It gets red color when it ripes and mesocarp remains yellow.

Origin and Distribution

Commiphora genus have around 165 species in the world, its origin is believed to be Africa and Asia. This plant is widely distributed in the tropical regions of Africa, Madagascar, Asia, Australia, Pacific islands, India, Bangladesh, Pakistan and other countries. Generally four species of Guggal occur in India. They are *Commiphora agollochoa* Engl; *Commiphora stocksiana* Engl, *Commiphora mukul* Hook ex stock and *Commiphora berryi* (Aen.) Engl. Guggal gum is derived from *C. mukul* which is distributed in the states like Gujarat, Rajasthan, Tamil Nadu, and Karnataka.

Cultivation

Guggal is the plant of arid zone which thrives well in arid subtropical to tropical climate. Growth is faster in the soils which have moisture retaining capacity. Rainfall in the guggul growing tract may average between 100mm and 50 mm. It prefers warm, dry climate for a good yield of oleo-gum-resin. Before rainy season land is prepared well in advance. 2-3 ploughing are done to fertile soil. Seeds are sown in the pits made at a distance of 3 to 4 m in rows. Guggal can be propagated by seeds and vegetatively through stem-cuttings. Air layering has also been successful in this plant. Semi hard wood cuttings are taken and cut end is treated with IBA or NAA growth regulator solutions are planted in nursery beds during June-July. The beds are irrigated lightly after planting and regularly thereafter. Cutting sprout in 10 to 15 days, becomes ready for planting in the main field after 10 to 12 months, during the main rainy season. Seed propagation is not very useful method because seed germination is slow and is also very poor due to the presence of the hard seed coat. Seed are mechanically scarified with sand paper and are kept under running water for 24 hours. The seedlings may be raised in polythene bags during the Kharif season and after hardening, may be planted in the main field.

Light irrigation during summer season support good growth of the plants, this crop has not shown good response to fertilizers, except to the lower level or nitrogen fertilization. Hence, urea or ammonium sulphate at 25/50 g/bush is given twice a year before irrigation. 'Interculture is confined to one weeding and hoeing in the early stages of growth.

Harvesting

Guggal plant gets normal height and girth after 8 to 10 years of growth, when they are ready for tapping. For tapping the gum, which is present in the balsam canals in the phloem, bark-deep, *i.e.*, a shallow, incision is made on the bark. If the incision is too deep, plant either dies or yields little resin in the following years. Period of December to March is ideal for tapping. If tapping is successful then exudation starts after about 15 to 20 days from date of incision and continues for nearly 30 to 45 days. A piece of polythene sheet can be spread on ground to collect exuded gum. A maximum of about 500 g of gum has been obtained from the plant; Plant which produced gum on incision usually dies ultimately showing progressive drying and darkening of incised branch. Drying process slowly extends from incised branched to other parts of plant and within 8-10 months, has been isolated and its pathogenicity is being studied.

Yield

Usually the incision is made after November but before April. The resin is collected at an interval of 10 to 15 days. Weather conditions influence the success of obtaining gum. The yield of gum-resin from each plant is about 700 to 900 g.

Diseases

Guggal is attacked by a leaf eating caterpillar *Euproctis lanata* Walker; white fly *Bemisia tabaci,* which sucks the sap of leaves as a result of which the leaves turn yellow and finally drop off. The rooted cuttings also suffer from termite attacks in the early stages of growth. The diseases noticed on this crop are leaf-spot caused by *Cercospora* spp.

Medicinal and Economic Importance

- Gum resin is used as incense, as fixative in perfumery, in medicine, such as substitute for African bdellium (Commiphora myrrha). It is a common adulterant of myrrh.
- Guggal has application in indigenous medicine against rheumatism, obesity, neurological and urinary disorders, tonsillitis and arthritis.
- In hay fever inhalation of fumes from burnt guggal is recommended. It is also used in chronic bronchitis.
- It is useful as stomachic and carminative, stimulating and antiseptic. It is used to improve digestion.
- Guggal gum can decrease elevated lipid levels.
- Antibacterial properties of essential oil (22.58 per cent), resin chloroform extract (33.96 per cent, yield on dry weight), and sesquiterpenoids, curzerene (8.74 per cent), furanoeudesma-1,3-diene (11.66 per cent), lindestrene (4.08 per cent), curzerenone (10.97 per cent), furanodien-6-one (2.04 per cent), 3-methyl-4,5-dihydrofuranodien-6-one (0.26 per cent), isolated from oleo-gum-resin of *C.mukul* bark were evaluated against 19 bacteria.
- Essential oil, chloroform extract and sesquiterpenoids were as effective as kanamycin 9 control) in inhibiting growth of Gram positive and Gram negative bacteria.

70

Cordia dichotoma Forst (Gunda)

Botanical Name: *Cordia dichotoma*. Forst

Family: Boraginaceae

Local Names

- ✰ **Gujarati:** Mota Gunda, Raj Gunda
- ✰ **Hindi:** Lasora, Gondi
- ✰ **English:** Indian Cherry, Bhokar
- ✰ **Sanskrit:** Bahu varah, Slesmataka
- ✰ **Tamil:** Naru villi
- ✰ **Telugu:** Banka Nakkeru, Batuka

Introduction

Mota Gunda is small tree found in deciduous forest areas. It is common in arid and semi arid areas to sub tropical areas. It is much benefited if water is provided. It is a good coppicer. A tree is of 9-10 m tall, with ash colored or blackish – brown rough and fissured bark. Leaves are 7-18.5 x 5-13 cm, broadly ovate, elliptic – oblong or suborbicular, coriaceous, glabrous, nerves 3-6 pairs, basal nerves are 3, rarely 5 nerves are seen. Flowers are creamy yellow or some times white in axillary and terminal cymes. Fruits are Drupe type 2 to 2.5 cm across, ovoid or rounded glabrous, mucilaginous, bright yellow with pinkish tinge when ripe.

Origin and Distribution

Mota Gunda or Vad Gunda is distributed in semi arid and arid regions of the

world like Pakistan, Myanmar, Sri Lanka, India, Japan etc. In India *Codia dicotama* is distributed in Maharastra, Gujarat, Kerala, Tamil Nadu, Madya Pradesh *etc.* like states. In Gujarat Mota Gunada is grown in most of the districts like Junagadh, jamanagar, Rajkot, Dang *etc.*

Cultivation

On variety of soil this plant can be grown. Generally it is propagated by seeds. The seeds germinate easily. One year old polypot raised seedlings are good for planting. Seeds are planted at a spacing of 7 x 7cm. Weeding is done to keep plant clear. Watering the plant in 15 days is advisable. Plant grows well to artificial watering and manures. Flowering starts in 6-7 years. How ever, commercial production of fruits can be expected after 10 to 12 years.

Harvesting

After 10 to 12 years, when fruits are fully matured harvesting is done during March-April. The fruits are picked by hands before they are ripe and then marketed. Unripe fruits are useful for making pickles.

Yield

Each tree yields about 50 kg of fruits every year which means total fruit production per hectare is 8000. They can be sold in the market at a rate of Rs. 3/- per kg. Thus net returns after 10 to 12 years are 20,000.

Medicinal and Economic Importance

- Bark is bitter, astringent, acrid after digestion, constipating, anthelmintic, cooling and depurative which is useful in fever, dyspepsia, diarrhea, burning sensation, vitiated conditions of Kapha, pitta, helminthiasis, leprosy and skin diseases
- Leaves are aphrodisiac, and are useful in gonorrhoea and ophthalmodynia.
- The fruits are sweet, cooling, emollient, anthelmintic, purgative, vulnerary, diuretic, expectorant, aphrodisiac, depurative and febrifuge whioch are mostly used in vitiated conditions of vata, pitta, ulcers, leprosy, skin diseases, hyperdipsia, burning sensation, bronchitis, dry cough, pectoral diseases, strangury, urethralgia urethritis, chronic fever, arthralalgia, pharyngopathy.
- This plant is also used to get relief from splenopathy and ring worm.

71

Crocus sativus Linn. (Keshar)

Botanical Name: *Crocus sativus* Linn.

Family: Iridaceae

Local Names

- **Gujarati:** Eshar
- **Sanskrit:** Keshrar
- **Hindi:** Keshar
- **Tamil:** Kungumapu
- **English:** Saffron
- **Telugu:** Kunkumapuvu

Introduction

This is perennial herb, which is bulbous. Leaves are long, linear and channeled. Flowers are deep violet in colour. Flowers are funnel shaped; each flower is stalkless, and appearing in autumn with leaves. Petals, narrow elliptic, equal, fused below into long slender corolla tuve. Styles are orange while stigmas are brick red in colour. Approximately 400 hours of work and 15000 flowers are needed to gather one kilogram of dried saffron. This is because each flower provides only three stigma. This species contain world's most important flower.

Distribution

Crocus sativus probably originated from Greece and Western Asia where

some wild species occur which are possibly related. Then it spread in east ward in Kashmir in India. This plant is distributed mainly in cold regions where there is high amount of moisture remains.

Cultivation

The plant prefers a well-drained sandy or loamy soil that is free from clay. It grows well on calcareous soils, and on hot sheltered stony banks. Plants are very frosts hardy. Plants produce less saffron when grown on rich soils. It produces also in poor summer. Saffron has been cultivated for over 4,000 years for its stigmas. 'Cashmirianus' comes from Kashmir and has large high quality corms. It yields about 27 kilos of rich orange stigmas per hectare.

Seed propagation is good method. Seeds are sown in the spring in a cold temperature around 10-20°C. Germination can take 1 - 6 months at 18°C. Seedling should not be transplated in first year as in first year; seeds should be provided proper liquid food. Small bulbs are divided once the when plants have died down, then planting is done into 2 - 3 bulbs per 8cm sized pot. They are grown for 2 years in green house then. Finally during late summer they should be transplated in their main position. It takes 3 years for plants to flower from seed. Clumps are devided in late summer after the plant has died down. The bulbs can be planted out direct into their permanent positions.

Harvesting and Yield

When plants start flowering, harvesting should be done, for the flowering period may last only 15 days. The triple stigmas are picked by hand daily just as the flower opens. Then flowers are allowed to dry. An annual yield of 8-10 pounds of dried saffron per acre is obtained in an established planting.

Storage

After harvesting, storage of saffron is essential, Storage is done in sealed tin containers. The final product is a compressed, highly aromatic, reddish brown strands which are about 1 inch long. True saffron has a spicy, pungent and bitter taste and a tenacious odor.

Chemical Constituents

This plant contains the major volatile components of the oil are safranal, isophorene, glucoside, corcin, crocetin, pcro-crocin essential oil, also contain β- and R-carotene and lycopene.

Medicinal and Economic Importance

- ☆ Saffron, marketed as powder or as "hay" performs the function of a spice by adding its faint, delicate and pleasing flavour to foods, which can be used in cream, cottage cheese, bouillabaisse, mayonnaise, rice, cakes, sweetments, chicken, meat, liqueurs and cordials.
- ☆ A saffron tincture is used to flavour essence.
- ☆ For complains including chorea and hysterical affections it is also used.
- ☆ It is used for pain in eyes and top of head. Eyes appear as after weeping and in smoke.
- ☆ Ciliary neuralgia, pain from eyes to top of head.
- ☆ It is also used in theatened glaucoma, and when dark blood hangs down from the nose.
- ☆ It is also used in uterine haemorrhages with clots with ling strings, with pain in left breast.
- ☆ It can be used in crawling sensation in anus, abdomen and stomach.

72

Curcuma angustifolia Roxb. (Tikhur)

Botanical Name: *Curcuma angustifolia* Roxb.

Familly: Zingiberaceae

Local Names

- ☆ **Gujarati:** Tikhur
- ☆ **Hindi:** Tikhur, Tikora
- ☆ **English:** East Indian arrow root
- ☆ **Sanskrit:** Tawakwsra
- ☆ **Telugu:** Aararoot

Introduction

It is known asTikhur. Leaves are long. Instead or roots rhizomes are seen which stores food. Rhizomes are light yellow to little red in color. Leaves are green to light green.

Distribution

Tikhur grows wild in India and it is also cultivated in South India, Bihar, Bengal and Maharastra and Gujarat. In Gujarat Tikhur is found in wild condition in Girnar hills area and some other forests like Dang *etc.* It grows in arid and semi arid region.

Cultivation

The sandy clay soil with better water drainage conditions and climatic conditions in which temperatures ranges between 4° C to 40° C are suitable. The land is to be ploughed 2-3 times and cow dung is added to soil 50 tones per hectare.

Generally this plant is propagated through rhizomes. Rhizomes are being sown at 30 cm distance. Sprouted tubers are cut into pieces which contain buds (eye). These pieces are sown on the ridges made between drains. They should be planted in rows and at a distance of 15 to 20 cm. Depth should not be more than 5-10 cm. Irrigation is essential in the summer before monsoon but after monsoon irrigation should be continued at 15 to 20 days interval. Weeding and hoeing work is done at 15 to 20 days after rains.

Harvesting

This plant matures within 7 to 8 months, its tubers are to be removed from the soil in the months of January to February and dry them under shaded place. The bold tubers may be preserved for sowing in next crop.

Yield

The yield per hectare can be 40 to 50 tones. The present market value of rhizomes is Rs. 15.00 to 20.00 per kg, while mother rhizomes @ Rs. 250 per kg.

Medicinal and Economic Importance

- ☆ Rhizomes are source of starch.
- ☆ Rhizome extract is being sweet and nutritious.
- ☆ This plant is used to purify blood.

73

Curcuma aromatica Salisb. (Jangali Haldar)

Botanical Name: *Curcuma aromatica* Salisb.

Family: Zingiberaceae

Local Names

- ✰ **Gujarati:** Jangali haldar, Zedo ary
- ✰ **Hindi:** Jangali haldi
- ✰ **English:** Wild turmeric, Yellow Zedoary
- ✰ **Sanskrit:** Aranya harida
- ✰ **Tamil:** kastri-manjal
- ✰ **Telugu:** kasturi pasupa

Introduction

The rhizomes are light yellow in color which is internally orange. Wild turmeric is some times used as a substitute for turmeric (*C.longa*) but not as a condiment. Leaves are long containing reticulate venation. Leaves are elliptic or lanceolate-oblong, caudate-acuminate, 30-60 cm long, petioles as long or even longer, bracts are ovate, recurved, more or less tinged with red or pink, flowers pink, lip yellow, obovate, deflexed, subentire or obscurely three lobed. Fruits are dehiscent, globose, three valved capsules type.

Distribution

Wild turmeric is distributed through out in India, Cultivated chiefly in Bengal and Travancore. It is seen to be grown in Gujarat, Maharastra, Rajasthan Southern

and eastern part of India. In Gujarat especially in forest of Girnar, Dang and Barda turmeric is seen abundant.

Cultivation

It can be grown in tropical diverse conditions from sea level to moderately high hills and under a moderate temperature where maximum temperature remains around 30 to 35 c and precipitation rate of 150 to 200 cm or more. Plant can be grown in different soil from light black to sandy loam and red soils to clay loams where drainage is well and humus content is high. Land is prepared during early monsoon. Deep ploughing makes the land with fine tilth. Bed is prepared as 1 to 1.5 m width and 15 to 20 cm high. Crop is planted in April to May time. Rhizomes are splitted weighing 30 to 40g for planting. Firm yard manner and cattle manner is useful. For one hactre 2500 kg of rhizomes are required.

Harvesting

Crop becomes ready for harvest in 7-8 months. Land is ploughed and rhizomes are gathered by hand picking, cleaned of mud *etc.*

Yield

47500 tones of dry rhizomes valued at Rs. 234 to 754 were exported in 1982-83.

Medicinal and Economic Importance

- ☆ Rhizomes are highly useful. They contain d-camphene, d-camphor, sesquiterpenes, sesquiterpene alcohols, acids and some unidentified residues. A sample from India contains 9.4 per cent essential oil, showing high amount of arcurcumene, β-curcumene and xanthorhizol. 2-4 methyl phenol is also present.
- ☆ Rhizomes are bitter, carminative, appetizer and tonic.
- ☆ Rhizomes are also used in combination with astringents and aromatics for bruises, sprains, high cough, bronchitis, cough, and leucoderma and skin eruptions.
- ☆ Some times wild turmeric is used as a substitute for *Curcuma longa* but not as condiments.
- ☆ Turmeric is an auspicious article in all religious observances in Hindu households. It is a normal constituent of condiments curry powders and prepared mustard
- ☆ An antiperiodic alternative, mixed with warm milk is said to be beneficial in common cold.
- ☆ Juice of the rhizome is applied in skin affections. Externally it is applied in indolent ulcers, and a paste made from powdered rhizome along with life forms a remedy for inflamed joints.
- ☆ Decoction of rhizome is said to relieve the pain of purulent ophthalmia oil, has antiseptic properties.
- ☆ In small doses acts as a carminative, stomachic, appetizer and tonic. In large doses, it appears to acts as an antispasmodic inhibiting excessive peristaltic movement of the intestines.
- ☆ It is used for dying wool, silk and unmornted cotton to which it imparts a yellow shade in an acid bath.
- ☆ It is used in dyeing cotton, sometimes in combination with other natural dyes like indigo and safflower to impart different shades. It is used as a colouring matter in pharmacy, confectionary and food industry.
- ☆ Turmeric paper is official reagent in British pharmacopoeia for testing alkalinity.
- ☆ It is stomachic, tonic and blood purifier.

74

Cymbopogon martini (Roxb.) (Palmarosa Grass)

Botanical Name: *Cymbopogon martini* (Roxb.)

Family: Poaceae

Local Names

- ☆ **Gujarati:** Sugandhit ghas
- ☆ **English:** palmarosa grass

Introduction

This plant is in the form of perennial grass which attains height upto three meters. It grows wild in our forest areas. After the first showers of monsoon, it starts sprouting and by the end of October-November, flowering and fruiting is complete and then plant dries.

There are two varieties of this grass namely Motia and Sofia. The commercial Rosha oil is obtained from motia variety by steam distillation. The oil of Sofia variety does not possess the minimum requirement of gernol content and thus is inferior and does not fetch good price. The oil is in much demand for perfumery and medicinal purposes. Large quantities of this oil are exported to various countries.

Origin and Distribution

This perennial grass is native to Southeast Asia, especially India and Pakistan. Mostly this grass is grown for its oil in throught world.

Cultivation

This plant needs well drained soil with irrigation facility in areas with average rainfall of 1500 m. It grows even in well drained clayey soils, but does not tolerate water logging. Loamy and sandy loams are the best soils.It can be propagated either by raising seedling in nursery amd then transplanting in the field, and or by planting grass slips, Nursery beds are prepared by the end of May. Raised beds are preferred, so as to avoid the risk of washing away the seeds at the time of flood watering. Farmyard manure may be added in good quantity to the field. The seeds are mixed with fine sand and sown at a distance of 15 to 20 cm. in lines. The seedlings should not be very close to each other. About 5 kg of seeds are sufficient for one hectare of plantation.

Regular watering in less quantity is done. For better growth of seedlings, urea is applied. After 3-4 weeks, the seedlings will be ready for transplanting in the field. At the end of June or in the first week of July, the transplanting of seedlings shall be done. If irrigation facility is available, transplanting may be carried out much in advance. The field is prepared before monsoon and kept ready for planting. The nursery seedlings are removed without disturbing the roots and are transplanted in the field in lines of 45 cm apart and the distance between plant in the lines is kept 30 cm. In case of root cutting method, the best oil yieling variety grass is propagated. The root slips from the known variety of grass are taken out and planted after the first showers of monsoon. Irrigation is applied for the growth of plant but additional watering induces decreases oil yield. For first two years there is no need of fertilizers but during the growing season, addition of fertilizers is good for grsss. Every year at the time of appearance of new leaves, mixture of Nitrogen, Phosphorous and Potash should be added.

Harvesting

Harvesting is done in first year when flowering is maximum in the month of October-November. Esssential oil is available in all parts of the plant namely inflorescence, leaves, stems etc. but is more in flowering plants. Generally the vegetative portion is cut abouve 10 to 15 cm from the ground level and entire plant is used for steam distillation of oil. In the subsequent years 2 to 3 crops are obtained in a year. Palmarosa gives oil upto eight years period, but after four year, the percentage of oil declines. Therefore it is advisable to keep the crop upto four years only and replace the same with new crop later on.

Yield

An oil yield of about 220-250 kg/ha is obtained from second year onwards from an irrigated crop.

Chemical Composition

Plant contains geraniol, geranyl acetate, linalool, geranyl octanoate, geranyl butyrate, dihemiacetal bismonoterpenoid, cymbodiacetal.

Medicinal and Economic Importance

- The plant is acrid, bitter, thermogenic, appetizer, carminative, digestive, cardiotonic, depurative, galactagogue, diuretic, sudorific and febrifuge and is used in vitiated conditions of kapha and vata.
- This plant is also used in vitiated conditions of pita and kapha, neuralgia, brohnchitis, cough, catarrh, helminthiasis, anorexia, dyspepsia, colic, cardiac debility, leprosy, skin diseases, aglactia, strangury, epileptic fits in children, pharyngopathy and fever.

75

Cyperus scariosus R.Br. (Nagar Motha)

Botanical Name: *Cyperus Scariosus* R.Br.

Family: Cyperaceae

Local Names

- ✰ **Hindi:** Nagar motha
- ✰ **Gujarati:** Nagar motha
- ✰ **English:** Nut grass, Cyperiol
- ✰ **Sanskrit:** Nar gmusta

Introduction

The plant grows wild in damp conditions.Nagar motha a perennial grass, slender, glabrous ridge 0.5-1.0 m high with under ground, Sympodial rhizome from which clustered clums shoot out. Rhizomes have length up to 20 cm hardening into wing roots, Leaves are in three rows. Consisting of a closed sheath and narrow blades. Spikelets in solitary or umbelled heads or spikes or linear or oblong spikelets with one or more leafy, involucrate bracts and with bracteoles under secondary division of the inflorescence. Rachilla is persistent. Glumes are distichous, where lowest of two remain empty and above remain sexual. Fruits are trigonous or triquetrous- nut type. Plant starts flowering in 6-8 months, when there occurs increase in temperature and humidity during summer and rainy season plant growth and rhizome development become faster. In the field plants have erect shoots and long pointed drooping leaves.

Distribution

The plant occurs in damp situations of Uttar Pradesh, Madhya Pradesh Punjab, Gujarat and other southern parts of India. About 600 species of *Cyperus* are known to grow in India. Of which *Cyperus rotundus* and *Cyperus scariosus* are very useful as aromatic and medicinal plant. They occur in damp places, road side ditches, river banks and marshy lands. In the Gujarat Nagar motha is grown in Junagadh particularly in Girnar hills areas, in Barda hills or Porbandar and some other parts of the state like Dang, Panchmahal *etc.*

Cultivation

Medium kind of black soil, marshy lands are suitable for the growth of plant along with hot climate condition. Although plant may be grown on low and medium fertile soils. Field is required to be tilled 2-3 times with plough and leter made plain after mixing 15 tones of cow dung manure per hactre. Plant is highly affected by shade because under prolonged shading the development of underground rhizome is adversely affected and may cause death of shoots. Generally plant is propagated by roots. In starting of monsoon roots are sown at 15 x 15 cm distances. Per hactre approximately 2, 00,000 plants can be grown of land. Plants become ready for transplanting after 15-20 days. After monsoon irrigation is given at an interval of 15 to 20 days. It will keep roots moist and help in plant development. Weeding and hoeing activities are done in September to October; during this season soil around the plants needs good hoeing.

Harvesting

Crop matures in 18 months. In December-January, when leaves start becoming yellow, roots may be removed, cleaned, dried under shade and are stored. Per hactre 12 to 15 quintals of dry roots can be obtained. Roots contain 0.3 to 0.4 per cent oil.

Yield

Root provides important oil. It is possible to harvest 200-250 quintals semi dry roots from one hectare land from natural growing conditions. Under cultivation rhizome yield may go high up to 350 to 500 quintals as per plant population under wild conditions are very low and the plants are found growing scattered. Market value of the oil is Rs. 35 to 60 per kg.

Medicinal and Economic Importance

- ✫ Roots are very useful in medicine. Rhizomes are used for washing hairs
- ✫ Oil is obtained from tubers (rhizomes) which are used by perfumers as fixatives.
- ✫ It forms good substitute for patchouli oil in soaps and other perfumes.
- ✫ Oil is also used as hair tonic.
- ✫ Rhizomes are used for commercial extraction of Cyperus oil and as such in dhoop, hawan samgries, agarbatties etc. Oil is also used for flavouring of tobacco.

76

Datura sp. (Daturo)

Botanical Name: *Datura* sp.

Family: Solanaceae

Local Names

- ✰ **Gujarati:** Daturo, Datura
- ✰ **Hindi:** Sadahdhatura
- ✰ **English:** Thorn apple, Jimson weed, Stink weed, Mad apple, Thorn apple
- ✰ **Sanskrit:** Dhatturah, Dhusturah
- ✰ **Tamil:** Ummattai

Introduction

Datura is genus of poisonous, annual or perennial herbs or shrubs. There are almost 15 species found in the world from which 10 spices are found in India. Some of them like *D.innoxia*, and *D.stramonium*, *D.metel* are very common and they are very useful drug plants.

1. *Datura stramonium*

This species is a glabrous or farinose annual, usually gets 90 cm height. In the rich soil it can reach upto 180 cm or more. Stem is erect with spreading branches. Leaves are pale green in colour. They are obovate or triangular-ovate, 12.5 to 15 cm long and irregularly toothed. Flowers are 7.5 to 20 cm long, white or violet in colour. The capsule is erect, 7 x 3.5 cm, ovoid, thickly covered with sharp spines and dehiscing into 4 valves. The seeds are numerous and reniform. Its chromosome number is 12 (2n) so it is diploid.

2. *Datura innoxia*

This species is bushy, coarse and annual attaining a height of 90 to 120 cm. The leaves are dark-green, ovate, often somewhat cardate about 12.5 x 7.5 cm. The flowers are white and fragrant and about 7.5 cm long. The fruits are about ovate-conical, 5 cm long and exposing a long central column bearing numerous light- brown seeds and emits a heavy, narcotic odour.

3. *Datura metel*

D.metel is subglabrous, spreading herb sometimes becoming shrubby. Leaves are triangular-ovate in out line and unequal at the base. The flowers are 17.5 cm long some times double or triple, they are white, violet, and reddish –purple on the out side and white inside. Fruits are globose, tuberculate or muricate, borne on a short thick peduncle; the capsule dehisces irregularly exposing a mass of closely-packed, light brown, flat seeds, which nearly fill the interior. Two improved varieties of D. metel have been evolved by RRL, Jammu. They are,

(a) PRL-Purple

This is an early flowering type variety in which entire plant is purple and there are seen dark-green leaves. The yield is about 240-290 q of green herb and 18-24 q/ha of seeds. The alkaloid content varies from 0.24 to 0.36 in the leaves and 0.12 to 0.19 in the seeds.

(b) RRL-Green

This is late flowering type variety in which plant is green and there are seen light green leaves. The yield is 210 q of green herb and 1500-2000 q/ha of seeds. The alkaloid content varies from 0.24 to 0.28 per cent in leaves and 0.098 to 0.128 in the seeds.

Origin and Distribution

The plant is distributed throughout tropical and warm temperate regions of the world. It is distributed in India, Africa, Asia, Europe, Mexico, South America and USA. Generally it is believed to be native of Caspian region of Europe, North America and Indian subcontinent.

Datura stramonium is believed to be a native of South America and it is grown on the hills throughout India up to an altitude of 2400 m. It is commercially cultivated in the USA and Europe with a view to obtain a drug of uniform potency. In India it is cultivated in the hills of Uttar Pradesh, Himachal Pradesh, Kashmir Valley and some parts of Karnataka.

Datura innoxia is a native of Mexico; it is also found growing in the Western Himalayas, the hilly regions of the western parts of the Deccan Peninsula, and a few other places in India. *Datura metal* is the native of Asia or Africa. It occurs throughout India and is occasionally grown in gardens. *Datura metel* and *Datura innoxia* are found mainly distributed in the states of Punjab, Hariyana, Rajasthan, Gujarat, Andhra Pradesh, Maharastra, Karnataka and Tamil Nadu.

Cultivation

Datura can be sown in the variety of soils but prefer alkaline rich or neutral, clay, loamy soil also, those tending to saline alkaline reaction, rich in organic matter, for their successful growth and yield. Clayey, acidic or water logged soils do not suit this crop. This plant is sensitive to frost and sheltered situations. Open areas with bright sunshine are therefore preferred for their cultivation. Generally Datura is grown in warmer parts of the world. It can be grown upto an altitude of 2400 meter. Plant can't withstand in high rain fall or high temperature. Where annual rain fall is 100cm and temperature range between 10-15 c in winter and 27-28°C in May-June are ideal for its cultivation. Season of sowing varies for different species of Datura.

- ☆ *D.innoxia* is sown as winter or spring crop, seeds are sown in October or in late January.
- ☆ *D.metel* is grown in summer and seeds are sown in March.
- ☆ *D.stramonium* is grown in temperate areas like Kashmir valley. Seeds are sown in March-April and in November in the North Indian Plains.

Land is ploughed 2-3 times followed by planking. Weeds and stubble are removed and FYM with a basal dose of fertilizers is incorporated into the soil during the preparation. About 2 kg of seeds are needed to raise seedlings for planting one hactre.

Generally seed germination is very low and irregular due to the presence of an inhibitor. Seed germination can be enhanced by soaking the seeds overnight in water and washing them repeatedly 2 to 3 times with fresh water before sowing. It can also be enhanced by freezing and thawing. Seeds are sown in rows 45-60cm apart and covered with soil. Seeds start germinating within a fortnight and in a months time, the germination is complete. Weeding and thinning is done when plants are 10-12 cm high, keeping a plant to plant distance of almost 30-45 cm. Seedlings are transplanted when they are 8-12 cm tall and possess four leaves, at a distance of 30-45 cm in rows which are 45-60cm apart. For getting higher yield spacing of 75 x 75 cm is recommended. Generally transplantation is done in December to January.

Fertilizers and Irrigation

Datura grows quickly with the application of fertilizers. Nitrogen, phosphate and potash are useful to growth of Datura. FYM is also essential. When there is no rains first irrigation is provided after sowing or transplanting immediately. Then irrigations are given at intervals of 8-10 days depending upon the weather conditions, till the final harvest. Proper weeding and hoelings are necessary for growth of plant.

Harvesting

Harvesting method is different for different species of Datura. They are explained below.

D.stramonium: Crop can be harvested in 6-7 months after sowing. Entire plant is cut when the fruits are mature but green, and partially dried in the sun or in the shade. Leaves are stripped and separately dried. The seeds are shaken off from the capsules when the fruits begin to burst.

D. innoxia: Crop is harvested in 6-8 months after sowing for final harvest. If plants are raised in October or late January harvesting is done in 5-6 months or one month after flowering which is in April. Fruits when fully grown or when still green are plucked by hand 2-3 times before the final harvest. Final harvesting is done with the help of sickle by cutting down whole the plant. This harvested fruits are spread out under the sun until they open. Open fruits are threshed with a stick to separate the seeds, which are dried in the sun and packed in the gunny bags when fully dried.

D.metel: First harvest is obtained in July, in which all the tender branches along with the leaves are cut. Herbage is dried in the shade immediately. The harvested plants regenerate and two more harvests are taken. Second harvest is taken in late August and third one is taken in October. Plant is cut from the base in the last harvesting, if a ratoon crop is not desired. Besides foliage, a sizeable number of fruits are also harvested as and when they ripen.

Diseases

No major pests or diseases are found in India. However, thrips act as vectors in transmitting the mosaic virus, which can be controlled by spraying a suitable insecticide like Metasystox.

Yield

D.stramonium gives output of 11.2 to 17 q/ha of leaves and 8-9 q/ha of seeds, *D. innoxia* gives yield of 1200-1700 kg/ha of dried seeds with an alkaloid content varying from 0.2 to 0.35 per cent, and *D. metel* provides yield of 240-290 q/ha of green herb and a yield of 210 q/ha of green herb and 15-20 q/ha of seeds can be obtained from the variety, RRL -green.

Medicinal and Economic Importance

- Leaves flowers roots and seeds are used for medicinal purposes.
- Alkaloids of pharmaceutical importance extracted from these plants at present are hyoscyamine, hyoscine and neteloidin. They are narcotic, poisonous when used in large doses and medicinal in small doses.
- *D.stramonium* is antispasmodic and it is used in asthma and Parkinson's disease. Dried leaves are used to treat asthma by smoking it in cigar.
- Hysociamine hydrobromine is used as a pre anaesthetic medicine in surgery, in childbirth, in ophthalmology and prevention of motion sickness. This hydrobromide is isolated from the plant datura.
- Stramonium is obtained from *D.stramonium* dry leaves and flowers. It has disagreeable odour and a bitter, unpleasant taste. It contains 0.3 to 0.5 per cent of tropane alkaloids, hyomine and atropine in small quantity.
- Stramonium is narcotic, antispasmodic and anodyne which is used to relieve the spasms of bronchitis or asthma.
- Flowers are used to make poultice which can be applied to wounds to reduce pain.
- Decoction of flowers and roots has been used as a sedative to calm patients during setting of fractures.
- Poultice is also obtained from the leaves which are useful in checks inflammation of breasts caused by excessive formation of milk.
- Seeds of *D.innoxia* are utilized as source of hyoscine
- Alkaloids are obtained from different parts of plant like, leaves (0.41-0.45 per cent), stem (0.25-0.26 per cent), roots (0.21 per cent), fruits (0.46 per cent) and seeds (0.19 per cent).

77

Dendropthoe falcata Linn.f. (Vando)

Botanical Name: *Dendropthoe falcata* Linn f.

Synonym: *Loranthus falcatus* Linn./*Loranthus longiflorus* Desr.

Family: Loranthaceae

Local Names

- ☆ **Gujarati:** Vando
- ☆ **Sanskrit:** Vrksadani
- ☆ **Hindi:** Banda
- ☆ **Tamil:** Pulluruvi
- ☆ **English:** Mistletoe
- ☆ **Telugu:** Badanika, Jiddu

Introduction

This plant is parasite plant which grows on stem of the other plant and gets nutrients from the host plant. Bark is smooth and grey in colour. Leaves are usually opposite type and thick. They are variable in shape from ovate to linear-oblong. Mid rib of the plant is prominent. Secondary nerves are obscure. Flowers are orange red or scarlet in short spreading, stout, axillary, racemes or calyx inflorescence. Flowers are bisexual. Fruits are ovoid or oblong berries type and they are crowned by cup shaped calyx present inside the flower.

Distribution

This plant is very common throughout in India, occurs as parasite plant on the other plant and mostly found in tropical and sub tropical regions of the world.

Cultivation

This plant is cultivated by two methods (1.) Seed propagation and (2.) Vegetative propagation. Although this plant is found to be grown naturally in moist forests and arid and semi arid forest.

Chemical Constituents

This plant contains cytokinins and casein hydrolysate, quercitrin, kaempferol, quercetagetin, quercetin, myricetin, an acylxyloside of quercetin, hyperoside, myricitrin, meratin and rutin.

Medicinal and Economic Importance

- This plant is cooling, bitter, astringent, aphrodisiac, narcotic and diuretic.
- It is used in pulmonary tuberculosis, asthma, menstrual disorders, swellings, wounds, ulcers, strangury, renual and vesical calculi and vitiated condition of *kapha* and *pitta.*

78

Desmodium gangeticum (L.) DC. (Shali-parni)

Botanical Name: *Desmodium gangeticum* (L.)DC.

Family: Fabaceae-pepillionaceae

Local Names

- ☆ **Gujarati:** shal parni, shal pun, Prasna parni
- ☆ **Hindi:** shal van, shal pani
- ☆ **English:** Salwan, sap parni
- ☆ **Tamil:** Orila, Pulladi
- ☆ **Telugu:** Gitanaram, Kolakuporna
- ☆ **Sanskrit:** Anshumati

Introduction

D.gangeticum is an annual or perennial, sub erect diffusely branched, under shrub found almost throughout India. It is found at 2000m in Himalayas as well. Leaves are unifoliate, alternate, stipulate and ovate-acute. Flowers are small, pink in colour borne in terminal elongate racemes, with companulate calyx, papilionaceous corolla is found. Stamens are diadelphous and sessile, ovule are many in the ovary. Style is filiform and incurved and stigma, capitate. Fruit is compressed and moniliform. Each fruit has 6-8 seeds. Several variable forms are met within the species.

Distribution

Desmodium is a large genus of perennial or annual herbs or shrubs found throughout tropical and subtropical regions of the world. About 38 species are found in India. Among them D.gangeticum is most useful as drug. It is seen at high elevation at 2000 meters in Himalayas. It is distributed in Jammu and Kashmir to Kanyakumari. It can be seen in Madhya Pradesh, Gujarat, Uttar Pradesh, Bihar and most of the other states of India. In Gujarat it is widely seen in Porbandar and Junagadh in Girnar hills and Barda hills.

Medicinal and Economic Importance

- ☆ Shal parni is used in Ayurveda as one drug of the Dashmoola. This species of Desmodium is utilized in the formation of Dhyavanaprasam and dhanvantharam tailam.
- ☆ Shaliparni is used as effective cardio tonic with diuretic and laxative actions.
- ☆ Roots of this plant are useful as antipyretic, expectorant, alterative and diuretic actions as they contain pterocarpenoids, gangetin, gangetinin

and desmodin. Aerial parts of this species contain tryptamine derivatives. Alkaloids from aerial parts have hypotensive and anticholinesterase activity and they act as stimulants of central nervous system.

- ☆ This species is employed as a nervine tonic.
- ☆ It is useful in burning sensation, fever, dysentery, and thirst and vomiting and allays difficulty in breathing.
- ☆ *D.pulchellum* is used against diarrhea and haemorrhage, *D.velutinum* is found to be useful against cancer, *D.heterocarpum* is useful in anti inflammatory action and *D.motorum* is used as aphrodisiac.
- ☆ Plant is useful in curing hazy vision. *D.triflorum* is effective as antidysentric and galactogogue which can be used against wounds and abscesses.

79

Dioscorea sp. (Vidari Kandh)

Botanical Name: *Dioscorea* sp.

Family: Diascoreaceae

Local Names

- ✰ **Gujarati:** Vidari kandh
- ✰ **Hindi:** Chupri alu, khamalu
- ✰ **English:** Greater yam, Asiatic yam,
- ✰ **Sanskrit:** Alukam
- ✰ **Tamil:** Perumvallikilanku, kappan kaccil
- ✰ **Telugu:** chedupadudumpa

Introduction

Dioscoria is twining annual herb plant which has around 600 species distributed worldwide, among them 60species are found in India. Stem of yam is weak, it is rope like structure which climb by twining. It usually grows to a length of several meters, which have to twine around some support. Variation is seen in shape and size of leaves ranging from 50 to 200 sq cm. tip of leaf is pointed. The lamina is simple except the species belonging to the section Lasio phyton (*D.hispida* and *D.dumetorum*) which have trifoliate compound leaves. Lamina surface of *D.cayenensis* is dark and glossy, veins remains distinct. The main veins are radiated from the base of the lamina and the other veins are reticulate. *D.rotundata* leaf veins contain variable amount of anthocyanin and petioles are enlarged at two extremes. Leaf colour is usually green except in *D.alata.* in which purplish anthocyanin pigments mark the young leaves and older leaves are quite larger and light green in colour.

Petiole may be winged or spined. Petiole may bear a set of prominent spines as in *D.alata* and *D.esculanta* or may be dilated into ear like auricles as *D.bulbifera*. Base of petiole is swollen. Phyllotaxy may be opposite or alternate on the lower parts of the stem and ossposite on the top. Flowers are dioecious but monoecious and hermaphrodite flowers are also seen. Generally flowers produced in early phase of the plant that is during June in West Africa. Female flowers are produced later. Roots are fibrous type and they remain within top 30 cm of soil in upper layer. Roots are developed early in their life cycle. But in sprouted tuber in storage, numerous short, stout roots arise at the base and once planted, those roots form a massive structure. These roots elongate rapidly and become the feeding roots.

Fruits are trilocular, dehiscent capsule type; junction of the locule is extended out into flattened wings. They are 1-3 cm long. When fruits become dried, dried fruit dehisces along the 3 sutures, separating the locules and making it easy for dispersal of the seeds. The plant dies at the end of the season, but the fruit is undehisced and remains so for several weeks. Seeds are small, light and possess a flattened wing like structure for its easy dispersal of wind. Some important species are explained below. Tuber formation in yam is the onset of meristematic activity at the junction of the stem. *i.e.* at the base of the vine and the hypocotyls region. Most of the yam tubers exhibit strong positive geotropism. In a structure that originates from the hypocotyls region, the transition zone between the stem and the root as a lateral outgrowth and hence, it is considered that yam tuber is neither a root nor a stem structure, not a root wit root cap. It lacks any scale leaves, buds or eyes from a massive corm like structure located at the base of vine. The corm is formed very early and later the tuber arises from it. Hence the corm is considered as the head of the tuber which has several buds presents and end of the dormancy period of the stored tubers, the sprouting will occur from the head end of the tuber, near the point of attachment.

Description of the Various Types of Tubers in Yam

Species	*Tuber Type*
D. alata	Shape is variable, tuber branching is seen, mostly cylindrical, brown to black colour, tuber watery in taxure. Tubers possess a thin layer of fiber below the skin.
D.rotundata	Tuber shape is cylindrical, skin is thick, smooth and brown, proviedes good protection in storage, flesh is white, starch grains are large and ovoid.
D. cayenesis	Cylindrical, skin is thick, brown and smooth, provides good protection in storage. tuber flash is yellow so it is called as yellow yam.
D. opposita	Tubers are very long and thin and hence very difficult to harvest. It also produces aerial tubers in leaf axils
D.bulbifera	Bulbil is the major storage organ, which is used as setts in the cultivation of this species. Bulbils are produced on the aerial parts of the plant. The underground tuber is small in both Asian and African varieties.
D.dumetorum	Tubers are bitter and poisonous, because of high content of alkaloids, dioscorine or its derivatives.
D.trifida	Tubers flash may be white, yellow, pink or purple. They are palatable.
D.hispida	Tubers are toxic so it is known as intoxicating yam. They are made edible after detowicating through running water or salt-water steeping.

Some of the species of yam is explained below.

1. ***D. deltodea*:** Rhizome is superficial, horizontal, tuberous, digitate and chestnut brown in colour. Vines are glabrous, left twining. Stem bears alternate petiolate leaves. Lamina is long and widely cordate. Flowers are borne in axillary spikes, male spikes, 8-40 cm long with 6 stamens. Female spikes are 15 cm long and 3.5 cm broad. Fruits get matured in 90 days. Capsules have 4-6 seeds that are winged.
2. ***D.composita*:** Tubers are large, white and deep rooted. Buds are confined to the crown portion. The vines are right twinning and nearly glabrous. Leaves are alternate and have long petioles, membranous or coriaceous lamina abruptly acute or auspidate acuminate, shallowly or deeply cordate. The fasciculate inflorescence is single or branched with 2-3 sessile male flowers, with fertile stamens. Female flowers have bifid stigma, stigma receptivity lasts for 30 hours. Fruits mature in 110 days. Tubers are large, white and deep rooted.
3. ***D.floribunda:*** Vines are glabrous and left twining alternate leaves which are borne on slender stems with broadly ovate, shallowly or deeply cordate coriaceous lamina with thick and firm petioles. Leaves are variegated to

varying degrees. Male flowers are solitary or in pairs. Female flowers have divariate bifid stigma. Capsules are obovate with winged seeds. Fruits mature in 120 days. Flowers are insect pollinated. Ants are main agent.

4. ***D.bulbifera*:** It is known as aerial yam or potato yam. It is characterized by the production of large number of bulbils on each plant which are edible. Bulbils are small, round. They are characterized by the swellings in the axils of the leaves. Thorne and pickles are found in many species. Tubers are source of tannins, alkaloids, and saponions which render them unpalatable or poisonous. Other medicinally useful species are *D.pentaphylla, D.opposityfolia,, D.hispida etc.*

Origin and Distribution

Dioscorea alata spreads from Burma to India, Malaysia, Sumatra and other parts of South-East Asia. Later, it spreads towards the West, to Madagascar, East Africa and other parts of the tropics such as West Africa, West Indies, Pacific Island, and tropical Asia and also to East Africa. After 1500 AD, its cultivation was expanded to west coast of Africa and widely known as Lisbon yam and grows as main food for ship slaves. It is most cultivated species of yam in India and it is grown as a subsidiary food. *D.rotundata* had been introduced from West Africa to India in 1976 through IITA, Nigeria in the form of seeds and now it is greatly established. It grows extensively in West Indies and some extent in East Africa. It has spread from Senegal to Ethiopia and to Uganda and Angola to Zimbabve.

D. bulbifera is mostly available in West Indies, South Pacific Islands and also to some extent in India. It is the only edible species, native to both Asia and Africa. *D. esculanta* is cultivated in China since the second century. It is spread widely in South East Asia, Pacific Island and West Indies. In India wild types are found. Madhya Pradesh, Orrisa, Uttar Pradesh, West Bengal, Tamil Nadu and Kerala are the states which grow this species. During sixteeth century, *D.cayenesis* was introduded from West Africa to West Indies. *D.dumetorum* is available only in tropical Africa in both cultivated and wild forms an widely cultivated throughout west Africa 15 N to 15 S. *D. trifida* is believed to be originated in South America and its cultivation is restricted to West Indies. *D.hispida* extends from Western India to Malaysia and Papua New Guinea. Cultivated types are poisonous as the wild ones. D. opposite is extensively cultivated in China and Japan and used for traditional medicinal purposes in China. *D.maxicana, D.composita, D.floribunda* are commercially grown in Maxico.

Cultivation

Dioscorea can be grown in several types of soils. Light or sandy soils require good irrigation and fertilization, whereas heavy clay soils restrict tuber growth and harvest and often create water logging. Medium loam and deep soils provide good yield. Some spicies like *D.floribunda* and *D.composita* are more suited to the tropics; *D.deltoida* is reported to be a suitable species for temperate locations. It can grow well in the temperate regions of Kashmir and Himachal Pradesh. The field should be ploughed and harrowed several times, leveled properly nd drainage channels should be made. Since yams have a high requirement of organic matter for good tuber formation, a sufficient quantity (20-25 t/ha) of FYM is incorporated at the time of land preparation. 45 x 30 cm for a 1-year old crop and 60 x 45 cm for 2 year old crop in *D.floribunda* is found to be optimum under irrigated conditions.

Deep furrows are prepared at 60 cm distances in which sprouted tubers are planted. After sprouting is completed, earthing up is done, utilizing the soil from the ridges.

Propagation

Yams can be propagated by (a) Tuber pieces (b) seeds and (c) by stem cuttings. It can also be propagated *in vitro* conditions.

(a) Tuber pieces: *Dioscorea* grows best by rhizome or tuber method. Tubers are prepared by cutting 50-60 pieces of rhizome. If pieces are small then yield is less. In fact 3 types of pieces are utilized for planting (1). Stems or crowns (2). Middle portions or Medians and (3). Tips or distal ends. Crown produce new shoots within 30 days of planting, while others take nearly 100days to sprout. Crown gives less diosganin compared to the median and the tips, hence the latter can be used for the extraction of alkaloids and the former can be used for propagation. Although median and tips can also be useful for planting. Only healthy tubers are selected to avoid the rooting or tubers. Healthy tubers must then be dipped in benlate fungicide (0.3 per cent) for 5 minutes followed by dusting the cut ends with 0.3 per cent benlate powder before planting or storage. The benlate treated tuber pieces should be kept in raised beds in the shade, covered with sand and watered daily. After 30 days sand may be removed and the sprouted crowns taken out and planted in the field. The median and tip portion are again covered with sand and watered regularly. Good time for planting tubers is February-March or June-July.

(b) By seeds: Seed propagation is also useful in some species of *Dioscorea.* Seeds contain wide membranous wing that can be removed without affecting germination. The seeds can be sown either in raised beds (there should be mixture of loamy soils and FYM.) in the shade or in polythene bags (filled with sand soil and FYM in 1:1:1).Planting depth should not be more than 1.25 cm and frequent watering of the beds is essential. The seeds germinate within three weeks and are ready for transplanting in 3-4 months. Seedlings are supported immediately. Just before start of the rain we should transplant seedlings to the field. Flowering time in Dioscorea is August to September. So female and male plant should be grown closer to each other that we can get seeds easily.

(c) By Stem cuttings: Stem cutting is also useful method. Vines are raised from 50-100g tuber pieces in the green house. One or two month old vines, are taken and cut into single node cuttings, each with one leaf. Before planting cutting should be treated with 1 ppm 2, 4-D and 0.1 per cent benlate for 4 hours. The beds are watered regularly. After rooting, the cuttings are transplanted to polythene bags and produce about ten leaves in a period of two months. *D.floriblunda* can also be done by air and ground layering.

(d) ***In vitro* propagation:** Both single nodes and apices explants are reported to produce full fleged plants after 90-100 days. Single nod leaf cuttings of *D.floribunda* when used as explants; give 100 per cent rooting when grown *in vitro*.

Irrigation and Fertilizers

During summer months frequent irrigation is needed. 4 to 5 days interval in hot condition is good while in winter 7-8 days gap for irrigation is good for better growth of plant. This plant grows quickly by application of fertilizers. Nitrogen, Phosphorus and potassium are useful for its growth. A fertilizer dose of 300kg N, 150 kg P and 150 kg K are useful. The entire quantity of P should given as a basal dose while N and K are given in four equal split doses at bimonthly internals commencing from 2 months after sprouting. For getting good tubers yield and diosgenin content the application of S, Ca, and Mg are useful.

Harvesting

Harvesting depends on the types of species. Time varies for different species. In *D.deltoidea*, harvesting should be done after three years to get optimum yield from the crop with maximum diosgenin content. Similarly *D.floribunda* should be kept at least for two years in the field for viable yields. Tubers are harvested generally in the month of February to March when tubers are in dormant condition. Tubers give maximum yield of diosgenin at that time. Harvesting can be done by manual labour with pickaxes.

Yield

Average yield varies by species to species. An average yield of 15 to 20 t/ha of fresh tubers can be obtained during the first year, and up to 40 to 50 t/ha during the second year. Diosgenin content of the tubers tends to increase on an average, from 2.5 to 3 per cent in 1st year and 3-3.5 in 2nd year. On the basis of prevailing market conditions with dry tubers a net profit of about Rs.30000 per hactre can be obtained in one year crop while Rs. 60000 per hactre can be obtained in two year old crop. Yield depends on type of soil.

Diseases and Control

No major diseases have been recorded to affect this plant. Rotting affects the crops of D.floribunda tuber pieces during storage in sand beds. This can be controlled by treating tubers with 0.3 per cent benlate solution. Leaf spot has also been reported in *D.composita* which is caused by *Drechelera sorokiniana*, can be controlled by spraying benlate solution. (0.1 per cent).

Medicinal and Economic Importance

- ✰ It is steroidal drug which alone constitute about 6 per cent of the total production of pharmaceuticals.
- ✰ It has antifertility activity and used in family planning programme of the developing countries.

- Very useful Diosgenin is obtained from the rhizomes of various species of Dioscorea. It is the major base chemical for several steroid hormones including sex hormones, cortisones, other corticosteroids and the active ingredient in the oral contraceptive pill.
- The other important sapogenins found from *dioscorea* are yamogenin, botogenin and kryptogenin.
- Minor sapogenins like pannogenin and tigogenin are also found in certain cases.
- It is estimated that Maxico is world's highest producer of diosgenin, producing about 750 tonnes annually. Indian pharmaceutical companies of India are the largest buyer of Diosgenin.

80

Diplocyclos palmatus (L.) Jeffrey (Shiv Lingi Ni Vel)

Botanical Name: *Diplocyclos palmatus* (L.) Jeffrey

Synonym: *Bryonia laciniosa* Linn.

Family: Cucurbitaceae

Local Names

- ☆ **Gujarati:** Shiv lingi
- ☆ **Tamil:** Sivalingakkay, Aiviralkkovai
- ☆ **Hindi:** Shiva lingi, Ishavar lingi.
- ☆ **Telugu:** Lingadauda
- ☆ **Sanskrit:** Lingini

Introduction

This plant is slender much branched tendril climber. From a thick permanent root stock, tendrils 2-fid; Leaves are simple, alternate, membranous, 5-lobed, scabrid above, pale and smooth beneath deeply cordate at the base, margins are sinuate, sometimes they are subserrate. Flowers are yellow in colour, unisexual, males in small fascicles of 3-6. Females are solitary or a few. Fruits are subsessile globose and are smooth berry brick red when they ripe with vertical lines. Seeds are yellowish brown in colour.

Origin and Distribution

This plant is found throughout India. It is found on hedges and bushes upto

1,200 meter elevation. This is very common in dry and semi dry forest in India, eastern and central tropical Africa from Mozambique northwards to the Sudan and Congo Republics, also in tropical Asia, Malesia including the Philippinesis, and tropical Australasia. Origin of the plant is still topic of uncertainty.

Cultivation

This plant is propagated through seeds or vegetative means.

Medicinal and Economic Importance

- ✰ The plant is acrid, thermogenic, anti-inflammatory, foetid, alterant, depurative and tonic.
- ✰ This is also useful in vitiated conditions of *Piita*, cough, flatulence, skin diseases, inflammations and general debility.

81

Eclipta alba L. (Bhangaro)

Botanical Name: *Eclipta alba* L.

Synonym: *Eclipta prostrate* L, *Eclipta erecta* L.

Family: Asteraceae

Local Names

- ☆ **Gujarati:** Bhangaro
- ☆ **Sanskrit:** Bhrangarajah, Tekarajah
- ☆ **Hindi:** Bhamgra
- ☆ **Tamil:** Kayyantakara, Kalkesi
- ☆ **English:** Trailing eclipta
- ☆ **Telugu:** Galagara, Guntagalijeru

Introduction

Bhangra is perennial herb, it is erect or prostrate and gets height of 30-40 cm. Stems are purple or green colour, bristly, thickened at the nodes. Leaves are opposite, subsessile, lanceolate-oblong, denticulate, hirsute on both sides. Flowers are white in colour and it is arranged in axillary or terminal head. Generally bisexual flowers are present. Achene 3-angled which is slightly flattened.

Distribution

This plant is distributed throughout India. It is mostly found in waste lands

and waste grounds. It also wildly grows on both the sides of the road. It is common weed of tropical climate. It comes up even on the poor soils.

Cultivation

It is propagated by seeds. The seeds are collected in advance and kept carefully preserved. The area is ploughed and continuous furrows are made at 25 cm apart. Then the seeds are sown in the furrows continuously, at the beginning of monsoon. The seeds germinate soon. Weeds are removed and leaves are collected in December or January.

Harvesting

The leaves are plucked and dried.

Yield

According to estimation, 2000 kg. of leaves are available from one hectare.

Chemical Constituents

This plant contains thiophene derivatives. The co-occurrence of mono-, di- and trithiophene acetylenes together with α-terthienyl in this species is noteworthy. The petroleum ether extract of aerial parts contains a terthienyl aldehyde, ecliptal besides stigmasterol and β-sitosterol, the aerial parts also contain 2-an-geloyloxy methylene-5′-dithiophene, 5′-isovaleryloxy methylene-2-dithiophene. The roots are very rich in thiophene acetylenes. They contain the dithiphene derivatives 5′-seneocioyl oxymetylene-2 dithiopene, and 5′-tigloy-loxymethylene-2-dithiophene in addition to 2-5-thiophene.

Medicinal and Economic Importance

- ☆ This plant is bitter, acrid, thermogenic, alterative, anti-inflammatory, anthelmintic, anodyne, vulnerary, ophthalmic, digestive, carminative, haematinic, diuretic, aphrodisiac, trichogenous, deobstruant, depurative and febrifuge.
- ☆ It is useful in the hepatosplenomeglay, elephantiasis, inflammations, vitiated conditions of vata, gastropathy, anorexia, helminthiasis, skin diseases, wounds, ulcers, ophthalmopathy, debility, hypertension, strangury, leprosy, pruritus, fever, jaundice, odontalgia, otalgia and cephalalgia.
- ☆ It is good for blackening and strengthening of the hair, for stopping haemorrhages and fluxes, and for strengthening the gums.
- ☆ The seeds are good for increasing sexual vigour.

82

Emblica officinalis Gaertn. (Ambala)

Botanical Name: *Emblica officinalis* Gaertn.

Synonym: *Phylanthus emblica* L.

Family: Euphorebiaceae

Local Names

- ☆ **Gujarati:** Amala, ambala
- ☆ **Hindi:** Amala
- ☆ **English:** Ambala, Indian gooseberry
- ☆ **Sanskrit:** Adiphala, Amaloko, Dhauri
- ☆ **Tamil:** Nelli
- ☆ **Telugu:** Usirikai

Introduction

This plant has many forms like herb, shrub and trees. Some of the species are often cultivated in green houses for their graceful foliage. Word phylanthus is derived from Greek word phyllon which means leaf and anthos (flower) which can be understood as bearing of flowers on the leaves. Embica is medium sized deciduous tree but grows quite tall reaching a height of 19 meters or even more. Tree is referred as evergreen tree by most of the people. Stem is mottled and gray color bark is seen. Leaves are feathery, small, fine and delicate. Tree has a peculiarity of shedding its twings along with leaves attached. Flowers are greenish, very small and are borne in clusters often on the naked portion of the twings below the leaves.

The tree flowers from March to May and fruits ripe from November to February. The fruit is nearly stalking less, smooth, yellowish, green and fleshy. The fruit is divided into six segments through pale linear grooves. The fruit is shiny and the size varies from small seeds embedded in the flesh. Aonla is hardy plant and does not require much attention.

Origin and Distribution

Origin of amala plant is tropical southeastern part of India. Plant is distributed throughout India. It is also grown in Myanmar, Sri Lanka, Puerto Rico, African Countries and many parts of south and Latin America. In India it grows in Karla, Tamil Nadu, Karnataka, Madhya Pradesh, Maharastra, Orissa and also in the plains of North India. It is seen naturally growing in mixed deciduous forests of most of the states of India and flowers and fruits even in hill tracts up to an elevation of 1000m from sea level.

Varieties of Aonla

Many varieties have developed in India. Some of them are described here.

1. **Amrit (NA 6):** This is a selection from 'chakaiya' and has mid season maturity. Fruits are most attractive, shining, medium to large sized. Generally they are 35-37 gm in weight, they contain very low amount of fibre content. They are very useful for candy making owing to low fibre content.

2. **Neelom (NA 7):** This is a selection from 'Francis' and has seasonal meturing. Fruits are medium to large sized containing weight of 35-37gm with conical apex and free from necrosis. Fibre content is little higher than NA-6. It is most precocious and prolific bearer (9-7 female flower/ branchlet). An ideal variety for preparation of processed products and has great commercial promise.
3. **Narendra Aonla 9:** This variety of Aonla is early maturing and is the selection of Banarasi. Fruits are medium to large sized, roundish at the stylar end and flattened. Fruits are pink tinged colored during early stage. This is good bearer and most suitable for dehydration and pickling.
4. **Anand 1:** This is a seedling selection from local seedling population. Trees of these varieties are tall with upright growth habit. It is moderate bearer having 1-2 female flowers per branchlet. Fruits are small with sligh rough and thick skin. Fleshy part of fruit is hard which contains fibers.
5. **Anand 2:** This is variety with tall tree, upright growth habit. It is a good bearer. Fruits are medium to small with rough skin. Female flowers are 2-2.5 per branchlet. This variety has medium keeping quality.
6. **Anand 3:** This variety is moderate bearer with yellowish green fruits. It contains 6-7 segments. It is developed at GAU, Anand.
7. **Luxmi 52:** It is recently developed at CISH, Lakhnow.
8. **Balwant:** This variety is the selection from Banarasi variety of Aonla. It is early variety with better salt tolerance. Fruits are big to medium in size containing weight of 40-50 gm. They are round and flat in structure. The flesh is fibreless.

Cultivation

Amla grows well under tropical or humid tropical condition. A well distributed rain fall is required for proper flowering and fruiting. Generally plant grows well in sandy loam-clay soils in India. It requires a deep soil. Presence of concrete layer at less than 1 meter depth may stop tree growth after 10-12 years. This can be cultivated in soils from slightly acidic to slightly alkaline or saline in reaction. Plant can be propagated by seed or it may also be propagated by Vegetative propagation.

Seed Propagation

When Amala fruits ripe seed become brown colored in November to December. Fruits are sun dried for removal of seeds. Fruits are also cut with the help of stone to obtain seeds. Some of the seeds may be embryo less and floating them on water just before sowing may be useful to separate seeds from fruits. Germination of seed is rapid and it is 100 per cent. Seedlings are raised for rootstock in field or in pots.

Vegetative Propagation

Vegetative propagation is done by inarching, budding or layering. The inarching is similar to that followed in mango and is done in the months of June and July with 25-30 percent success. There is a drawback with inarching that a large number of grafts may dry up soon after the season is excised from the mother plant. Further, a heavy loss may also occur when inarched grafts are transplanted in field, due to separation of scion from the stock at the point of union. Budding is the most successful technique in the multiplication of Amla. at the point of union. Budding is the most successful technique in the multiplication of Amla. The bud take as well as survival, both are much better as a result of patch budding in May. Three different budding methods namely, shield; patch and fork have been tried with good success in the months of June to September. Bud take of 100 per cent is obtained with the fork and patch methods and the maximum success and survival after transplanting occurs in September budding. Budding methods can be successfully used to rejuvenate seedling trees or old bearing trees of gooseberry. Primary branches of such trees are headed back and cut portions are treated with coal tar. The selected new shoots arising from these stubs are budded in June to September. The technique of air layering has been tried with success at Kerala Agricultural University. Use of moss as a media is found to be best for air layering.

Grafted of budded plants or layers should be planted at the distance of 8-10 m. The planting is done in June July. Pits are prepared of 1 x 1x 1m size. In each pit 20-30 kg of compost mixed with red soil is filled in. Plants are planted in the centre of this pits. This tree needs sharp light and is sensitive to forst and drought. Under suitable environmental condition the tree coppices well and pollards moderately well. Growth of plant is fast in starting but later it becomes very slow.

Harvesting

Harvesting is done in the 10 to 11 year of planting. How ever it continues to yield well for about 50 years under well nourished conditions. Fruits are plucked by hands. Fruits should not be droped on ground because they can be destroyed and their quality can be degraded.

Diseases

No specific diseases have been noticed.

Yield

Per tree around 100 to 150 kg of fruits are produced. Within one hectare if there are 150-160 trees are there then they will give yield whole life.

Medicinal and Economic Importance

- Almost all the parts of alma tree are of great economic importance. Fruits are very good sources of vitamins and minerals and hence used as a constituent of food and medicines.
- Vitamin c obtained from fruits of amla is more useful than vitamin c obtained from capsule.
- Amala are one of the three constituents of the well known Indian preparation Triphala. (Triphla is obtained from amla, Terminalia bellerica and terminalia chebula) Triphla is laxative and it used in the treatment of enlarged liver, piles, stomach complaints, pain in eyes etc.

- Fruits of Emblica are good liver tonic, raw fruits are cooling and mild laxative.
- Fermented liquor is made from the fruit is useful in indigestion, anaemia, jaundice, certain heart complaints, cod in nose and for promoting urination. Dried froots are used in diarrhea and dysentery.
- Even pickled fruits are used in Indian medicines. *Chyavanpras* is obtained from the fruits of the Emblica. This is used for growth, vigour, andgeneral health.
- Oil prepared from the fruits is said to have the property of promoting the hair growth.
- It is largely used in making inks and dyes and for hair shampoos. Fruits along with bark are used for tannins and polyphenols.
- Several alkaloids, flavanoids, quinines, steroids and terpenoids are also isolated from Amla plants.
- Amala candy, preserve, jam, pickle etc. are some of the commonly made products. Fruits acts as detergent,
- Bark and leaves are rich in tannin (21-22 per cent). Wood is also used for poles and agricultural implements.

83

Embelia ribes Burm.f. (Vav Ding)

Botanical Name: *Embelia ribes* Burm.f.

Family: Myrsinaceae

Local Names

- ☆ **Gujarati:** Vav ding
- ☆ **Hindi:** Vay vidang
- ☆ **English:** Emblica
- ☆ **Sanskrit:** Vidangah
- ☆ **Tamil:** Vayu-vilamga
- ☆ **Telugu:** Vidanga, Vatuvidanga

Introduction

Vav ding is very common plant in India. It is grown throughout in India., It is also grown in the areas up to 1,500 m elevation in hilly regions. It is a large scandent shrub which has long, slender, flexible branches. The bark is studded with lenticels. Leaves are simple alternate elliptic-lanceolate and short. They are glabrous on both the sides, shiny above, silvery beneath gland dotted, glandular pits near the midrib beneath. Flowers are white or greenish white. Inflorescence is axillary, lax panicled racemes. Fruits are globular dull red to nearly black berries, longitudinally striated with short slender persistent pedicel and style base. Seeds are single globose, hollowed at the base, white spotted, testa are membranous, albumen pitted.

Distribution

This plant is seen to be distributed in tropical and sub tropical countries like Japan, India, Pakistan, Bangladesh, Sri Lanka etc. The plant is endogenous to India and found throughout in India upto an altitude of 5,000 feet. In India it is grown in Gujarat, Maharashtra, Punjab, Haryana, Bihar and most of the other states.

Cultivation

Light and sandy soil with good drainage conditions is suitable for its cultivation. Dry and temperate climate is good for its better growth. Cow dung provides good growth to the plant. Generally 7-8 tones of decomposed cow dung manure is applied in the field per acre. Plants are propagated through seeds and vegetative method. The seeds are sown in nursery during May-June to prepare the plants. The seeds sprout in nearly 30-45 days. After 6 months of sowing plants are ready and should be transplanted in the field at a distance of 2 x2 feet in ditches of 1 x 1x 1 feet prepared in advance. Irrigation should be given at irregular intervals of 15-15 days. Weeding and hoeing activities are given according to need.

Harvesting

Plant provides fruits after two years generally in August to September. Fruits are ripen in November to January. The ripped fruits are to be plucked by hands. Then they are dried under the sun and then filled in the bags for storage. Roots are also harvested. For that whole plant is uprooted after two years of age and is separated from plant. Roots are cleaned and dried.

Yield

From the second year onwards nearly 25-30 quintals dry fruits per hectare can be obtained. In the market these dry fruits are sold at the rate of Rs. 80.00 to Rs. 100 per kg.

Medicinal and Econonomic Importance

- ☆ Generally whole plant is used for medicinal value. Roots, leaves and fruits are used.
- ☆ Roots are acrid, astringent, thermogenic and stomachic, and are useful in vitiated conditions of *vata*, odontalgia, colic, flatulence and dyspepsia.
- ☆ Fruits are also acrid, astringent, bitter, thermogenic, anthelmintic, depurative, brain tonic, digestive, carminative, stomachic, diuretic, contraceptive, rejuvenating, alterant, stimulant, alexeteric, laxative,

anodyne, vulnerry, febrifuge and tonic, and are useful in vitiated conditions of *kapha* and vata, helminthiasis, skin diseases, leprosy, pruritus, nervous debility, amentia, dyspepsia, flatulence, colic, constipation, strangury, tumors, asthma, bronchitis, dental caries, odontalgia, hemicrania, dyspnoea, cardiopathy, psychopathy, ring worm infection, fever, emaciation and general deability.

☆ Leaves are used as astringent, thermogenic, demulcent and depurative, and are useful in prutitus, skin diseases and leprosy.

84

Ephedra sp. (Som Vel)

Botanical Name: *Ephedra gerardiana* Wall.

Family: Ephedraceae

Local Names

- **Gujarati:** Som vel
- **Hindi:** Som lata, Phock, Khanda
- **English:** Ephedrine plant

Introduction

Ephedra is small erect shrub with dark green, striated, cylindrical branches, which are arising in whorls. The branches are green to dark green, and curved. Nodes and inter nodes are seen clearly. Internodes are 1-4 cm long and 1-2 mm in diameter. The fruit is ovoid, red, sweet and edible. In each fruit 1 to 2 or more seeds are enclosed by succulent bracts. There are so many species of Ephedra like *E. intermedia, E.major, E.foliata etc.*

Ephedra intermedia

It is densely branched, erect or prostrate shrub, met with in Kanawar and to lesser extent in Kashmir.

Ephedra major

An upright, rarely ascending, densely branched shrub, twigs of which closely resemble with *E.gerardiana.*

E.foliata

A tall, scandent shrub bearing edible fruits found in plains of Punjab and Rajasthan.

Distribution

It is grown in Russia, UK, Kenya and Australia. A few areas of Northern Himalayas like higher altitudes of Kashmir, Punjab, Gujarat, Himachal Pradesh and Uttar Pradesh are suitable for its cultivation. 4 species of ephedra occur in India.

Cultivation

Seed propagation is good method. Seeds have poor viability. Therefore they are collected fresh from the fruits which are ripe from June to July. The seeds, when not removed from the fruits germinate viviparously. The seeds are sown in the nursery beds in September to October. By the end of cold season, when germination is over weeding is done. Seedlings will be ready for transplanting in the field by the onset of monsoon. The field is thoroughly ploughed or raked up and unwanted vegetative matter is removed. After providing for drainage channels, the seedlings are brought from nursery and transplanted in the field. In June after one or two pre monsoon showers. At the spacing of 1 X 1 m plants are planted. After one month of planting first weeding is carried out. Second and third wee dings are also important in first year.

Harvesting

Harvesting is done from second year and every year afterwards. The crop can be retained up to 5 years, after which the crop is renewed once again. October –November is good time for harvesting when alkaloid contents are higher at that time. The first harvest is taken in the second year and later on every year. The aerial branches are cut flush to the ground and the material is dried completely under shade, before dispatching the same for marketing.

Yield

Every year 2000kg drug/ha is obtained from ephedra. Drug is sold at Rs. 15/-kg in market. Net return per hectare after 2nd year to 5 th year is 15, 000 Rs.

Medicinal and Economic Importance

- This plant is widely used in the treatment of bronchial asthma.
- Drug ephedrine is obtained from the plant which is utilized as effective cardiac stimulant.
- Another drug pseudo ephedrine is also obtained from this plant which is used for different therapeutical uses.
- Tincture of ephedra is a valuable cardiac stimulant, especially during cases of diphtheria and pneumonia.
- Decoction of stems and roots is used against rheumatism and syphilis. It is known remedy for treatment of bronchial asthma, for which the drug is administered orally or intravenously.
- The drug has also been used in havey fevers, and rashes of allergic origin.
- Nasal sprays containing the drug are used against sinusitis and inflammation of mucous membrane.
- Overdoses of ephedrine can cause nausea, sweating and skin inflammations.
- Rhizomes have large knobs of the size of a foot base and are used as fuel by Tibetans.

85

Eranthemum roseum Nees. (Dashmuli)

Botanical Name: *Eranthamum roseum* Nees.

Synonym: *Daedalacanthus roseus*

Family: Acanthaceae

Local Names

- ☆ **Gujarati:** Dashmuli
- ☆ **Sanskrit:** Dashmulah
- ☆ **Hindi:** Dashmula

Introduction

It is a perennial herb growing up to 1-2 meters high. Leaves are oppositely arranged, oblong lance-like with 10-20 cm length. The leaf stalk is 1-3 cm long. Flowers are blue or violet occuring either singly in the leaf axils or in a spike 10-15 cm length.

Flowers are terminal and found at the end of branches. Flower contains 5 petals and two stamens emerging from the flower tube. The flowers have a strong fragrance. Flowering is occurred in November to April. It is commonly found on shady forests of the hills of Western Ghats.

Distribution

This plant is distributed throught India and semi arid regions of the world. This plant is highly useful in Ayurvedic system of Medicine. In Gujarat, this plant is found in Gir forest, Sasan, across the river, and very common in river banks.

Medicinal and Economic Importance

- Roots are used in rheumatism.
- It is also used in the fever and leucorrhoea.
- It is cultivated in the garden for their attractive foliage and flowers.

86

Erythrina indica Lamk. (Pandervo)

Botanical Name: *Erithrina indica* Lamk.

Family: Papilionaceae

Local Names

- ✰ **Gujarati:** Panderavo, Panervo
- ✰ **Hindi:** Pangli, Bombay pangara
- ✰ **English:** Indian coral tree
- ✰ **Sanskrit:** Rakta puspa
- ✰ **Tamil:** Muruka, Kavir
- ✰ **Telugu:** Badida

Introduction

Erithrina indica is the tree which is found in sub tropical to tropical regions of the world. There is legend about this plant that this tree is the part of Loard Indra. The name of the tree is taken from Greek word *Erithros* or *Erithrinos* which means red color type, Indiga means endogenous to India. There are three leaflets on the stalk, in which middle one is believed to be of loard Vishnu, right one is of Brahma and left is of loard Shiva. So this plant became famous because it concluded *trimurties.* This tree is medium sized or long tree which some time extends up to 15 meters or so. Branches have small spines of Prickles. When they are small until then it occurs. Leaves are larger, containing long stalk and on the stalk three leaflets are seen. Leaflets are broad, and ovoid, sometimes length and width are same in size. Upper side is acute, in the base there is round type. Last leaflets are bigger than

lateral leaf. Flowering time is early month of March to May when leaves are not there. Flowers are arranged in spike inflorescence, they are arranged in clusters. Flowers are shiny red in colour. Calyx is red upper layer or protective layer. Calyx is polypetalous type. One calyx is larger in size and it is called standard. Pods are large, in the starting they are green in colour and finally they obtain grayish black. Broader at the apex and thinner in the base portions. Seeds are 10 to 12 in fruit. Seed are hairy. Fruits remain on the tree until leaves are there.

Distribution

Erythrina is a genus of trees or shrubs rarely herbs, usually armed with spines, wodely This tree is distributed in the tropical and sub tropical regions, coastal and sub coastal areas of the world. This tree is distributed in Japan, Russia, India, Pakistan, Sri Lanka *etc.* In India tree is found in the Gujarat, Maharashtra, Bengal, Madhya Pradesh *etc.* In Gujarat this plant is seen to be distributed in Junagadh, Porbandar, Jamanagar *etc.*

Cultivation

Light to medium soil can be utilized for cultivation of *erythrina.* Although it grows well on fertile soil. Propagation is done by vegetative propagation method and by seed propagation method. Generally a seed are grown in nursery beds and after one year when plant gets proper size it is transplanted in the field in the beginning of rainy season. 1-1.5 meter long and 10cm wide branches are planted in the particular method that basal portion remain inside the soil and upper region remain open into air. This specific method is used for maintaining polarity characteristics of plant. In the plants liquid moves in upward direction so we should always rise cuttings in such a manner that polarity is maintained. Some times seed propagation gives good results but vegetative propagation is more useful than seed propagation.

Medicinal and Economic Importance

- ☆ This tree is grown as it provides protection to the plants as wind breakers.
- ☆ Bark is having antipyretic, astringent, tonic, anathematics, anti inflammatory and sedative property.
- ☆ In conjunctivitis bark is applied externally.
- ☆ In India this is grown to support other plants as Nagar vela (piper bettle), Vitis, champa and some other plants.
- ☆ This plant belongs to laguminoceae family so it fertile soil by adding nitrogen, because of this character, plant helps to grow tea and some other crops.

87

Eucalyptus sp. (Nil Giri)

Botanical Name: *Eucalyptus citriodora* Hook,

Synonyms: *E.globulus* Lobill., *E.teriticornis* Sm.

Family: Myrtaceae

Local Names

- ☆ **Gujarati:** Nil giri
- ☆ **Hindi:** Nil giri
- ☆ **English:** Blue grass, Eucalypt
- ☆ **Sanskrit:** Tailparrah
- ☆ **Tamil:** Yukkalimaram.

Introduction

E.citriodora is tall, graceful, shaft like tree about 40 m high and the foliage are strongly lemon-scented. It is perennial trees Leaves are of 5 distinct types. 1- Cotyledonary, 2-Seedling, 3-Juvenile which are opposite for 4 to 5 pairs with a few peltate, adult leaves alternate, 4-Intermediate and 5-Adult leaves which are lanceolate up to 15 cm long and 3 cm broad, acuminate, with venation finely marked, lateral; Veins are numerous, oblique, parallel and widely spreading. Inflorescence is usually axillary, corymbose panicle; umbels- 3-5 flowered on terate 5-7 mm long peduncles. Buds pedicellate, Calyx is tube hemispherical to cylindrical. Anthers are adnate.

E.globulus is an evergreen glabrous tree, attaining a gigantic height of 55m, highly aromatic. Bark is blue grey. Juvenile leaves opposite for a large number of pairs, adult leaves are alternate. Inflorescence is axillary, usually solitary, but occasionally in 3-flowered umbels, on a very short or a rudimentary peduncle. Buds sessile, operculum flattened-hemispherical. Fruits are sessilec or almost globular to broadly conical 4-ribbed warty.

Origin and Distribution

Almost 700 species of Eucalyptus are found. Some species possess medicinal value or volatile oils having fragrances varying from camphor, thymol, peppermint to rose and lemon grass and are native to Australia and large number of these contain essential oil in their leaves. French botanist L'Heritier in 1778 first time discovered this plant and called it as Eucalyptus. Few species are commercially exploited in India. They may be categorized in three groups.

1. Medicinal oil producing species or cineole rich essential oil containing species.
2. Perfumery oil producing species or citronellal rich essential oil containing species.
3. Phellandrene rich essential oil containing species

E.globulus, E.outraliana and *E.smithii* are species which are useful as it provides medicinal oil. This species is generally produced in China, Spain, Portugal, South Africa, Brazil, Australia and India. Perfumery oil producing species are *E.citriodora.* India, China and Portugal are the main producing countries. Phellandrene rich essential oil containing species are *E.dives, E.dives* var. A and *E.australina* var. B. Austrailia is the only country which produces this species.

Cultivation

E. citriodora is cultivated in medium loam soil, well drained, with adequate water holding capacity. *Euacalyptus globules* is medium to light loam soils, which is rich in organic matter with good water holding capacity. *E.Citriodora* grows well up to an altitude of 600 m in subtropical areas as well as in mediterranean type of climate with moderate summer temperatures and mild winters. In India, Southern plateau of Karnataka, Maharastra, hills of Tamil Nadu and Kerala are suitable for cultivation. When cultivated in North Indian plains the tree thrives well but suffers from a number of diseases. Generally dry and cool regions are most suited for good growth. Propagation is done by seeds. Seeds are germinated at moderate temperature at both high and low temperatures seeds do not germinate. At these two extreme temperature, if germination takes place, seedlings do not survive. High and frequent rainfall checks the growth and even death of the plant. Best time of sowing is February or September. Generally sowing is done directly or seedlings are rinsed first in nursery. After 4-14 days germination occurs depending upon atmospheric humidity and temperature. Seedlings are planted in the field after 10-16 weeks at a height of 20-30 cm in rainy seasons of August and September. If seedlings are placed at 90 x 75 cm apart, then they can produce good amount of herb and oil.

Harvesting

After six or seven months of planting *E.citriodora* is harvested. When trees are pollarded at a height of 1-1.25m and after that subsequent harvest are obtained every four months. Generally three harvests are obtained and in fertile land with good irrigation and climate 4-5 harvests are obtained. The pronounced shoot potentially is a very advantageous character for regular supply of leaves.

Processing and Storage

On an average 200-250 kg/ha of oil can be obtained in southern India. The oil can be distilled in direct fired field distillation unit or stills which are operated by a steam boiler from the leaves of *E.ciltriodora* or *E.globulus.*Leaves separated from the branches are distilled fresh immediately. Distillation of leaves takes generally 3 to 4 hours. The essential oil in fresh leaves of *E.citriodora* is 1-2 per cent and *E.globulus* is 0.7-1 per cent. Oil of the Eucalyptus should preferably be stored in Aluminium or S.S.containers preferably at low temperature. The oil of Species *E.citriodora* should not be stored for longer period and required more precaution in storage than other species.

Yield

Old tree about 6 to 8 years gives 30-60 kg of leaves. They should not be stored for more than 3-4 days, because of possible decomposition and attack by insect pests and diseases. The leaves should be dried and spread out in thin layers on clean yards for a couple of days. The dried leaves are then packed in polythene bags for longer periods of storage. Rutin contain in leaves vary from leaf to leaf and from plant to plant. The rutin content is maximum about 20 per cent in young leaves. The content decreases gradually and tips and decreases gradually in mature leaves up to 3 per cent. The July-August harvest of 3 year old plants yields the maximum rutin content, when compared to the other months of the year. Generally rutin content decreases slightly with the aging of the plant.

Medicinal and Economic Importance

- ✰ Eucalyptus oil is acrid and bitter. It is reputed astringent, thermogenic, antiseptic, decodorant, stimulant, carminative, and digestive, cardio tonic, diuretic, expectorant, insect repellent, rubefacient and antipyretic.
- ✰ It is useful in healing flatulence, halitosis, worm infestation, cardiac debility, tuberculosis, chronic cough, asthma, bronchitis, pyorrhea, burns, and dyspepsia and skin diseases.
- ✰ Any internal consumption of the oil in excess will cause cardiac debility, vomiting and diarrhea.
- ✰ *E.macrorhycha and E.Youmani* provide large scale rutin in Australia and New Zealand. Rutin content in these species is of the order of 8-10 per cent. Rutin is used in the treat ment of capillary fragility, retinitis and possibly, rheumatic fever of haemorrhagic conditions.
- ✰ Generally capillary fragility occurs in a significant number of cases of high blood pressure, diabetes and other conditions. It is also utilized in prevention of apoplexy and retinal hemorrhage, in those cases where capillary fragility is concerned.
- ✰ It can be used for cases of coronary thrombosis and for the purification of blood. Rutin also protects against the harmful effects of X-rays, indicating that it may be of use to persons exposed to dangerous atomic radiations. Under certain conditions, rutins can protect animals against histamine shock.
- ✰ This species is reported to improve the texture and fertility of land. It is envisaged that in Karnataka, areas around Bangalore, Hassan and Chikmagalur, which are less useful for other crop production, can be profitably utilized for the cultivation of the species.
- ✰ *E.citriodora* provides essential oil which is used in perfumery; it is also source of citronella for manufacture of citronellol and menthol. A small proportion of oil added to germicides and disinfectants made from other Eucalyptus oils greatly improves their odor. Wood pulp is suitable for preparing writing and printing paper.

88

Euphorbia hirta L. (Dudheli)

Botanical Name: *Euphorbia hirta* L.

Family: Euphorbiaceae

Local Names

- ☆ **Gujarati:** Dudheli
- ☆ **English:** Pill bearing spurge, Asthma plant
- ☆ **Hindi:** Dudhi
- ☆ **Tamil:** Amampatchai arisi
- ☆ **Telugu:** Reddinanabrolu

Introduction

This plant is erect and annual or perennial plant growing to 50 cm, with pointed oval leaves and clusters of small flowers. This plant is also known as Australian asthma weed, snake weed, Cat's hair etc. This plant is small annual herb, ascending or erect upto 50 cm high, with round stems covered with yellowish hairs. Leaves are small, dark green upper side and pale on the lower side. Flowers are minute, whitis, in small stalked clusteres. Fruitsare small, hairy, 3-angled, wrinkled, light reddish brown seeds.

Origin and Distribution

This plant is native to India and Australia, pill bearing spurge is now widespread throughout the tropics. It comes up in tropical climate. It is a common weed of plantations and agriculatural fields in arid and semi arid areas.

Cultivation

This plant is propagated by seeds and vegetative method. Seeds are collected in advance and kept ready before monsoon. The area is ploughed and cleared of all rank growth. Then seeds are directly sown in the field in parallel spaced at 30 cm apart at the beginning of monsoon. Ten-to twelve kg of seed is required to cover one hectare area. The seeds germinate soon and plants establish with little care. One weeding with soil working is carried out in the second month after sowing of seeds.

Harvesting and Yield

The plants are uprooted in the month of February/March/April after the seeds are fallen on the ground. It is estimated that about 1000 kg of dried drug of *Euphorbia* would be obtained from one hectare.

Chemical Composition

Pill bearing spurge contains flavonoids, terpenoids, alkanes, phenolic acids, shikimic acid and choline. The letter two constituents may by partly responsible for the antispasmodic action of this plant.

Medicinal and Economic Importance

- ☆ A specific treatment for bronchial asthma, pill bearing spurge relaxes the bronchial tubes and eases breathing.

- Mildy sedative and expectorant, it is also taken for bronchitis and other respiratory tract conditions.
- It is most often used along with other anti asthmatic herbs, notably gum plant and lobe.
- In the Anglo American tradition pill bearing spurge is taken to treat intestinal amoebiasis.

89

Evolulus alsinoides L. (Shankh-pushpi)

Botanical Name: *Evolvulous alsinoides* L.

Family: Convolvulaceae

Local Names

- ☆ **Gujarati:** Shankhapushpi
- ☆ **Sanskrit:** Vishnukranta, Shankhapushpi
- ☆ **Hindi:** Shankhahuli

Introduction

This plant is perennial herb with prostrate branches and small elliptic to oblong, lanceolate, obtuse, mucronte leaves, flowers are mostly in upper axils. Corollas are blue in colour and rotate and broad funnel shaped, capsule is globose.

Distribution

It is a common weed growing wildly in open grassy places throughout India. Mostly found in arid and semi arid regions of the world.

Cultivation

This plant is cultivated by seeds and vegetative method.

Chemical Composition

The plant contains an alkaloid shankhapushpine. Fresh plant contains volatile oil and potassium chloride. It also contains a yellow neutral fat, an organic acid and

saline substances. Three alkaloids evolvine, betaine, and an unidentified compound have been isolated.

Medicinal and Economic Importance

- ☆ This plant is used in chronic bronchitis.
- ☆ It is also used in general weakness.
- ☆ This plant can be used in fever, nervous debility, loss of memory, also in syphilis and scrofula.
- ☆ It is used as Rasayan.

90

Ferula assa-foetida Linn. (Hing)

Botanical Name: *Ferula assafoetida* Linn.

Family: Apiaceae

Local Names

- ☆ **Gujarati:** Hing
- ☆ **Sanskrit:** Hingu
- ☆ **Hindi:** Himg
- ☆ **Tamil:** Perunk yam
- ☆ **English:** Assafoetida
- ☆ **Telugu:** Inguva

Introduction

This plant is herbaceous perennial plant, which has fleshy, massive and carrot shaped root with one or more forks. Stem gets height of 1.8-3 meter. It is solid with membranous leaf sheaths. Leaves are 45 cm long and radicle, they are shiny, coriaceous with pinnatified segments and petiole looks channeled. Flowers are 10-20 in the main and 5-6 in the partial umbels. Fruits are reddish brown in colour, they are flat and thin. The English and scientific name for Hing is Asafoetida. This name is derived from the Persian 'aza' (for resin), and the Latin 'foetidus' (for stinking).

Origin and Distribution

The origin of this plant is believed to be Iran. This is wild plant in Punjab,

Kashmir and Afghanistan. It grows in arid and semi arid regions of the world. It grows throughout arid and semi arid regions.

Cultivation

This plant is cultivated by seeds propagation or by vegetative propagation method. Succeeds in most soils. Prefers a deep fertile soil in a sunny position. This species is not hardy in the colder areas of the country; it tolerates temperatures ranges -5 and -10°c. Plants have a long taproot and are intolerant of root disturbance. They should be planted into their final positions as soon as possible. The whole plant, especially when bruised, has an unpleasant smell like stale fish.

Seed are best sown as soon as the seeds ripe in a greenhouse in autumn. Otherwise they are sown in April in a greenhouse. Seedlings are pricked out from individual pot as soon as they are large enought to handle. Because plant doesn't like root disturbance they are planted in small condition. Give the plants protection mulch for at least their first winter outdoors.

Harvesting

The resin is extracted after the plant is about four years old. The older the plant, the more resin it produces. The time to start harvesting the resin from the succulent stem and the root is just before flowering, in the months of March and April. An incision is made in the upper part of the root or lower part of the stem and the exuding gum is collected. Several incisions can be made in the root/stem till there is no more oozing of gum. This process can continue up to three months. A single plant can yield up to 1 kilogram of resin.

Chemical Constituents

This plant provides gum resin. This gum resin contains coumarins, 5-hydroxy-umbelliprenin, 8-hydroxyum-belliprenin, 9-hydroxyumbelliprenin, 8-acetoxy-5hydroxyumbel-liprenin, assafoetidin, ferocolicin, asacoumarin A and B, farnesiferol A, -B ana –C and the disulphides, asadisulphide and sec-butylpropenyl disulphide.

Medicinal and Economic Importance

- ☆ The oleo resin is bitter, acrid, carminative, antispasmodic, expectorant, anthelmintic, diuretic, laxative, nervine tonic, digestive, sedative and emmenagogue.
- ☆ It is used in flatulent colic, dyspepsia, asthma, hysteria, constipation, chronic bronchitis, whooping cough, epilepsy, psychopathy, hepatopathy and vitiated conditions of kapha and vata.

91

Ficus benghalensis Linn. (Vad)

Botanical Name: *Ficus benghalensis* Linn.

Family: Moraceae

Local Names

- ☆ **Gujarati:** Vad, Vadlo
- ☆ **Hindi:** Bad, Bargad
- ☆ **English:** The banyan tree
- ☆ **Sanskrit:** Vata
- ☆ **Tamil:** Ala, Peral, Alam, Al
- ☆ **Telugu:** Marri, Vati

Introduction

It is common tree in whole India. No village would be in India without any tree of banyan. So many legends are related with this tree. Loard Buddha obtained his enlightenment after long meditation under a Banyan tree. According to Mahabhagavata, Savitri worshiped Banyan for saving her husband *Satyavan* from *Yamadharma.* Hindus believe that Lord ***Mahavushnu*** sleeps on the Banyan leaf at the end of creation when the earth completely submerged under water during jalapralaya. Lord Srikrishna is known as *Vatapatrasai* as he slept on leaf during his childhood. The popular lore is that the name "Banyan tree" derived from Hindi word *Baniya* which means merchants. In old days the members of Indian merchant class used to carryout trade in Persian Gulf, with pearl, dimonds and textile products. *Baniyas* frequently sat under this tree and they discussed about business there.

It is large tree having aerial roots and large leaves. Plant is milky. Leaves are simple, alternate and broadly elliptic to ovate, subcordate at base, at the apex leaf is blunt. Inflorescence is sessile hypanthodium and figs are monoecious (2 cm in diameter). There are 4-5 bracts, copular; 6 mm shortly connate, obtuse, persistent, tepals are 3-5, shortly connate, glabrous. The figs are axillary, globose, turning from green to dark brick red on ripening. The male and female flowers are seen, with oblong anther and unequal sessile female flowers. Ovary is obovoid with erect style, tapering and sterile flowers similar to female flowers, pedicellate.

Distribution

No village in India is without banyan tree. It is distributed in semi topical and tropical regions of the world. It is grown in Gujarat, Maharastra, Madhya Pradesh, Bengal and other states of India. This tree is grown in throughout the forest tracts of India mainly in deciduous and semi evergreen forests and planted extensively, in avenue plantations for shade in villages and roadsides. It can be seen from sea level to about 1,300 m and in other countries like Burma, Bangladesh, Pakistan, Sri Lanka and some Asian countries.

Cultivation

Banyan tree grow on different kind of soils including shallow and stony sites or even on rocky crevices. It also thrives on saline soils. It requires a rainfall of 500-4000 mm. Sub tropical climate, well drained medium to heavy loamy soils provides a good growth to the plant. Seed propagation is useful method, natural regeneration is coming up from seeds left over by bids, monkeys etc. Seeds are germinated on buildings or on walls etc. and grow into trees. After that they produce epiphytes, which rapidly send roots to the ground and engulfing the supporting tree finally. Seedlings germinate on the ground naturally too.

But if we want to grow this plant artificially, dried seeds are sown on beds or in polythene bags. These seeds germinate poorly, and hence should be sown thickly in seedbeds and seedlings are pricked out into polybags after about one month. During the monsoon time seedlings are grown in bags or beds for one year. Seedlings should be kept moist and in shade. Then they are planted in the field in 30 x 30 x 30 m of pits. Vegetative propagation is also done by cutting long mature stem. Long stem cuttings should be used for planting during January to March season which is believed to be dormant season. Plant should be given water regularly, rather than planting only during monsoon season. Plant grows very quickly by watering, mulching, soil working, weeding and pruning regularly.

Medicinal and Economic Importance

- ☆ Roots leaves and stem is used widely. Latex is obtained externally for joint pains, rheumatism and lumbago.
- ☆ Bark is astringent and cooling.
- ☆ A decoction of buds in milk curves haemorhages and a paste of leaves promote suppuration on wounds.
- ☆ Banyan tree is used in treating skin diseases with the parts like stem bark, root bark, leaves, new shoot buds, milky latex and aerial roots. Prop roots are good in checking external as well as internal bleeding in hoemoptysis, menorrhorgia and ulcers.
- ☆ An infusion of bark cures dysentery, diarrhea, leucorrhoea, nervous problem etc. It is also used in treatment of diabetes.
- ☆ Banyan tooth brushes are very famous. Some people use young buds of Banyan tree as toothbrush.
- ☆ Figs are useful food for wild life and avian population. Some tribal people eat fruits at the time of scarcity.
- ☆ The tender leaves are lopped for fodder to goats; Cattle eat the wooly tender prop roots and washer men for boiling the clothes.

- ☆ Dried roots are also used as fuel wood while dried leaves are used to make paper plates and cups.
- ☆ Wood obtained from this plant is highly useful for making pulp. Old wood is used for making yokes, well curbs, and cart shafts and for general carpentry.

92

Ficus carica L.

(Anjir)

Botanical Name: *Ficus carica* L.

Family: Moraceae

Local Names

- ☆ **Gujarati:** Anjir
- ☆ **Hindi:** Anjir
- ☆ **English:** Common fig
- ☆ **Tamil:** Anjir
- ☆ **Telugu:** Simaiyatti

Introduction

It is small deciduous tree, reaching 5-8 m in height. Aerial roots are found in this species. Fig is warm temperature fruit crop. It is grown in wide range of soils and is known to do well on heavy clays. Leaves are coriaceous, broadly ovate entire or 3-5 lobed, pubescent beneth.They are 7.5- 24.5 x 5-16 cm in size. Fruits are receptacles type, 1.2 – 3.5 cm across, axillary, ovoid-oblong, thiny pubescent.

Origin and Distribution

This tree is widely distributed in all over the tropical and sub tropical regions of the world. The native of this plant is Carica in Asia Minor.In India this tree is known to be distributed in most of the states like Gujarat, Maharashtra, Madhya Pradesh, Bengal etc. This tree is very common in India.

Cultivation

Fig is cultivated in wide range of soils but it grows well on heavy clays, rich loams, and light sandy soils, with good drainage. It grows best in light early rains. For fruit development and maturation light early rains is very useful. This plant can tolerate more moisture when the crop is grown for use of fresh fruit, but when it is raised for drying; a warm, moderately dry, fairly sandy soil with a considerable amount of lime is required.

Plant is propagated by cuttings, layering and propagated by *gotee* and seed are also practiced. Patch budding cleft or bark grafting are also possible, but propagation by cuttings is the most common method followed. Cutting should be 20-30 cm long and 1-2cm in diameter, this cuttings are taken for mature wood 2-3 years old and planted at 30 cm apart each way in nursery beds at the beginning of rains. After a year, they are transplanted in orchards. Different spacing are adopted depending on local conditions But in normal case spacing of 3 m in between the lines and 2.5 m from plant to plant in the lines is recommended. Good irrigation is very useful for the growth of the plant. After the planting soil is regularly weeded and worked up. Farm yard manure is also applied.

Harvesting

Harvesting is done in two to three years after planting. In proper conditions trees continue to bear for 12-15 years, after which they show a marked decline in the yield. Normally, the trees bear two crops in July-October and January to May. How ever fruits of the first season are sour in taste, therefore only second crops of good quality which is gathered for marketing. Generally fruits are allowed to fall on the ground when they are ripening. For the green fruits they are picked up from the tree by hand or by bamboo. Especially dried fruits are selected which are fallen on the ground.

Storage

Figs are arranged in wooden racks and turned over daily for 5 to 7 days to ensure drying. They are pressed flat to economies packing space and to improve market appearance. Before packing they are dipped in boiling boiling salt solution of about 3 per cent to render them soft and to improve the taste First grade figs possess semi transparent rosy skin, while second grade figs are dark and less attractive. Frozen figs can be held for several months in good conditions.

Yield

When tree becomes ready for fruits it can give 180-350 fruits per year. However a minimum of 1 kg dried fruits per tree can be expected. So from the 1 hactre of land we can get 1000 kg of dried fruits in 6 th years and onwards. Figs are sold at the rate of Rs.50 per kg in the market. Till 6 to 15 years yearly expenditure is Rs. 13,000 to 14,000 per hactre and net return per hactre is Rs. 52,000 per year.

Pruning and Notching

To obtain proper height and shape, trees are pruned every year. In Maharastra light pruning is given in July when trees are headed back to 1.2m from capacity of trees. While in UP heavy pruning is done when trees are headed back to 30 - 45 cm height from the ground level. Pruning is done to induce the growth of the fruit bearing wood and so increase the yield of fruits. Notching is specific method to increase or stimulate laterals on vigorous upright branches. Notching is one method to achieve the desired effect. This consists in giving slanting cuts in the bark, a little above the buds, and removing a slice of the bark. The depth and width of the notches vary according to the size of the branch. Restricting the root growth by planting in 50 cm diameter pot, sunk in the soil, gives good yield.

Diseases and Control

In India no big diseases and pests are recorded on fig plant but in other countries large number of diseases is seen. Some of the diseases recorded on fig are explained here.

1. **Leaf rust:** Rusty pustules are found on the under side of the leaf. The leaves drop off prematurely and in severe cases affect the yield of fruits. Spraying

with Bordeaux mixture or dusting with sulpher has been recommended.

2. **Stem border:** Adult beetle feeds on the bark, leaves and fruits of the tree but it does not cause big defect to the fig. The grub however eats the inner bark and xylem tissue, and kills the branches of the entire plant. As a control measure, the trunk of the tree is protected by paper painted with coal tar or with wire gauge. (1/16 inch mesh). The grub is killed by injecting kerosene or chloroform reosote mixture into hole.

Medicinal and Economic Importance

- ☆ A major portion of fig production is consumed as dry fruit. Fresh and dried fruits are used for its laxative properties.
- ☆ It is demulcent, emollient and highly nutritive.
- ☆ Figs are also utilized in the form of syrup and confection.
- ☆ Tree of this plant gives latex which contains caoutchouc, resin, albumin, cerin, sugar and malic acid.
- ☆ Fig latex is used as an anthelmintic.
- ☆ Fruits or figs are useful in the prevention of nutritional anaemias.
- ☆ Leaves of this plant are used as fodder.

93

Garcinia indica Choisy. (Kokam)

Botanical Name: *Garcinia indica* Choisy.

Family: Clusiaceae

Local Names

- **Gujarati:** Kokam
- **Sanskrit:** Amlavetasa, Tintidika, Vrksamla
- **Hindi:** Kokam
- **Tamil:** Murgal
- **English:** Kokam, Butter tree

Introduction

This is slender tree which is 4-10 meter tall. Wood grayish white in colour, bark looks light brown. Bark is thin and smooth. Blaze is bright yellow. Branches are drooping. They are cylindrical or sub cylindrical, slender, and hairless. Leaves are opposite and they are 5-10 x 2-5 cm and inverted egg-shaped-oblong to elliptic lanceolate. Each leaf becomes narrow at the base, rounded or shortly acuminate at apex, shiny, dark green, lateral nerves are 7-18 pairs, prominent. Male and female flowers are separated, fleshy and stalked, orange –yellow in colour. Male flowers are 3-8 in cluster, about 1.2 cm long. Female flowers are solitary, terminal. Berries are spherical, 3-4 cm across, wine brown or purple or pinkish orange, with persistent calyx lobes. Fruit pulp are white with red tinge, acidic, fleshy. Seeds are 5-8, flat, smooth, shiny and brown.

Origin and Distribution

This plant is native of Western part of India. It grows well under semi arid condition. It is widely found in the evergreen forests in Maharashtra, Goa, Karnataka, Kerala, South Gujarat, Assam and West Bengal like states of India.

Cultivation

This plant is cultivated by two methods: (1) By seeds propagation and (2)By vegetative propagation. It can be grown in a variety of soil and in different agro-climatic conditions. It can be cultivated by soft wood grafting and planted in the month of July-August. Weeding and thinning of the plants may be done as and when required usually after 15-20 days. This plant is normally grown as rainfed crop. Hence regular irrigations is not required.

Fertilizer

Organic manures like, Farm Yard Manure (FYM), Vermicompost, Green Manure etc. is used as per requirement of the species. To prevent diseases, biopesticides could be prepared (either single or mixture) from Neem (kernel, seeds and leaves), Chitrakmool, Dhatura, Cow's urine etc.

Harvesting and Yield

Harvesting is done in March-April. Fruits and bark are separated and dried in shade.

Chemical Constituents

The fruit rind contains a polyisoprenylated phenolic pigment, garcinol and its isomer isogarcinol, along with hydroxycitric acid, cyaniding-3-glucoside and cyaniding-3- sambubioside, L-leucine and DNP L-leucine hydrochloride have been reported from the leaves.

Medicinal and Economic Importance

- The seeds are bitter which are used as a remedy in scrofulous diseases.
- This plant is also used in dysentery, mucous, diarrhea and externally applied for exocoriations, chaps, fissures of lips and as a substitute for spermaceti.
- The ripe fruit is anthelmintic and cardiotonic.
- Root bark, fruits and seed oil are used to cure piles, abdominal disorders, mouth diseases, cardiac diseases and worm infestation.

94

Gardenia gummifera Linn f. (Nadi-hingu)

Botanical Name: *Gardenia gummifera* Linn f.

Family: Rubiaceae

Local Names

- ☆ **Gujarati:** Nali hingu, Nadi hingu
- ☆ **Sanskrit:** Nadi hingu, Gandharaj
- ☆ **Hindi:** Nadi-hingu
- ☆ **Tamil:** Kambi, Vella pavetta, Thikka malli
- ☆ **Telugu:** Bikki, Sirubikki, Manchibikki, Tellamanga

Introduction

It is small tree or large shrub which gets height of 3-7 meter. Branches are crooked, twisted and brittle. Thin, smooth, grayish brown bark is seen. Wood is yellowish white in colour. Leaves are opposite, inverted egg-shaped to oblong, 4-8 x 2-4 cm, base rounded to slightly heart shaped, apex is obtusely acute. Margins are entire, leathery, nearly stalklees. Lateral nerves are in 10-16 pairs, distinct and with a dot-like gland known as domatia at the axil of the each nerve. Flowers are bisexual, solitary and axillary. Fruits are berry type, ellipsoid to oblong and they are about 4 x 3 cm, smooth. They are not prominently ribbed. Seeds are minute and many.

The leaf buds and young shoots of this species yield resinous exudation, known in commerce as Dikkamali or cumbi gum. Resin is secreted freely in the form of tears.

Shoots and buds are broken off with the tears of resin attached and marketed either in this form or after agglutination into cakes or irregular masses. Resin is transparent, greenish, yellowish, with sharp pungent test and peculiar offensive odour.

Distribution

This plant is very common in the dry and deciduous forest. It is found in the degraded slopes and in the mountainous areas. 6-8 cm cutting of semi hard wood

Cultivation

Nadihingu requires laterite type of soil or clay loam soil for better growth. Stem cutting method is generally used for propagation of the plant. Cuttings are planted in nursery bed during November-December in the nursery beds. Rooting occurs after 20-30 days. Regular watering is essential; in the shade plant grows rapidly. When plant in the nursery gets proper height it is transffered to the main field.

Chemical Constituents

This plant contains resin, volatile-oil, colouring matter-gardenin etc.

Medicinal and Economic Importance

☆ Resin obtained from the leaf buds is used in curing wounds, indigestion, gas trouble, piles, chronic coughs, neuropathy, anorexia, colic, foul ulcers, intestinal worms (especially round worms),

- It is also used in cardiac debility, leprosy, skin diseases, intermittent fever, enlargement of the spleen and obesity.
- The plant is one of the ingredients of a veterinary herbal antiseptic ointement, which is effective in healing wounds, fungal infections and other skin lesions in cattle. The ointment showed prominent antimicrobial effect.

95

Gloriosa superba Linn. (Vachh Na)

Botanical Name: *Gloriosa superba*

Family: Liliaceae

Local Names

- ✰ **Gujarati:** Vachha nag
- ✰ **Hindi:** Kali hari, Languli
- ✰ **Sanskrit:** Langali, Agni shikha
- ✰ **English:** Glory lily, Gloriosa lily, Tigers claws.
- ✰ **Tamil:** Kalapaikkilanku, Nabhikkodi
- ✰ **Telugu:** Advi nabhi

Introduction

The name of this plant Gloriosa superba has came from Latin word gloriosus which means flowers. It is climbing herb which is perennial growing between 3.5 to 6 m in length, but it is trained at 1.5 meter above ground level. Roots are tuberous and vines are tall, weak stemmed that support themselves by means of cirrhosed tips. The leaves are ovate, lanceolate, acuminate, the tips spirally twisted to serve as tendrils. The flowers are large and solitary. Inflorescence is laxcorymbose. Fruits are of capsule type, each fruit contain numerous seeds. Seeds are warty and dorsally compressed.

Origin and Distribution

This plant is believed to be native of tropical Asia and Africa. The plant grows

throughout in India. from the North West Himalayas to Assam and the Deccan peninsula, extending up to an elevation of 2,120 m, all along the Western Ghats (Karnataka). In world it grows in Madagascar, Sri Lanka, Indochina and on the adjacent islands.

Cultivation

This tropical plant can grow in humid and warm areas. Rain fall around 373 cm is good for its growth. Plant can not tolerate continuous moisture stress; however, it needs frequent irrigations up to the flowering in dry periods. Plant prefers sandy loam soils on the acidic side, The soil should have pH between 6-7, well drained soil is useful. Propagation is done by three methods. (1) Propagation by tubers (2) Propagation by seeds and (3) propagation by rhizomes.

(1) Propagation by Tubers

Biforked tubers are produced during the growing season of the plant. This tuber has only one growing bud. Tubers should be handled carefully, because they are brittle and liable to break easily if the growing bud is subjected to any kind of damage, the tuber will fail to sprout. The vigor of the vine and its flowering and fruiting depends on the size of the tuber. Weight of each tuber should not be less than 50-60 g. Plants raised from smaller tubers do not produce flowers during their first year. Large tubers are divided into two by breaking them in the middle. The tuber which is dormant starts sprouting from the month of May.

(2) Propagation by Seeds

Seed propagation is also good method. Seeds take three to four years for flowering. Except for experimental purposes, seed propagation is not favored by the growers.

(3) Propagation by Rhizomes

V-shaped rhizomes are used for commercial cultivation of *Gloreosa superba.*

All the grasses are removed and field is ploughed two to three times until it is brought to a fine teeth. Field must be leveled and drainage arrangement should be done. FYM is applied in the field or compost manure can also be applied depending upon soil type. Generally one foot deep furrows are prepared at spacing of 45-60 cm. Treated tubers are planted at a depth of 6-8 cm, keeping a plant to plant distance of 30-45 cm depending upon soil type. Nitrogen, K20, P2O5 are very useful for growth of this plant. During sprouting time plant requires frequent irrigation because it helps the soil to keep surface soft, which helps easy sprouting and emergence of the growing tip outside the soil. Until flowering is over irrigation should be withheld. In the initial stages weeding is also necessary for the growth of the plant.

Harvesting

Flowering time of this plant is September to October. The fruit requires about 105 to 110 days from the set to reach maturity. The right stage of harvest is hen this capsules starts turning light green from dark green and skin of the fruit shows shrunken appearance and becomes light in weight. At this stage when pressed the pod gives a crinkling sound. After picking, the capsules should be kept in the shade for 7 to 10 days to facilitate the capsules to open up, displaying deep orange yellow coloured seeds. The seeds and pericarp are separated manually and they are dried for a week in the shed, by spreading them uniformly over any clean, dry floor or any platform specially erected for the purpose. At the later stage seeds are moved into sunlight till they dry completely. The dried seeds are then packed in moisture proof containers and stored until exported or extracted for the alkaloids.

Yield

Yield of the seeds differ greatly, depending upon the vigour and age of the plant, which in turn depend on the size of the tuber. The yield in the initial stages will be low. But by time to time it increases gradually. After three years, we can get 200-250 kg/ha of dried seeds from well managed farm. Yield of pericarp or husk is about 75 per cent of the seeds.

Diseases and Control

Plant is affected by following fungal and pests diseases.

1. **Leaf blight:** This disease is caused by the fungus *Curvularia lunata*. The spores of the fungus are airborne and attack the foliage of the plant. Leaves turn into light yellow in colour in the beginnng but later there are produced small spots on the leaves. This desieases is seen mostly when there is monsoon season. Diseases can be controlled by spraying 0.3 per cent Dithane M-45.
2. **Tuber rot or basal stem rotting and wilting:** This disease is caused by *sclerotium* fungi which damages to the underground tubers lead to death of the plant. In the initial stages infected tubers start becoming soft and the foliage exhibits a yellow appearance and in advanced stages whole

tuber gets infected giving the appearance of a discoloured mass and the plant dies off. Drenching the soil with Bavistin @ 0.2 per cent or Cuprassol near the root zone of the plant has been observed to control the disease.

3. **Lily caterpillar:** This disease can be marked with red and black bands on the surface of the plant. They lay a mass of eggs on the undersurface of a tender leaf or in the crown of the growing shoots. This disease can be over come by spraying metacid at a concentration of 0.2 per cent at fortnightly intervals.
4. **Green caterpillar:** Caterpillar pests are green coloured which attack the leaves and flowers of the plant. It completely eats away flowers causing severe damage to the crop.

Medicinal and Economic Importance

- Seeds, tubers, leaves are mainly utilized for medicinal value.
- Tubers are used in medicines as anthelmintic and leaf juice is reported to kill lice in hair. Its tubers are used as a tonic, antiperiodic, antihelminthic and also against snakebites and scorpion stings.
- Drug obtained from the plant is used as gasterointestinal irritant and may cause vomiting and purging. It is also used for some times promoting labour pains and, conversely, also as an abortifacient. It is considered useful in colic, chronic ulcers, piles and gonorrhoea.
- It is used in local applications against parasitic skin diseases and as a cataplasm in urological pains.
- Leaves when applied in the form of a paste to the forehead and neck, they are reported to cure asthma in children. The leaf juice is used against head lice.
- Alkaloids colchicines and gloricine are isolated from the plant in which colchicines is highly useful in treatment of gout, common disorder in the temperate parts of the world. Gout is caused because of deposition of microcrystal of uric acid in the joints, which is attributed to a defective regulatory mechanism for endogenous passive synthesis. Colchicines interrupt the cycle of new crystal deposition, which seems to be essential for the continuance of acute gout.
- These alkaloids are used as poly ploidizing agents in polyploidy breeding in crop research. In the old times tubers are used to get colchicines but seeds also provide colchicines.

96

Gmelina arborea L. (Shivan)

Botanical Name: *Gmelina arborea* L.

Family: Verbinaceae

Local Names

- ✰ **Gujarati:** shevan, shivan, savan
- ✰ **Hindi:** Gamhar, Kambhari
- ✰ **Sanskrit:** Asweta, Bhadra, Gambhari
- ✰ **English:** Gumhar,
- ✰ **Tamil:** perungumpil, Ummi-thekku
- ✰ **Telugu:** Gumartak, Kummadi

Introduction

Savan is desiduous tree which is scattered throughout in India. This tree is medium in size (20 meter or more) and attains girth of 2 meter. Bark is smooth and whitis grey in colour. Flowers are in terminal panicles, and brownish yellow in colour. Fruits are drupe type, fleshy, ovoid and 2 seeded. Leaves are simple type, opposite, broadly ovate, cordate and glandular.

Distribution

This tree is distributed in deciduous forest of tropical and sub tropical semi arid areas. In India it is widly distributed perticulary in Maharastra, Madhya Pradesh, Gujarat and other warm states of country.

Cultivation

This plant is grown in fertile moist valley with good drainage. It can grow upto 1600m altitude in the hilly areas. It is a light demander and moderately frost hardy. It is never gregarious in nature.

Clay soil may be avoided. Seed propagation is good method for its cultivation. Direct sowing, transplanting of nursery raised container seedlings or planting of stumps have been successful. Six months old seedlings raised in polypots give excellent results. Althouygh seeds have poor viability. That's why seeds are sown directly after their collection. Seeds are planted at 2 x 2 m spacing, The plants are kept free by weeding. Plant is fast growing but irrigation is also useful for the growth of the plant. Good fertilizers help plant to give good yield.

Harvesting

Generally harvesting is done in three to four years of planting. Root portion can be harvested after 4 years. Plant is uprooted by digging around the plants. The roots dried and dispatched for marketing. Bark is removed and dried. Then bark is taken for marketing.

Yield

From one hactre of land each year 500 trees are exploided. Area is replanted, thus covering one ha. over a period of 5 years. It is believed that each plant gives 2 kg of roots and 500 gms. of stem bark. A production of 1000kg of roots and 250 kg of the stem bark would be available from 500 plants. The drug obtained from the plant is sold at the rate of Rs. 20-30 per kg. Total expenditure in 6 th year is around 5000 while net return will be around 20000 Rs. in 6th year and onwards.

Medicinal and Economic Importance

- ☆ Root, stem, drupes, flowers, bark are used in the medicinal treatments.
- ☆ The root is bitter tonic, stomachic, laxative a galactogogue. It is prescribed in indigestion, fevers and anasarca in the form of infusion or decoction.
- ☆ The root is one of the ingredients of the ayurvedic drug Dashmoola. Root powder is also used for gout.
- ☆ The drupes which are sweetish and bitter, are used as an ingredient of refregeirant decoction for fevers and bilious affections. Fruit is best tonic
- ☆ The tender leaves are demulcent. A poultices of the leaves are used in head ache in fevers.
- ☆ The leaf juice is used as a wash for foul ulcers.
- ☆ Flowers are used for blood diseases. Bark is a bitter tonic and stomachic and is considered useful in fever and indigestion.
- ☆ These leaves are sometimes used as feed for eri-silkworms. Leaves are also used for the fodder for cattle.

97

Gymnema sylvestre (Retz.) R. Br. (Madhu Nasini)

Botanical Name: *Gymnema sylvestre* (Retz.) R.Br.

Family: Asclepiadaceae

Local Names

- ☆ **Gujarati:** Madhunashini
- ☆ **Sanskrit:** Meshashringi, Madhunasini
- ☆ **Hindi:** Gudmar, Meradingi
- ☆ **Tamil:** Adigam
- ☆ **English:** Periploca of the woods
- ☆ **Telugu:** Podapatra

Introduction

It is climber which is large, much branched, woody and has pubescent young parts. Leaves are simple, elliptic or ovate. They are more or less pubescent on the both sides. Base is rounded or cordata. Small flowers are there, they are yellow coloured and arranged in umbellate cymes. Fruits are slender, follicles upto 7.5 cm long.

Origin and Distribution

This herb is native to tropical forests of southern and central India. It is found in dry and semi dry forest, arid regions of the world and also grown upto 600 meter. It is seen coming up on all types of soils. It can grow in moist localities of semi arid zones also. It tolerates over head shade to some extent.

Cultivation

This plant is cultivated by seeds and vegetative method.. Semi hard wood is good for stem cuttings. Stem cutting is planted in polypot in nursery in the month of October. The cuttings start sprouting within 15-20 days. The plants grow and by the end of May, they will attain 30-45 cm of height and will be ready for transplanting in the field. Afterwords, furrows are prepared at one meter distance. At the beginning of monsoon in the month of June, plants are transplanted. In the month of July/August, one weeding is carried out and. In the first year, the leaves are not harvested, but efforts are made to maintain the bushy form of the plants. From 2nd year, leaves can be harvested every year.

Harvesting

Leaves can be harvested from 2nd year and onwards. The leaves can be plucked in two or three pluckings. Leaves are dried under shade for 2 to 3 days and are packed in polybags

Yield

The leaves (dried) may fetch a rate of Rs.10/- per kg. An yield of 4000 kg of leaves can be expected from one hectare of plantation.

Chemical Constituents

This plant is anti sweet. Gymnema saponins I-V were isolated from the plant. An aqueous extract of the leaves containes the saponins gymnemic acid III, IV, V. They have been characterized as 21-0-2-methtylcrotonyl, 21-0-benzoyl, and 29-0-benzoyl gymnemagenin 3-0-glucuronides, respectively. Gymnemic acid is also isolated. The leaves contain dammarane-type saponins, gymnemasides I-VII and gypenoside XXVII.

Medicinal and Economic Importance

- The plant is bitter, astringent, acrid, thermogenic, anti-inflammatory, anodyne, digestive, liver tonic, emetic, diuretic, stomachic, stimulant, anthelmintic, alexipharmic, laxative, cardiotonic, expectorant, antipyretic and uterine.
- It is tonic, which is useful in linflammations, hepatosplenomegaly, dyspepsia, constipation, jaundice, halminthiasis.
- It is also applied in the cardiopathy, cough, asthma, brtonchitis, intermittent fever, amenorrhoea, vitiated conditions of vata, conjunctivitis and leucoderma.
- The fresh leaves are chewed for its property of paralyzing the sense of taste of sweet and bitter substances for sometime.

98

Hemidesmus indicus (Linn.) R.Br. (Anant Mul)

Botanical Name: *Hemidesmus indicus* (Linn.)R.Br.

Family: Asclepiadaceae

Local Names

- ☆ **Gujarati:** Anant mul
- ☆ **Hindi:** Anant mul, Magrabu
- ☆ **English:** Indian sarasaparilla
- ☆ **Sanskrit:** Anant mullah, Sariba
- ☆ **Tamil:** Nannari, Saribam
- ☆ **Telugu:** Sugandipala

Introduction

It grows throughout in India. Anant mul is found in tropical and sub tropical areas of the India. It is perennial, twinning or prostrate, wiry shrub with woody root stock and numerous slender, terete stems having thickened nodes. Leaves are simple, opposite, very variable from elliptic-oblong to linear lanceolate, variegated with white above, silvery white and pubescent at the lower side. Flowers are 10 cm in length, they are greenish purple in colour, crowded and arranged in subsessile cymes in the opposite leaf axils. Fruits are slender follicles, cylindrical, they are 10 cm long, tapering to a point at the apex, Seeds are flattened, black, ovate oblong and silvery white in colour. The root is tuberous and dark brown in colour, tortuous with transversely cracked and longitudinally fissured bark. It has a strong central vasculature and a pleasant smell and taste.

Distribution

Anant mul is distributed throughout in sub tropical and tropical regions of the world. It is seen in Pakistan, Sri Lanka, Myanmar, Burma etc. like countries. In India this plant is found to be distributed in the semi desert and desert areas like Gujarat, Madhya Pradesh, Uttar Pradesh, Bengal and some other states of the country.

Cultivation

Although it is naturally produced in some of the areas of the Gujarat like in Girnar Hills (Junagadh) and Barda Hills (Porbandar). Plant does not need any specific farming method for its cultivation but it is spread with seeds. Seed propagation and vegetative propagation are the two methods for its cultivation.

Harvesting

Whole plant is pulled by hand or it can be uprooted with falcket. Plant is dried. and stored.

Medicinal and Economic Importance

- ☆ Stem is very useful as it is bitter, diaphoretic and laxative, are useful in inflammations, cerebropathy, hepatopathy, nephropathy, syphilis, metropathy, leucoderma, odontalgi, cough and asthma.
- ☆ The latex is good for conjunctivitis. Root, stem and leaves are utilized in medicines.
- ☆ Roots are bitter, sweet, astringent, aromatic, refrigerant, emollient, depurative, aphrodisiac, carminative, appetizer, anthelmintic, alterant, demulcent, diaphoretic, febrifuge, expectorant and tonic.
- ☆ They are useful in vitiated conditions of pitta, burning sensation, leucoderma, leprosy, skin diseases, pruritus, asthma, bronchitis, hyperdispsia, ophthalmopathy, hemicrania, epileptic fits, dyspepsia, helminthiasis, diarrhea, dysentery, haemorrhoids, strangury, leucorrhoea, syphilis, abscess, arthralgia, fever and general debility.
- ☆ Other constitutes present in the roots are: β-sitosterol, α-and β-amyrins, lupeol, tetracyclic triterpene alcohols, small amount of resin acids, fatty acids, tannins, saponins, a glycoside and a ketone.
- ☆ Leaves are used in vomiting, wounds and leucoderma. The air dried roots yield essential oil containing P-methoxy salicylic aldehyde as the major constituent. The aroma of the drug is attributed to this aldehyde.

99

Hibiscus rosa-sinensis L. (Jasud)

Botanical Name: *Hibiscus rosa sinensis* L.

Family: Malvaceae

Local Names

- ☆ **Gujarati:** Jashud
- ☆ **Tamil:** Smparuthi
- ☆ **Hindi:** Jasut, Jasum
- ☆ **Telugu:** Java pushpamu dasana
- ☆ **Sanskrit:** Japa, Java, Rudra pushpam
- ☆ **English:** Shoe flower, Rose of China, Chinese hibiscus

Introduction

This is evergreen, woody, glabrous and showy shrub which gets height of 5-8 feet. Leaves are bright green in colour, ovate, entire till below. They look coarsely toothed above. Flowers are solitary, axillary, bellshaped, with pistil and stamens projecting from the centre. Fruits are capsules type and reddish, they are many seeded. This plant is very common in gardens.

Origin and Distribution

Origin of the plant is unknown. Although some people think that origin of plant is Southeast Asia and China. It is commonly cultivated as garden shrub because it looks so beautiful in the garden. It is cultivated from the sea level to 500 meters.

Cultivation

It prefers a well-drained humus rich fertile soil in a warm, sheltered position in full sun. It is not very frost-tolerant and needs to be grown in essentially frost-free areas. It might succeed outdoors in the very mildest areas of the country if given a very sheltered warm position. Alternatively, it might be possible to grow the plant as a tender annual by starting it off early in a warm greenhouse. If well-grown it can flower and set seed in its first year. This plant is cultivated by two common methods *i.e.*through seeds and vegetative propagation.

Seed are sown early in spring in a warm greenhouse. Germination is usually rapid. Seedlings are grown in individual pot. If plants are grown as annuals, they should planted out into their permanent positions in early summer and protect them with a frame or cloche until they are growing away well. If they are grown as a perennial, then it is better to grow them in the greenhouse for their first year and to plant them out in early summer of the following year. Cuttings of half-ripe wood are sown in July or August in a frame or outside in the nursery.

Chemical Constituents

Taraxeryl acetate, beta-sitosterol, campesterol, stigmasterol, cholesterol, erogosterol, lipids, citric, tartaric and oxalic acids, fructose, glucose, sucrose, flavonoids and flavonoid glycosides. Hibiscetin, cyaniding and cyanin glucosides and alkenes are found.

Medicinal and Economic Importance

- To induce abortion, ease menstrual cramps and to help in childbirth.
- It is used to treat headache.
- A preparation from the leaves is used to treat postpartum relapse sickness.
- It is also used to treat boils, sores and inflammations.
- It is very good medicine for hairs.

100

Holarrhena antidysenterica Wall. (Kutaj)

Botanical Name: *Holarrhena antidysenterica* Wall.

Synonym: *Holarrhena pubescence* (Butch-Ham.) Wallich ex Don.

Family: Apocynaceae

Local Names

- ☆ **Gujarati:** Kutaj
- ☆ **Hindi:** Kutaj
- ☆ **English:** Kutaj
- ☆ **Tamil:** Veppalei
- ☆ **Sankrit:** Kutajah
- ☆ **Telugu:** Tedlapala

Introduction

This plant is small, medium sized deciduous tree. Leaves are opposite, short petioled, ovate to oblong, elliptic oblong, membranous, pale green. Flowers are white in colour, fragrant, in terminal or sub axillary many flowered corymbose cymes. Fruit is of two slender, elongate parallel, follicular mericarps, 8-16 inches long. Each follicle enclose many seeds. Flowering time is March to May.

Distribution

This plant is more or less throughout distributed in India upto 1000 ft.This tree us vey common in the Indian forest.

Cultivation

Plant is easily propagated by seed propagation method. Before propagation, seeds are kept in cotton and allowed to soak for 24 hours in water. Then seeds are sown in polybags or in nusery beds. Generally seeds take one week for germination. When seedlings get height of 5 cm it can be transplanted in the main field after one year. 30 cm cube pits are prepared with 6 x 6 meter spacing for planting.

Medicinal and Economic Importance

- This plant is used as astringent, febrifuge which is used in dysentery, diarrhea, fever and intestinal worms.
- The plant is also useful in asthma, chronic bronchitis, boils and ulcers.

101

Holostemma annularium (Roxb.) K. Schumn (Chirvel)

Botanical Name: *Holostema annulare* (Roxb.) K. Schumn

Synonyms: *Holostema ada-kodien* Schultes

Family: Asclepiadaceae

Local Names

- ☆ **Gujarati:** Khirdodi
- ☆ **Sanskrit:** Jivanti, Arkapushpi
- ☆ **Hindi:** Chhirvel
- ☆ **Telugu:** Polagurugu
- ☆ **English:** Jivanti

Introduction

This is handsome, laticiferous, twining shrub. Flowers are large and conspicuous. Leaves are simple, opposite, cordate type. Flowers are purple colour and they are arranged in axillary umbellate cymes. Fruits are thick follicles. Length is 9-10 cm. They are cylindrical in shape. Point of the fruit is blunt. Roots are irregularly twisted, pretty long upto a meter or more, they are thick and cylindrical.

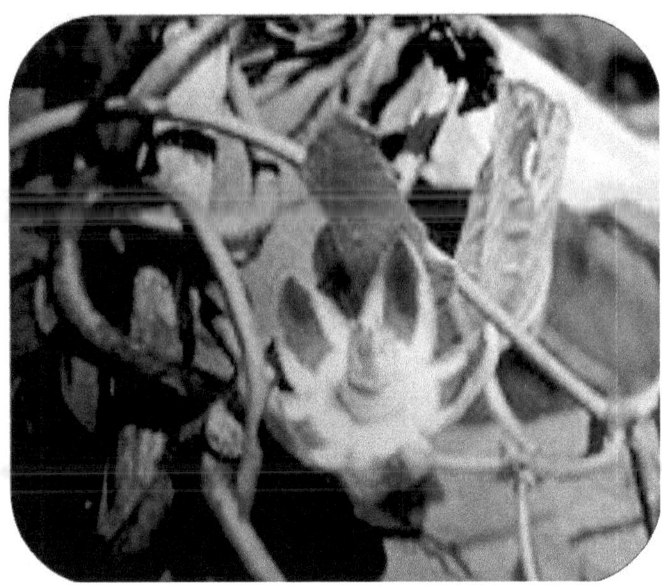

Distribution

This shrub is found throughout in India. They are generally found in open hedges and forests. This plant is very common in the arid and semi arid regions of the world.

Chemical Constituents

Root contains many important chemical constituents. It contained moisture, protein, sugar, starch, fiber and ash. The ash contains calcium and phosphate.

Medicinal and Economic Importance

- ✰ The roots are sweet, refrigerant, ophthalmic, emollient, alterant, tonic, stimulant, aphrodisiac, expectorant and galactagogue
- ✰ They are used in ophthalmopathy, orchids, cough, burning sensation, stomachalgia, consumption, fever and tridosa.
- ✰ Leaves, flowers and fruits are eaten as vegetables.

102

Juglans regia Linn. (Akhrot)

Botanical Name: *Juglans regia* Linn.

Family: Junglandaceae

Local Names

- **Gujarati:** Akhrot
- **English:** Common walnut,Persian walnut, European walnut
- **Sanskrit:** Aksotah
- **Tamil:** Akrottu
- **Hindi:** Akhrot
- **Telugu:** Akrotu

Introduction

This plant is very large deciduous monoecious tree with tomentose shoots. Longitudinally fissured bark is seen. Bark is grey coloured. Leaves are alternate, imparipinnate. Leaflets are entire and aromatic in nature. Flowers are yellowish green in colour. They are very small. Male flowers are arranged in pendulous slender catkins inflorescence. Female flowers are arranged in 1-3 flowered terminal catkins. Fruits are drupes type. Each fruit has 5 cm length. Exocarp of the fruit is leathery. Fruits are woody wrinkled. Hard endocarp encloses 4 lobes. They are corrugated, oily. Seeds are edible.

Origin and Distribution

This plant is cultivated mainly in the Himalayas and Khasia Hills. It is the native of region stretching from the Balkans eastward to the Himalayas and southwest China. The plant is distributed in the Europe, America, Roman, Greece, Bulgaria like countries in the world.

Cultivation

This plant is cultivated by two general methods (i). By seeds and (ii). By cuttings. Various methods are used for preparing an area for a walnut orchard, all giving equally good results, when measured by profitable crop production. Since trees are deep-rooted, soil should be fertile, well-drained, alluvial, 2 m or more deep, of medium loam to sandy or silt loam texture, and free of alkali salts, especially excessive boron. Seedling trees show great variation as to hardiness, type of fruit and fruitfulness.

Persian walnuts are planted in the orchard from 10 to 20 m each way; however, many spacings are in use depending on the variety and the cultivation methods to be used. Intercropping young walnut orchard with trees of a different species may be useful, at least for the first 5–10 years. Intercropping may be difficult because of irrigation, spraying, and use of equipment for cultivation of the intercrop. Holes should be dug amply wide to accomodate roots, a few meter deeper than the roots, and planted no deeper than they were in the nursery. Roots should never be allowed to dry out during the transplanting process. Topsoil should be used to fill hole when planting tree, and firmly tamped around roots. Do not transplant when soil is wet. Nut trees must have tops reduced or cut back, either before or after planting, usually to about 1.5–2 m from ground level. Lower buds should be suppressed so the upper ones will be forced to grow and make the framework of the tree.

Newly planted trees should be staked, either with a single stake driven close to the tree and tying it to the stake, or driving three stakes equidistant, fastening tree to each with stout cord so as not to injure bark. After trees are planted, they should be watched, and watered during every severe dry spell until they have become established. When irrigated, total of 2 1/2–5 acre feet of water per acre should be applied through-out the year, including normal rainfall. The modified central leader system of training young walnuts is recommended for western orchards, in which 4 or 5 main framework branches spaced both vertically and horizontally are developed; the first branch should be started no lower than 2 m from the ground.

Chemical Constituents

This plant contains a globulin, juglansin. The juglansin has been isolated from the edible kernel. The nitrogen distribution of the globulin is as follows. Basic N, 5.41; non-basic N, 11.51; humin N, 0.15; and amide N, 1.78. The globulin contains cystine and tryptophane.

Medicinal and Economic Importance

- ✰ The leaves and bark are antiscorbutic and detergent, and are useful in herpes, eczema, scrofula and syphilis. Leaves are astringent, tonic and anthelmintic.
- ✰ Fruits are sweet, emollient, thermogenic, aphrodisiac, tonic and carminative. They are used as alternate in rheumatism.
- ✰ The expressed oil of the fruit is considered useful against tapeworm and is used to strengthen and lubricate the muscles.
- ✰ The kernels are said to posses aphrodisiac properties and are recommended in colic and dysentery.

103

Jasminum grandiflorum L. (Champo)

Botanical Name: *Jasminum grandiflorum* L.

Family: Oleaceae

Local Names

- ✰ **Gujarati:** Chambeli, Malti
- ✰ **Hindi:** Chameli
- ✰ **English:** jasmine, Catalonium jasmine
- ✰ **Sanskrit:** Jati, Chetaki, Malati
- ✰ **Tamil:** Pinchi, Koti malligai
- ✰ **Telugu:** Jati, Jaji, Malti

Introduction

Jasmine is an evergreen profusely branched shrub attaining a height of 2m, if allowed to grow and given support can climb to height of 10-12m. It is a large glabrous shrub erect while young and climbing or scrambling when older. Leaves are opposite, imparipinnately compound, up to 7 cm long, the rachis flattened or winged, the leaflets are 5-7, elliptic, round elliptic or oval, the terminal leaflet are mostly ovate lanceolate and acuminate. Flowers are white in colour and they are arranged in axillary terminal long cymes. Bracts are foliaceous ovate, bracteoles are linear. Calyx is glabrous, five lobed, subulate. Corrolla has 5 lobes. They are star shaped; elliptic or obovate, carpels are two. The corolla tube encloses tow stamens borne on short slender filaments and fruit is a berry type. Generally plant can be

identified with the help of bushes of 1.5-2 m diameter. Full grown plant form a oval shape look with leaves and branches both having dark green appearance.

Origin and Distribution

Jasmine is next to the rose in odour. In India white blossoms of the jasmine is used since the time immemorial. It was used for ceremonial purpose and for scenting of hair oils and ointments. *J. grandiflorum* was originated in Indo-Burma region in the foothills of Himalayas. The plant was introduced in Morocco, Egypt, S.Africa, Syria, Algeria, United Arub Republic, India and China for commercial plantation. The plant was brought to North Africa and Spain, apparently by the conquering Moors, for the name Jasmine is of Arabic origin (Yasmin). It is found in France, Egypt, Morocco, Algeria and Italy, India, Pakistan, Sri Lanka *etc.* In India jasmine is grown in the Tamil Nadu, Karnataka, Andhra Pradesh, Maharastra, Gujarat *etc.* states in about 8000 hectares.

Cultivation

Plant suits high humidity and soil moisture promotes good growth from newly planted cutting. Similar condition also favours plant growth which is found to be maximum during raining season. Although flowering starts fewer first years of plantation but the economic yield is obtained only in the third year till 15-20 years. It is propagated by layering and stem cuttings in India, in France by grafting of *J.grandiflorum* on stocks of *J.officinalis* which is a hardy plant resistant to frost, and in Italy by rooted stem cuttings. For layering, mature shoots, which are one year old, are lowered in soil 10-15 cm deep after giving a shallow cut in the portion to be lowered, roots in 90-120 days. Rooted layers are separated and kept in moist gunny bags in shade till these are planted. Ideal time of layering in India is June-December.

Also plantation is raised from cuttings, which are obtained in March from near mature wood. Each cutting should have 3 or 4 eyes and it should be 22-25 cm long. Cuttings are buried more than 5 cm deep and are spread 7 cm apart and are ready for transplanting in the field within 4 to 5 months, which takes place during March of the following year. The plants are spread 1m in rows, 1-5m apart. Pruning is done every year to get good development of fresh flowering shoots. Generally December and January is good time for pruning.

Harvesting

Harvesting is done in the second year of planting in April and May; hand plucking is done in the early morning because at that time flowers contain maximum essential oil. After 10'o clock oil content seems to be decreased. Normally, hand plucking of the blossom is done.

Yield

In India a yield of 7500-8000 kg per hectare has been reported. In France, during first year, a one hectare plantation produces 1,200-2000 kg of flowers. In the second and following years a yield of 3000-4000 kg of flowers are obtained. In Italy, the yield of flowers from the 3rd year onwards averages 4500 kg per hectare.

Storage

The oil should be stored in a cool place. When it becomes old it darkens in colour.

Grading and Processing

Enfleurage method is widely used in India for production of jasmine attars. In this method dehusked seeds of till (Sesame) and jasmine flowers are spread in thin layers one over the other for 10 to 12 hours daily. The exhausted flowers are replaced by fresh flowers and the process is repeated for 5 to 7 days. One kg of seeds need about 3 kg of fresh flowers. The perfumed seeds are distilled. Other method of extraction of jasmine perfume is solvent extraction which is more modern method and recovers particularly all the odouriferous constituents and is also economical. Hexane or petroleum ether is used as solvent for extraction. The solvent is recovered by vacuum distillation and the residue constituents the concrete which is purified by extraction with 95 per cent ethyl alcohol whereby jasmine absolute is obtained. Jasmine concrete is a yellowish brown waxy mass with an odour of jasmine flowers. The absolute on the other hand is a viscous, clear yellowish brown liquid possessing a delicate odour of fresh jasmine flowers.

Diseases and Prevention

This plant is affected by many pest and diseases. Leaf blight is caused by *Glomerella cingulata* which can be controlled by spraying of Bordeax mixture or organic fungicide.

Medicinal and Economic Importance

- ☆ In countries like India, bulks of the flowers are used in garlands and decorative bouquets for religious offerings to temples and small amount is used for production of attar and scented hair oil.

- ☆ Jasmine concrete and absolute is used in high grade perfumes, ranking next to the rose in the order of importance. Nearly all high quality perfumes contain at least a small amount of jasmine oil.
- ☆ The absolute gives the best results, it blends with any floral scent imparting smoothnes and elegance to perfume composition.
- ☆ Jasmine oil is used for perfuming expensive soaps, cosmetics mouth washes and dentofroces, bath salts, sachets and tobacco. It is also used in incense and fumigants.
- ☆ Alcoholic washings of concentrate are used in handkerchief perfumes.

104

Jatropha curcas L. (Ratan-Jyot)

Botanical Name: *Jatropa curcas*.Linn.

Family: Euphorbiaceae

Local Names

- ☆ **Gujarati:** Ratan Jyot
- ☆ **Hindi:** Safedaranda, Jangliarandi
- ☆ **Sanskrit:** Kananaeradna
- ☆ **Tamil:** kadalamanakku, Rattamanakku
- ☆ **Telugu:** Nepalmu, Peddanepalamu

Introduction

The genus Jatropha is tropical crop. Jatropa has around 170 species on the world. Generally it in the form of perennial herb. Jatropa word is derived from the Greek word Jatros and trophe. Jatros means doctor and trophe means nutrition. It is desiduous large shrub or small tree. It gets height of 3-5 meter with very smooth grey bark, which excudes whitish coloured watery latex, when it is cut. Leaves are alternate, green to pale green, broad, cordate, usually palmately 3 or 5 lobbed glabrous, unisexual; occasionally hermaphrodite flowers occur. Ten stamens are arranged in two distinct whorls of five each in a single column in the androecium, and in close proximity to each other. In the gymnosperm, the three slender styles are connate to about two third of their length, dilating to massive bifurcate stigma. The length of petiole ranges between 6-23 mm. The infloresce is formed in the leaf axils. Small yellowish green flowers are born in loose penicles of cymes. Fruits are

trilocular ellipsoidal. Fleshy exocarp is seen till the seeds remain matured. Generally they are produced in winter particularly in the monthe of October-December. Fruits of Jatropa are capsule types and they get metured in 57 to 60 days after anthesis. Seeds are ovoid and oblong and black in colour. They are similar to that of castor. After two of four months form the fertilization, when fruits (capsule type) become green to yellow, seeds become matured.

Origin and Distribution

The native of the plant is Mexico and Central America. But is cultivated in Africa, Latin America, India and South Eatern part of the Asia. In earlier days when its medicinal importance was not very much known at that time this plant was used for making fencing or shed. In India during 16 th centuary Portuguese Navigator introduced it first time. It is grown throughout in India including Andaman Island and generally grown as live fence/hedge for protection of farmer's field against damage which is done by animals and bulls. Tree grows fast in arid and semi arid condition.

It has good drought resisting capacity. It is suitable for sand dune stabilization and soil conservation areas. Being hardy, it can be used for the ecological and economic rehabilitation of wastelands in the tropical and sub tropical regions of the world. Bio diesel is obtained from the plant so it is highly economic plant which gives foreign exchange to the country too.

Cultivation

Jatropa can be cultivated in any kind of soils. Generally dry areas where annual rain fall is between 300-1000 mm can be very useful. The drier regions of the tropics are also useful. This plant is propagated by seeds and by branch cuttings. Direct seeding can be done in the field.

Seed Propagation

When seeds are mixed in soil naturally plant can also be produced. Seeds are collected in October to December however it varies from place to place depending upon the climate conditions. For cultivation purpose the seeds should be collected from superior trees in the month of October to December. For cultivation purpose the seeds should be freshly collected. Seeds may be cleaned by blowing or winnowing. Dry up the seeds before storing. Seeds are stored in the gunny bags.

Seed beds are adjusted that the nursery operations could be carried out easily without entry in the beds. The length of the beds may be up to 12 meters. Seeds are sown under pan stick nursery operation. One or two healthy and bold seeds may be sown in poly bags at a depth of 1.5 to 2.0 cm. Seeds are soaked in cold water to get good germination. Irrigation is very useful for the growth. FYM and vermicompost manure are utilized. Good time for sowing is February to March and September to October. Seeds are sown in furrows of 2 cm depth. After placing the seeds furrow should be covered with a thin layer of soil and pressed so as to embed the seeds. Deep sowing is not useful.

Vegetative Propagation

Cuttings of 2 to 3 cm should be taken from the base of the stem because it is healthy portion and gives good results.

Transplanting is done in the rainy season. During this time pits are prepared of 45 x 45 x 45 cm in the May to June. Firm Yard Manure is also applied. Dust is used to protect plant against termite. Seedlings are put in the center of the pit. The soil around the seedling should be pressed and irrigation should be given, if there are no rains, transplanting can be done in February to March. Two or three seeds can be sown in each pit.

Prooning, Hoeing and Weeding

Proonign is done to deveop more number of branches at a height of 40-60 cm. The plant will become bush rather then a medium tree. It will give vertical woody growth. Plant basin should be kept free from weeds. Two to three hoeing and weeding are enough.

Harvesting

Harvesting is done in the month of August to December in North Indian condition. Fruits can be harvested in two to four months after flowering.

Yield

Yield is obtained from 2nd year and onwards. From one acre plantation about 800-1000 kilograms yield can be obtained. About Rs. 10,000-12,000 can be earned from one acre plantation of this plant. Oil is also obtained form the plant. Mechanical processing technique is simple and feasible for farmers. Solvent extraction can provide 95 to 99 per cent of the total available oil.

Medicinal and Economic Importance

- ☆ Whole plant is medicinally and economically useful. Plant is used in traditional medicines and its effects for the cure of many diseases.
- ☆ Jatropine is obtained from the plant which contains anticancerous properties.
- ☆ Plant extract is used for wound healing, allergies, burns, cuts and wounds, inflammation, leprosy, leucoderma, scabies, small pox like dieses.
- ☆ Juice of leaves is used for plies. Plant juice is purjative.
- ☆ Seeds are good source of traditional medicines to treat arthritis, gout and jaundice. Stem is used for dental prolems like toothache, gum inflammation, gum bleeding, pyorrhea *etc.* Twing sap is used for wounds and ulcers.
- ☆ Water exctract of the plant is used for the treatment of HIV tumor.
- ☆ *Jatropa curcas* is source of Bio diesel which is highly useful. Oil is renewable source of energy, it is visible alternative to diesel, kerosene, LPG, furnace oil,. Coal and fuel wood. It is also used in resins, polish, paint, soap and candle industries. Jatropa contains nitrogen (4.4 per cent), Phosphorus (2.09) per cent, and potassium (1.68 per cent)
- ☆ Large scale cultivation fertilizes the bare lands and thus it is very useful in the low rain fall areas for farmers.
- ☆ Seed oil contains 21 per cent saturated fatty acids and 79 per cent unsaturated fatty acids.
- ☆ *J.carcas* oil is organic fertilizer which is used to make weste lands into useful lands and it does not impure soil like other chemical fertilizers.
- ☆ Plant is used as food for the animal and fishes, rats etc. Bio-fence is also made to protect other species of plants from jatropa.

105

Lavandula angustifolia Mill. (Lavender)

Botanical Name: *Lavandula angustifolia* Mill.

Synonyms: *Lavandula officinalis* Chaix, L.vera DC, *Lavandula lotifolia* Mill Spica DC

Family: Labiatae, Lamiaceae

Local Names

- ☆ **Gujarati:** Lavandar
- ☆ **Hindi:** English Lavadula
- ☆ **English:** Lavender, Spike lavender
- ☆ **Sanskrit:** Lavandin

Introduction

Lavender is a temperate plant and is hardy herbaceous bushy with straight woody branches 50-80 cm in height. There are 28 aromatic species of Lavandula but three species are of economic importance. (1) Lavender (2) Spike lavender and (3) Lavandin

1. **Lavender:** Lavender It is small perennial shrub. Leaves are opposite, long, narrow, lanceolate, light grayish green with a downy appearance. Inflorescence at the end of each slender stem where the densely packed layers of flowers seem to be in whorls, they are in fact so closely grouped that there are no leaves in the floral sector and they form a compact spike. Calyx is long, tubular, piloset and 5-lobed. Whole flower is mauve to violet shade tinged with light blue.

2. **Spike lavender:** Spike lavender is perennial shrub which gets height of 60-80 cm. Leaves are dimorphic, linear, lanceolate, narrowly elliptic to spathulate, much alternated at the base. Primary leaves of young shoot are up to 6.0 cm long, basal and axillary leaves are small and fastigiated. Peduncles typically branching, spreading, after very long spikes often interrupted, compact, rather slender. Bracts are linear to lanceolate, acute, tomentose, equal to calyx or slightly longer, medium nerve alone conspicuous. Bracteoles are linear, green or greenish calyx is up to 5 mm long, marginal teeth obtuse or rounded.
3. **Lavandin:** Lavandin is hybrid between lavender and spike lavender and has characters which are intermediate between the two species. Lavandin is a shrub which is 40-50 cm tall. Leaves are dimorphic, linear, lenceolate 4-5 cm long and narrowly spathulate, attenuated at the base. Basal leaves are smaller and tastigate densely grey. Spikes mostly at the interior end, compact to obovate, and green or grayish in color.

Plant grows well in month of March to April. Old clumps are regenerated quickly but to make it further faster heavy fertilization is required. Flowering time for this plant is April, May and June.

Origin and Distribution

Lavender is native to Europe bordering western half of the Mediterranean region extending to eastern coast of Spain, France, Italy and North Africa. It is cultivated in France, UK, USSR, Bulgaria, Italy, Hungary, Australia, China and India. Principally Bulgaria and the USSR is now almost certainly the world's largest producing region.

Cultivation

The plant can well tolerate temperate up to 26°C. Light promotes growth and improves oil quality. In India its cultivation has been found successful at attitude of 1666 m where summer temperature ranges from 25-35°C and winter temperature 0°C to -8°C. It grows well on sloppy lands thereby checking soil erosion to a great extent. It prefers open sunny soils for proper growth.

Lavandula angustifolia Mill.

Lavender is propagated through seeds and rooted cuttings. Seed propagation is very easy and economical. Seeds are planted in 1m wide raised nursery beds in rows 10-12 cm apart. Prior to planting, seeds are mixed with fine sand or ash as the seeds are very small and planted at a depth of 1-2 cm and covered with a mixture of FYM and sand. The proper time for plantation is October to November or March to April. Sprinkler irrigation is essential to keep the beds moist and seedlings should be pruned regularly. Seedlings of March April are ready for transplanting in autumn and those of October to November are ready in March to April. Propagation through cutting is also useful because genetic purity of the clones is maintained. Cuttings of 10 cm long are taken from one year old growth and planted in nursery beds and covered with a mixture of sand and well rotten compost. Cuttings are planted in the holes at a distance of 7.5 x 7.5 cm and fertilized regularly. Transplantation can be done in October to November or March to April. Cuttings or seedlings are planted at a distance of 30- 40 cm in rows 1.2-1.4 m apart.

Harvesting

Depending upon the exposure of the plantation, plants start flowering earlier in warmer and low altitude areas and later at higher slopes. Flowers are cut off with stem lengths not greater than 12 cm as there is no oil in the stems and leaves. However harvesting of lavender has been mechanized and at present, most of the lavender is harvested and chopped mechanically by specially designed tractor mounted harvesters. Generally in full bloom harvesting is done. Such harvesting is started with 50 percent blossom and 75-90 percent flowering. Harvesting is carried on dry, warm and sunny days.

Oil Extraction

Lavender in fresh conditions should not be stored for a longer time and immediately distilled. Flowers or flowering tops are distilled by stem distillation method. Hydro distillation method is not so useful as it causes ester saponification and thus lowering the ester content in the oil and may change its composition.

Storage

Oil should be stored in aluminium or stainless steel container and filled to its capacity, preferable in cool and dry places.

Yield

Fresh spikes are useful for oil. Per hectare 80-100 quintal of fresh spikes per year is obtained. In Bulgaria and USSR yield goes up to 100-150 kg oil per hectare has been obtained.

Medicinal and Economic Importance

- The lavender species can be exploited in number of ways. As a medicinal plant its dried inflorescence has traditionally been considered to be an antispasmodic, a carminative, a diuretic, a nervine, a stimulant and a tonic. Lavender has also been an insect repellent.
- Essential oil is obtained from lavender which is antiseptic, carminative and spasmodic activity. It is also used in cologne toilet waters, lotions and a wide range of high quality perfumes often with blending or stretching.
- Spike lavender oil is used to scent cheap soaps, polishes, detergents and liquid cleaners, although the more delicate French oil is also used in room sprays, deodorants, disinfacectants and insecticides.
- The major consumer of lavandin oil is the soap industry, but the oil also has many applications in men's colognes and numerous inexpensive perfumes, detergents, cleaning and washing up liquids, polishes, talcum powder and hair oil preparations.

106

Lawsonia inermis Linn. (Mehandi)

Botanical Name: *Lawsonia inermis* Linn.

Family: Lythraceae

Synonym: *Lawsonia alba* Lam.

Local Names

- ☆ **Gujarati:** Mehadi, Heena
- ☆ **Hindi:** Mehandi
- ☆ **English:** Henna, Egyptian priven, Cypress shrub
- ☆ **Sanskrit:** Medhini, Madayantika
- ☆ **Tamil:** Mailenati, Marutani
- ☆ **Telugu:** Goranta

Introduction

It is found in dry deciduous forests and widely cultivated as a hedge plant. It is a glabrous much branched deciduous shrub with 4 – gonous lateral branches often ending in spines. Leaves are simple and opposite, 2-2.5 cm long, elliptic, acute at either ends or tips, they are obtuse, minutely petioled, entire, and coriaceous. Flower is 0.5 cm in diameter, greenish-white, sweet scented, and is arranged in large corymbosely branched terminal panicles inflorescence. Some times flowers are looked like rose colored. Petiole is very short or absent. Fruits are globose capsule type, 0.5 cm in diameter, ultimetly one celled, irregularly breaking up; seeds are angular, smooth and pyramidal.

Distribution

This plant is distributed in whole tropical and sub tropical regions of the world. It can survive well in areas where rain fall is 1000-2000 mm. It is grown in Gujarat, Maharastra, Madhya Pradesh, Bihar and all the other states in India. It is observed in Sri Lanka, Pakistan, Burma and some other countries as well.

Cultivation

This plant can be grown in hot and arid areas although it can survive in any kind of soil. It grows well in clayey soils possessing good water holding capacity. It tolerates alkalinity in the soil to some extent. It can be propagated by seeds and also by branch cuttings. Propagation through cuttings gives good results. Furrows are prepared at one meter distance in the field. The 45 cm long cuttings are used for propagation and it should be planted at a distance of 50 cm distance at the beginning of monsoon. Cuttings are sprouted quickly. First weeding is carried out in the month of July. By the end of monsoon plant reaches at the height of 1-2 meter. Plant does not require artificial watering. However if water is available it is advisable to irrigate the field once in a month. Seeds are sown in nursery beds in the month of March to April. For one hectare of land 20 kg of seeds are required. Seeds start germinating in 10 to 15 days intervals. When plant gets height of 40-45 cm, these are to be transplanted in straight lines at 30 cm distances in the field. At this time the soils should be nicely moist.

Irrigation and Fertilization

Plant is affected by termites so to prevent this plant from them, the land is tilled 2-3 times with plough and the powdered oil cakes of neem mixed at the rate of 50-60 kg per hectare is added. It needs very less or no irrigation. Very less irrigation should be given to this crop; otherwise the color contents in the leaves will be affected.

Harvesting

Harvesting is done after 6 months of planting. Leaves are plucked by hands without damaging stems. In the first year one plucking and in the second and third year 2-3 pluckings are carried out. These leaves are collected and dried under shade or sent to distillation plant. A production of 500 kg in the first year, 1500 kg of air-dried leaves per hectare in the 2nd and subsequent years is expected.

Yield

The yield of oil is on an average 0.75 percent. Thus, the production of oil from the leaves is taken as 4 kg in the first year and 12 kg in the subsequent years.

Medicinal and Economic Importance

- ☆ Roots, leaves, flowers, seeds are used for medicinal values.
- ☆ Leaves are bitter, astringent, acrid, refrigerant vulnerary, diuretic, emeticm expectorant, anodyne, anti-inflammatory, constipating, depurative, liver tonic, haematinic, styptic febrifuge and trichogenous.
- ☆ Flowers are intellect promoting, cardiotonic, refrigerant, soporific, febrifuge and tonic. They are used in cephalagia, burning sensation, sardiopathy, amentia, insomnia, and fever.
- ☆ Decoction is used as gargle for relaxed sore throat. Essential oil is obtained from flowers by stem distillation. This oil is used in perfume industry. Herbal hair oil is also made up from this plant.
- ☆ Henna contains 25-33 per cent water soluble matter, aqueous solutions are orange in colour and show a green fluorescence. The principal colouring mater is lawsone, 2-hydroxy-I; 4-naphthaquinone which is present in dried leaves in a concentration of I-0-I-4 per cent. Behenic, arachidic, stearic, palmitic, oleic etc.

107

Lepidium sativum L. (Dodi)

Botanical Name: *Lepidium sativum* L.

Family: Cruciferae

Local Names

- ☆ **Gujarati** Dodi, Kharkhodi
- ☆ **Sanskrit** Asalika, Asalic
- ☆ **Hindi** Chandrashur, Dodi
- ☆ **Tamil** Alivirai
- ☆ **English** Garden cress,Water cress
- ☆ **Telugu** Aditya

Introduction

This plant is erect, herbaceous glabrous annual up to 45 cm in height. Leaves are entire or variously lobed or pinnatisect, the lower petiolate, the upper sessile. Flowers are white and small and they are arranged in small racemes. Fruits are obovate; they are in the form of small pods. Pod is notched at the apex with two seeds per pod.

Origin and Distribution

The plant has uncertain origin but according to some people Iran is the place from where this plant has originated. It is also believed to be naturalized in Britain. It is distributed in whole India.

Cultivation

This plant is cultivated by two methods (1).Seed propagation and (2). Vegetable propagation.

Chemical Composition

Seeds contain an alkaloid, glucotropaeloin, sinapin, sinapic acid, mucilaginous matter and uric acid. Saturated acids like stearic, arachidic behenic lignoceric, loeic and linolenic acids are isolated from the plant.The unsaponifiable mater contain β-sitosterol and α-tocopherol. The oil possesses anti oxidant properties. The optimum stabilizing concentration when used as an additive for linseed oil is ten percent.

Medicinal and Economic Importance

- ☆ The roots are bitter and acrid, and they are useful in secondary syphilis and tenesmus.

- ☆ Leaves are stimulant, diuretic and antibacterial and are useful in scorbutic diseases and hepatopathy.
- ☆ Seeds are bitter, thermogenic, depurative, rubefacient, galactagogue, emmenagogue, tonic, aphrodisiac, ophthalmic and diuretic.
- ☆ They are useful as poultices for sprains and in leprosy, skin diseases, dysentery, diarrhea, splenomegaly, dyspepsia, lumbago, ophthalmopathy, leucorrhoea, scurvy, seminal weakness, asthma, cough, hiccough, haemorrhoids and vitiated condition of *vata*.
- ☆ It can be asdministered to cause abortion.

108

Leptadenia reticulata W. and A. (Jivanti)

Botanical Name: *Leptodenia reticulata* W. and A.

Family: Asclepidaceae

Synonyms: *Gymnema aurantiacum* Wall. ex Hook.f.

Local Names

- ✰ **Gujarati:** Dodi, Nani dodi, khirkhodi
- ✰ **Hindi:** Dori
- ✰ **English:** Jivanti
- ✰ **Sanskrit:** Jivanti
- ✰ **Telugu:** Kalasa, Mukkutummudu
- ✰ **Tamil:** Palaikkodi

Introduction

This is twining shrub. The stem is with cork like deeply cracked bark, branches, numerous, younger ones and glabrous. The leaves are coriaceous, ovate, acute, glabrous above finely pubescent. Flowers are greenish yellow, and are arranged in lateral or subaxillary cymes inflorescence, often with small hairs. Fruits are follicles sub woody, 6.3-9 cm tapering. The seeds are 6 mm in length.

Distribution

This plant is a small genus of erect or twining shrubs distributed in tropical and sub tropical parts of Asia and Africa. Mainly two species are found in India.

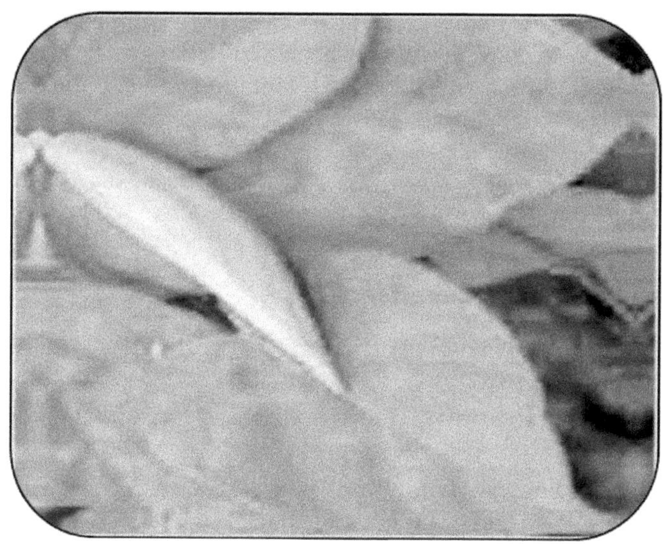

Leptadenia reticulata is found in sub Himalayan tracts of Punjab and U.P. and throughout the Deccan Peninsula up to an altitude of 900 m, particularly in hedges. In India plant is distributed in Punjab, Konkan, Poona and Gujarat. It is also found in the Western Peninsula.

Cultivation

It is mostly grown in hot humid or in semi dry climate. The climatic condition of Rajasthan and Gujarat are suitable for its cultivation. Generally sandy or black to medium black soil is useful. This plant is propagated through seeds and vegetative means.

Seed Propagation

The first necessity for sowing the seeds is to prepare a nursery in the first week of June. These seeds are soaked under dark and kept them for 24 hours. The seeds should be sown at one cm deep in polyethylene bags filed with the mixture of sandy soil, black soil and cow dung in 2:1:2 ratio.

By Layering Method

The month of July is suitable for planting through air layering. 0.5 to 1 cm thick green branches are selected. Node is selected in the branch. Two cuts are made at the interval of 5 cm distances. The bark between both cuts should be totally removed. On the barkless portion coating of siradit which is a plant hormone is made. Thereafter, burry and cover the branch along with node under wet black soil. Two Leptadenia reticulata creepers and their interwoving should not in anyway affect its productivity. The spacing between two plants should be 3 x 3 m.

Irrigation and Fertilizers

The use of carbonic manure is useful. Cow dung manure with carbonic manure is given to the plant for better growth. Irrigation is provided in summer at the intervals of 3-5 days and in the winter season 8 to 10 days intervals. For each plant 8-10 liters of water is sufficient. After transplantation and in initial days, irrigation is to be given at one day intervals.

Harvesting

Good crop is harvested in two years to three years. 1100 kg of above ground parts can be harvested.

Medicinal and Economic Importance

- ☆ Stem, leaves and roots are useful for medicinal values.
- ☆ Roots of the plants are useful in lactation and skin infection.
- ☆ Leaves and roots are used in skin diseases, inflammation of the skin and on wounds.
- ☆ Alcoholic extracts of roots and leaves show antibacterial activity against gram-positive and gram negative bacteria.

109

Limonia acidissima L.
(Kotha)

Botanical Name: *Limonia acidissima* L.

Family: Rutaceae

Local Names

- ✰ **Gujarati:** Kotha
- ✰ **Hindi:** Kotha, Bilin, Kavit, Kaitha, Katbel
- ✰ **English:** Wood apple, Elephant apple
- ✰ **Sanskrit:** Kapittha, Kapitthah
- ✰ **Tamil:** Vilankay, Maram
- ✰ **Telugu:** Tora elka, Velaga, Veagapandu

Introduction

It is big tree which often reaches up to height of 15-16 m in height and 60-120 cm in girth. Leaves are pinnate, they are 7-10 cm long, leaf lets are obovate or ovate. Flowers are polygamous in lax panicles inflorescence. Fruits are wood, globose and hard they are covered with white bloom. Pulp of fruit is aromatic. Seeds are numerous and small they are covered in pulp.

Distribution

This plant is distributed throughout in India. It can be cultivated from up to elevation of 500 m in western Himalayas and all plains in the India. It can be cultivated in almost all conditions and climates. It is grown in Jammu Kashmir to

Kanyakumari. Gujarat, Maharashtra, Madhya Pradesh are some of the states which are related with the production of wood apple.

Cultivation

There are two types of this species: one with small acidic fruits and the other with large sweet ones. It is propagated by seeds or by cutting or by layering. Seeds are used to raise seedlings in polypots in the nursery. After two years these raised seedlings are transplanted in the field at 8 m x 8m x 10m spacing. Growth can be increased by using fertilizers because this plant is slow growing plant. Weeding is done time to time to for raising of seedlings. Generally after 14 -15 years, plant start to flower and giving fruits.

Harvesting

Ripining period of fruit is November to March. Fruit are plucked from the trees and taken to market at that time. Fruits can remain without spoiling for long time.

Yield

Each tree gives 60-70 fruits per year which means it can give 9000 fruits per hectare. Each fruit can be sold at the rate of Rs.1/-. After 1st 15 years net return is Rs. 7,000 per year from this plant.

Medicinal and Economic Importance

- The pulp of raw fruit is useful in arresting secretion of bleeding in the treatment of dysentery, diarrhea and piles. Pulp of fruit is eaten as such or with sugar.
- It can be used for making sharbats in the same way as Bael. It is useful in preventing and curing scurvy and flatulence.
- The gum is excluded from by the stem has a soothing effect on the skin and mucous membrane.
- Gum is also useful in bowel complaints, dysentery and diarrhea, tenesmous, *etc.*
- Ripe fruits are aromatic, digestive and tonic. They are useful as refresher too.
- A mixture of the ripe pulp of the fruit, cardamom, honey and cumin seeds is very useful in preventing cancer of breast and uterus and cures sterility.
- The juice of the tender leaves is used in relieving bowel complains of children, urticaria, skin eruptions caused by biliousness.
- Wood apple is used for treating hiccups.
- Powder of leaves dried in shade, with equal quantity of sugar candy, is useful in spermatorrhoea or involuntary ejaculation, premature ejaculation and functional impotency. The gum is also found useful as aphrodisiac.

110

Mallotus philippensis (Lam.) Muell-Arg. (Rohini)

Botanical Name: *Mallotus Philippensis* (Lam.) Muell-Arg.

Family: Euphorbiaceae

Local Names

- ☆ **Gujarati:** Sundari, Japilo
- ☆ **Sanskrit:** Recanakah, Kampillakah
- ☆ **Hindi:** Sindur, Kamala, Rohini
- ☆ **Tamil:** Manjanai, Kunkumam, Kamala
- ☆ **English:** Kamala tree
- ☆ **Telugu:** Sundari, Vasanta, Kumkumamu

Introduction

This is very small tree about 10 m in height with grey or pale brown rough bark with irregular fissures. Leaves are alternate, longer than broad, three ribbed at the base ovate or ovate lanceolate, entire or shortly serrate, red glandular beneath. The transverse nervules are prominent. Flowers are dioecious and small, they are arranged in spikes, the males clustered sessile or very shortly pedicellate in erect long terminal spikes usually several together, the females nearly sessile in short spikes;

Fruits are globose type, three lobed capsules covered over with dense reddish brown glandular pubescence, seeds are black in colour they are smooth and sub globose.

Origin and Distribution

This small tree is distributed throughout India. It is mostly seen in evergreen and deciduous forests up to 1,500 meter. It is distributed from India, Sri Lanka and South China to the West Pacific and Australia. In Borneo collected in Sarawak and Sabah. The plant is endogenous to India.

Cultivation

Tree is frost hardy and drought resistant which falls to the ground in the beginning of the hot season and germinates in the ensuing rainy season. Seeds are susceptible to drought and insect attack. Where loose bare soil is available. Seeds get covered during rains and germination is facilitated. Artificial propagation is done by soeing fresh seeds by about April. Germination is rather uncertain and it is advisable to sow seeds quite close at 5 cm distance and to thin out during first rainy season. The more vigorous seedlings are ready for transplanting during first year. Smaller ones are kept for another year in the nursery. Line sowing with field crops has proved successful. Regular weeding and loosening of soil should be carried out during first two years and as often as may be nessasary afterwards. The tree also reproduces from root suckers. Although growth is slow.

Chemical Constituents

Plant contains kamlolenic, conjugated dienoic, linoleic, oleic, lauric, myristic, palmitic and stearic.

Medicinal and Economic Importance

- The glandular hairs, which are reddish brown in colour are acrid, thermogenic, purgative, digestive, lithontriptic, styptic, Vermifuge, alexipharmic and depurative.
- They are useful in vitiated conditions of kapha and vata.
- It is useful in the verminosis, constipation, flatulence, wounds ulcers, cough, renal and vesical calculi, hemorrhages, poisonous affections, scabies, ringworm herpes and other parasitic skin affections.

111

Manilkara hexandra Roxb. (Rayan)

Botanical Name: *Manilkara hexandra* Roxb.

Family: Sapotaceae

Local Names

- ✰ **Gujarati:** Rayan
- ✰ **Hindi:** Khiri, Khirni
- ✰ **Sanskrit:** Rajadanah, Rajadani
- ✰ **English:** Rayan, Bakula
- ✰ **Tamil:** Nanupala, Palai
- ✰ **Telugu:** Khirni, Mangipala

Introduction

It is a large evergreen tree with tough branches, generally attaining the heights of 10-12 m. The bark is grey and rough. Leaves are simple, alternate, stiff, leathery, 5-13 x 2-5 cm, obovate or oblong, usually rounded at the top and narrowing into petiole, glabrous on both sides, dark green and smooth above, pale beneath, nerves 12-20 pairs, not conspicuous, petiole is 1-1.5 cm. channeled above, they are glabrous. Flowers are star like white, solitary or 2-6 together in axils of the leaf. Calyx is rusty tomentose, 6-lobed, reflexed, ciliate, inner 3 narrower than outer 3, corolla, tubular, 18 lobed in 2 series of 6 and 12, stamens are 6. Fruits are berry, 2-2.5 cm, 1 seeded some times there are two seeds, elliptical or egg-shaped.

Distribution

Genus *Manilkara* is distributed mainly in tropics, three species of genus manilkara found in India, in which *Manilkara hexandra* and *Manilkara littoralis* are endogenous to India while *M.kauki* is exotic and species has introduced in gardens.This tree is found in evergreen forest and dry parts of the India. This tree is particularly found in Gujarat, Rajasthan and adjoining areas.

Cultivation

It is the plant of arid and semi arid climatic zones. In hilly areas, it is found along with water courses. It is drought resistant and have low needs. It grows well in light soil and alluvial soil. Seed propagation is good method. Two years old seedlings are raised in polypots (20 x 30 x 300) are best for raising plantations. At the beginning of monsoon the seeds are sown in polypots filled with soil. They are regularly watered on non rainy days. They germinate without any difficulty. They are kept free of weeds and watered regularly. When they become two year old, they attain a height of 80-90 cm. height and are ready for transplanting in the field. In the area to be planted pits of 45 x 45 x45 cm are dug at a spacing of 10m x 10m. After adding farm yard manure to each pit, the nursery raised seedlings are transplanted in the pits, at the beginning of monsoon. The plants establish soon and start growing. Regular weeding and soil workings are carried out. Application of fertilizer (50 grams of urea in 2 split doses per plant every year) helps plant to grow quickly. The plants are slow growing and they take 12-15 years before flowering and fruiting. Trees have long life almost more than 100 years.

Harvesting

Harvesting is done in 16th years and onwards because in this time tree becomes capable to give fruits. Maximum yield of fruits are obtained in 25-30 years of age. In the march fruits are plucked from the tree. Since all the fruits are not ripe at one time, several pickings of the fruits are necessary.

Yield

Each tree produces about 25kg. initiallty and reach up to 100kg. at the age of 30-35 years. An average production of 2500 kg of fruits can be expected from one ha. of plantation. The fruits can be sold at the rate of Rs. 10/- per kg. Total expenditure till 25 years and onwards reaches to Rs. 10,000 per hectare and net return is around Rs. 40,000 per hectare.

Medicinal and Economic Importance

- ☆ Fruits of the tree are edible.
- ☆ The tree gives gum which has commercial importance.
- ☆ The oil is also extracted from the seeds is used to adulterate ghee.
- ☆ Rayan seedlings are in much demand because they are used as stock for grafting sapota (Manil-*kara Zapota*) seedlings.
- ☆ Excellent walking sticks are made out of the branches of this tree.

112

Mentha arvensis L. (Pudina)

Botanical Name: *Mentha arvensis* L.

Family: Lamiaceae

Local Names

- ☆ **Gujarati:** Pudina, Pudino
- ☆ **Hindi:** Pudina
- ☆ **English:** Mint
- ☆ **Sanskrit:** Pudinah, Putinah
- ☆ **Tamil:** Putina, Podina, Putina
- ☆ **Telugu:** Pudina, Igaenglikara

Introduction

Mentha arvensis is a downy perennial plant with rootstock creeping along or just under the ground surface. It has a rigid branching, pubescent 60-90 cm tall stem. The leaves are lanceolate to oblong, 3.7-10 cm long, sharply toothed, sessile or short petiolated and hairy. Flowers are arranged in cymes which are usually sessile or rarely pedunclate. Flowers are purplish minute.Calyx is 2.5 to 3 mm long, narrowly deltoid, acuminate, corolla white to purple with 4-5 mm length. Plant grows quickly in the April May months when the temperature reaches around 30°c. When there is increase in the temperature growth resumes quickly. Fully developed plants cover the ground surface complety and give dark green look to entire area.

Origin and Distribution

In India, the plant has been introduced by Arabs and has been cultivated in the north Indian plains to a limited extent for more than 300 years. As a source of essential oil the plant was again introduced in Jammu and Kashmir in 1957. Scientific cultivation and distillation technologies of chamomile were developed by CIMAP, Lucknow, NBRI, Lucknow and RRL, Jammu. In India most of the states grow this plant.

Cultivation

In most part of the world Japanees mint is grown in sub temperate to temperate climate in northern latede which provide ample sunshine during most part of growing season. Crop suits subtropical and tropical climate with fairly cold winter and warm summer. High temperatures support vegetative growth and oil production, while low temperature supports the growth of underground stolen. It is long day plant which needs more light to grow. It can be grown in well drained medium or light soils which are rich in organic matter, contain high nutrients status, have good water holding capacity and with a pH range of 6-7.5. Water logged and heavy clay soils are to be avoided.

Plant grows well by vegetative propagation method. Terminal tips and above ground stems (Suckers) can be used for propagation. Usually planting material is obtained from the old plantation which has remained dormant in winter. However it is advisable to plant new crop in small area in August or beginning of September for stolon production because in nursery plantations sucker or stolen production is very high compared to the old plantation left for production of planting material.

Harvesting

Harvesting is an important operation which affects both the production and quality of oil. It is therefore important to harvest the crop at proper stage of plant growth. Early harvesting results in lower recovery and poor qualiy of oil, whereas delay in harvesting results in low production of oil due to the fall of leaves mainly lower leaves where oil is present in maximum concentration. Harvesting of Japanese mint is recommended at the time of initiation of flowering buds which occurs somewhere between 100-110 days after planting and again and 65-70 days after first harvest. Harvesting should be done on sunny days and irrigation should be withdrawn a week before harvesting. Rains at the time of harvest of excess soil moisture in the field have adverse effect on oil accumulation in the plant, resulting in low recovery of oil on distillation. On large plantations it is advisable to use tractor operated forage harvester or mowsers.

Yield

On an average 20-25 tones of herb per hectare is obtained which on distillation gives 125-150 kg oil/ha. With newly developed varieties higher yield of 200-250 oil/ha is obtained.

Storage

The crude oil is made free of moisture by treatment with dehydrating agents like sodium sulphate, filtered and stored in air tight containers of HDP and galvanized iron.

Medicinal and Economic Importance

- ☆ Fresh or dry herb is used for the distillation of oil. Japanese mint oil is generally used as a primary source or menthol. Menthol is used as a flavouring agent in tooth pastes, candies, chewing gums and mouthwashes *etc.*
- ☆ This is also used as an ingredient in a number of medicinal preparations like ointments, pain balms, cough syrups, cough lozenges and tabletes as well as a large numbers of beverages and other items like tobacco, cigarettes, confectionery, betel nut flavouring *etc.*
- ☆ Menthol is used as a soothing ingredient in a number of cosmetic preparations, cologue, deodorants, after shave lotions and perfume bases.
- ☆ The oil and menthol are also used in flavouring mouth fresheners, aerosols, polishes, lipistics and hair lotion. After isolation of menthol, demantholised oil is used as a flavour especially in toothpaste.

113

Merremia turpethum Linn. (Nashotar)

Botanical Name: *Merremia turpethum* Linn.

Synonym: *Convolulus turpethum* Linn.

Family: Convolulaceae

Local Names

- ☆ **Gujarati:** Nashotar
- ☆ **English:** Nashoti, Indian jalap, Turpeth
- ☆ **Hindi:** Nashotar
- ☆ **Tamil:** Shivadai, Kumbam

Introduction

This plant is perennial extensive climber with milky sap. Stems are narrowly winged and pubescent. Leaves are variable in shape. They vary from 4.5-14 x 3-13 cm, cordate or truncate at the base, pubescent on both sides. Flowers are solitary or in few flowered cymes. They are peduncles 1.5 to 9 cm long. Calyx is 15 to 20 mm long. Lobes are subequal, minutely pubescent, much enlarged in fruit. Corolla are 4 to 5 cm long and white in colour. Fruits are depressed globose; they are 15 to 20 mm in diameter. Seeds are dull black in colour, obscurely trigonous.

Distribution

This plant is widely distributed in arid and semi arid regions of the world. This plant is very common in the Barda hills on the river sides at the Junagadh, Sasan

in the Gir Forest in Gujrat. It is found all over India, upto an elevation of 900 m. It comes up on all types of soils.

Cultivation

This plant is not under cultivation anywhere in India. It can be propagated by seed. The area is ploughed thoroughly well and ridges/furrows are prepared at one meter distance. At the beginning of monsoon, the seeds are dibbled on the ridges continuously. The seeds soon germinate. For one hectare area 10 kg of seeds are sufficient. Effortrs are made to get the seedling at 1m x 0.5m. At the end of 2nd year (in April) the plants can be uprooted. Immediately after the harvesting, the field may be prepared.

Harvesting and Yield

The drug consists of root and stem. It consists of cylindrical pieces of 1.5-15 cm long and 1.5 cm in diameter, often with central woody portion removed by splitting the bark on one side. Its external surface is furrowed giving the drug a rope like appearance. The plants are uprooted and dried under sun before cutting into pieces. A production of 5,000 kg of drug can be expected from one hectare once in two years.

Medicinal and Economic Importance

- ☆ Roots of the plant has been used for long as a valuable purgative, known in commerce under the name of Turpeth Root or Indian Jalap.
- ☆ It is effective as jalap (*Exogonum purga*) and superior to rhubarb (*Rheum emodi*).
- ☆ White turpeth is preferred to black turpeth, because the black turpeth causes drastic purgation, vomiting, fainting and giddiness.

114

Mesua ferrea L. (Nag Keshar)

Botanical Name: *Mesua ferrea* L.

Family: Clusiaceae

Local Names

- ✰ **Gujarati:** Nag keshar
- ✰ **Sanskrit:** Nag puspah, Nag kesharah
- ✰ **Hindi:** Nag keshar
- ✰ **Tamil:** Nagappu, Nanku
- ✰ **English:** Mesua, Iron wood tree
- ✰ **Telugu:** Nageshvar, Gajapuspam

Introduction

This plant is medium sized to large handsome tree which is glabrous, evergreen. It gets height of 18-30 meter. Bark is reddish brown in colour which peels off in thin flakes. Leaves are simple, opposite, thick lanceolate, coriaceous, covered with waxy bloom underneath, become red when it gets young age. They are acute or acuminate. This tree is very common in Srilanka and declared as National tree of Sri-Lanka on 26th February-1986.

Origin and Distribution

This plant is distributed throughout India. It is found in evergreen forests upto 1,500 meter. Plant is distributed in Bengal, Assam, Eastern and Western ghat and also Andaman island. Origin of the tree is India.

Cultivation

This plant is cultivated by two methods (i).By seeds propagation and (ii). By vegetative propagation.

Chemical Composition

This plant contains polenske val Hehner val., palmitic, stearic oleic, and linoleic, palmitostearo-olein, dipalmoto-olein distearo-olein, stearo diolein, palmito-diolein, linoleodiolein, triolein.

Medicinal and Economic Importance

- Flowers are astringent, bitter, acrid, midly heating, anodyne, sudorific, digestive, carminative, constipating, anthelmintic, diuretic, alexipharmic, expectorant, stomachic, haemostatic, aphrodisiac, febrifuge and cardiotonic.
- They are useful in vitiated conditions of Pitta and Vata, asthma, cough, hiccough, halitosis,,eprosy, scabies, dermatopathy, pruritus, pharyngodynia, vomiting, dysentery, haemorrhoids, ulcers burning sensation of the feet, dipsia, impotency, leucorrhoea, haemoptysis, strangury, cephalagia, fever and cardiac debility.
- Seed oil is used in vitiated conditions of *vata* and skin diseases.

115

Michelia champaca L. (Peelo Champo)

Botanical Name: *Michelia champaca* L.

Family: Magnoliaceae

Local Names

- ☆ **Gujarati:** Rechampo, Peelo champo
- ☆ **Hindi:** Champa
- ☆ **English:** Golden yellow
- ☆ **Sanskrit:** Champak
- ☆ **Tamil:** Sampang
- ☆ **Telugu:** Sampangi

Introduction

Champa is evergreen medium sized tree which spread 2.5 to 3 m. It is tall and gets height of 30 m sometimes. It grows well in the humid area with plenty of rainfall. Generally it is 6-10 meter in height. Leaves are ovate to lanceolate, 25 cms long x 6.5 cms across, buds of the plant are silky. There are three well varieties of champa. They are distinguishing by their flowers. Flowering time is April to May. Generally flowers are white; sovereign red flowers 5 to 6.5 cms long and creamy light yellow. All these flowers have numerous petals. Fruits are produced during July-August. Seeds are available in late August.

Origin and Distribution

Champa is the native of India, Java, Philippines and Nepal. In India champa

is found in Bengal, Assam, Nilgiris and also in Western part of India like Gujarat, Maharastra and some other states. It is cultivated for its fragnant flowers and handsome foliage.

Cultivation

This plant is grown intemperate and sub temperate climate where annual precipitation is 150-250 cm. Tree can survive well in damp climate and shady conditions. This tree is propagated either from the seed or cuttings on sandy soils and planted at a distance of 5-6 m apart. After 4 to 5 years, 12-15 seedlings are planted in the field. Cuttings of root and shoot are also useful for propagation. Plant has strong roots which can reach up to deep soil layers. Too much water lagging is not useful to the plant because some time it leads to the death of plant.

Harvesting

Flowers are main yield of Champa plant. Flowers are picked up in April to May and again in September to October. They become brown in a few hours after being picked, hence to prevent its odour being impaired the oil must be prepared soon after picking. 375 to 425 weigh one kilogram and yield 640-670 ppm of an oil on fresh flower weight basis. In India, perfumers make champa attar mostly. Solvent exctraction method is used to prepare the oil using benzene as a solvent. The concrete yields by solvent extraction is 26 to27 per cent of steam volatile oils.

Yield

A champa tree yields about 50-100 flowers daily during the season, which lasts for about 80-100 days. From one kilogram of flowers 640-670 ppm of oil is obtained on fresh weight basis. Storage of oil is done in aluminum container under dry and cool condition.

Medicinal and Economic Importance

- ☆ Whole plant is used for medicinal and economic uses. Bark of Champa tree is considered as stimulant, diuretic and febrifuge.
- ☆ Dried roots and root bark are purgative and emmenagougue. Juice of the leaves is used in colic.
- ☆ Flowers and fruits are stimulant, antispasmodic, stomachic and diuretic and are considered useful in dyspepsia, fever and in renal diseases.
- ☆ The flowers oil is used as an application in cephalagia, opthalmia, gout and rheumatism.
- ☆ Fruits and seeds are considered useful for healing cracks in feet.
- ☆ Enflureraged flowers in sesame oil yield an excellent product often used as hairs oil and as a head coolant. The flowers yield dye as a base for other colours and for dyeing silk and cotton fabrics. The oil is highly esteemed in perfumery and used for the production of high grade perfumes.
- ☆ Champa oil has delightful velvety odour closely approximately to that of the flower and recalling the fragrance of tea, orange blossoms.
- ☆ It provides one of the most exquisite raw materials for perfumery, being used in some of the finest French creations.
- ☆ Genuine concrete is produced in very small quantities and commercial champa perfumes are mostly synthetic.
- ☆ This plant contains cineole, isoeugenol, phenyl ethyl alcohol, benzaldehyde and methyl anthranilate, benzyl alcohol, p-cresol and methyl ether.

116

Mitragyna parvifolia (Roxb.) Korth. (Kalam)

Botanical Name: *Mitragyna parvifolia* (Roxb.) Korth.

Family: Rubiaceae

Local Names

- ☆ **Gujarati:** Kalam
- ☆ **Hindi:** Kayim, Kadam Kamgi,
- ☆ **English:** Kaim
- ☆ **Sanskrit:** Vitanah
- ☆ **Tamil:** Katampai, Nirkkatampu, Sinakatampu
- ☆ **Telugu:** Nerkadamba

Introduction

It is large and deciduous tree. It has rounded crown and light grey smooth bark exfoliating in small scales. Leaves are simple, opposite, very variable, elliptic, suborbicular or obovate, rounded, acute or bluntly acuminate, glabrous on both sides, main nerves 6-8 pairs, stipules 13 by 5.8 mm, Flowers are greenish yellow, fragrant and in globose heads. Fruits are oblong capsules with blunt rounded tops and blunt ribs.

Distribution

This plant is distributed throughout in India. It is mainly found in deciduous and evergreen forests. This plant is also seen at the elevation of 1,200 m height in

Himalayas. This is widely seen in the Gujarat, Madhya Pradesh, Rajasthan, Bengal and also in other states.

Cultivation

This plant is grown in tropical and sub tropical regions but it can be grown in any kind of lands. This plant is suitable to rich soil but any medium rich soil is also used for its cultivation. Generally Kalam is propagated by two methods. (1). Seed propagation and (2) Vegetative Propagation.

Natural reproduction takes place by seeds scattered in the hot season. Natural reproduction sometimes comes up freely on abandoned cultivation and also on badly drained ground. Artificial reproduction, seedlings are raised in boxes and then they are transplanted when 2-3 months old. Trees grow moderately fast, mean annual girth increment being 0.7 inch.

Medicinal and Economic Importance

☆ Roots, leaves and bark of the trees are utilized for medicinal importance.

- The roots and bark are acrid, bitter, stomachic and febrifuge, and they are useful in gastropathy, colic and fever.
- The leaves are acrid, bitter, sweet, styptic, vulnerary, stomachic, anti-inflammatory, anodyne depurative and febrifuge.
- They are useful in vitiated conditions of *Vata* and *kapha*, internal and external haemorrhages, wounds, colic, flatulence, dyspepsia, inflammations, myalgia, skin diseases, leprosy, erysipelas and fever.
- Stem provides good wood which is used for house furniture.

117

Morinda citrifolia L. (Noni)

Botanical Name: *Morinda citrifolia* L.

Family: Rubiaceae

Local Names

- ☆ **Gujarati:** Noni
- ☆ **English:** Indian mulverry, East Indian mulberry, awl tree
- ☆ **Hindi:** Surangi, Bartindi
- ☆ **Sanskrit:** Ashyuka

Introduction

This plant is medium sized tree which gets height of 6-8 meter. Young twings are angular, slightly compressed and grooved. Leaves are opposite, stipuliferous. They are shining green above and pale below. Margins of the leaves are undulate type. Flowers are arranged in dense ovoid head inflorescence which is opposite to leaf. Flowers are white in colour then they become yellow. Fruits are ovoid including many drupelets. Numerous seeds are embedded in the pulp.

Origin and Distribution

The plant is native of Southeast Asia and now extensively distributed throughout India, the tropical Pacific and Caribbean. It grows wild and cultivated for its medicinal uses in South India and Viet Nam. It grows in relatively dry to moderate wet land, from sea level to about 1500 feet elevation. It tolerates salinity also.

Cultivation

This plant is cultivated by two methods (1).seed propagation and (2).Vegetative propagation. Noni seeds are reddish-brown, oblong-triangular, and have a conspicuous air chamber. They are buoyant and hydrophobic due to this air chamber and their durable, water repellant, fibrous seed coat. A single large noni fruit can contain well over 100 seeds. Only soft, ripened noni fruits should be chosen for seed collection. The seeds must be separated from the fibrous, clinging fruit flesh. Site should be selected where full or partial sun light reaches well. Heavy soils, compacted areas, and flood-prone sites are not useful for cultivation of noni plant. Young noni transplants do not grow well where winds are strong. Such conditions may exist along windward coasts. An appropriate interplant spacing for noni is 10–15 feet. At 12-foot spacing there are 290 plants per acre. Closer interplant distances result in plant crowding and may exacerbate certain pest and disease problems.

Seed Scarifying

Scarification of seed becomes helpful in reducing time for germination Hard seed coat is punctured or nicked for this purpose.

Seed Drying and Storage

Seeds should be cleaned properly and allowed to dry for two to three days in shade under room temperature seeds should be stored in air tight container.

Planting

Fresh noni seeds can be planted immediately after extraction from the fruit. Some growers soak the seeds until they start to germinate, then they are planted in containers, while others plant fresh seeds without presoaking treatment. The seedlings are grown for about nine months to a year before they are transplanted to the field. Some growers just plant fruit fragments containing seeds directly into the field soil. Noni seeds require some months for its germination. Cultivation of noni plants from stem cuttings (verticals or laterals) reduces the time required to obtain plants that are ready for transplanting. Cuttings from stems and branches will sprout roots readily under the proper conditions. A rooting compound may prove helpful in promoting rapid root establishment. Branch or stem is removed and checked for fresh sap flow from the wound. If the sap flows readily, cuttings can be made from these materials. Cut end of the freshly cut noni stem is inserted into a pot containing a general-purpose growth medium. Again, an artificial, pathogen-free medium is preferred to untreated agricultural field soil. Irrigation is required in proper amount.

Harvesting and Yield

Noni plants can begin to bear fruit about nine months after planting. Fruits can be harvested at this early stage, although they generally are small and few. Some farmers choose harvesting during the first and second years in favor of pruning back the branches instead. This pruning results in a bushy plant with more vertical and lateral branches and ultimately produces greater fruit yield. The expected yield from mature, healthy plants is 250–500 pounds per plant per year, depending on nutrition and plant spacing. However, yields at some locations can exceed 200 kg pounds per plant per year with good crop management. Fruits may be picked just before they begin to ripen fully and turn completely whitish- yellow on the tree. These fruits are suitable for shipping and will continue to ripen unless measures are taken to retard ripening. For processing locally, fresh fruits are picked when ripe, just before they fall naturally from the tree. Juice can be squeezed directly from the ripened fruits or allowed to drop from the fruits and ferment as in the traditional method.

Chemical Composition

Root bark contains morindon, morindin, morindadiol, soranjidiol, rubichloric acid, alizarin a-methlyl ether, and rubiadin 1-methyl ether.

Medicinal and Economic Importance

- ✰ Root bark has a beneficial effect in hypertension, osteodynia and lumbago.
- ✰ A decoction of the leaves taken by mouth in effective for fever, dysentery and diarrhea.
- ✰ A poultice of pounded fresh leaves cures furunculosis.
- ✰ The fruit, taken together with a little salt is stomachic, aperient, and active on dysentery, uterine haemorrhage, metrorrhoea, cough, coryza, oedema and neuralgia.

118

Morus alba L. (Shetur)

Botanical Name: *Morus alba* L.

Family: Moraceae

Local Names

- ☆ **Gujarati:** Shetur
- ☆ **English:** White mulberry, white-fruited mulberry
- ☆ **Hindi:** Tut, tutri, chinni
- ☆ **Tamil:** Musukette, Kambli
- ☆ **Telugu:** Reshme

Introduction

This is deciduous tree, 3-15 m high, bark or large stems are brown in colour. It is rough. Fissures are mostly vertical. Leaves are alternate, very variable in size and shape, usually broadly oval, 6-18 m long by 2-4.5 cm wide, apex acuminate; Rounded or cordate shape. Margin is irregularly dentate or incised-lobate, glabrous or slightly pubescent along the nurve 3, lateral nerves are forked near the margin; Petiole is usually is 1.8 to 2.5 cm long. Flowers are monoecious or dioecious and green in colour. Stalk is frouped. Fruits are aggregate, constiting of all the ovaries of the catkin forming a crustaceous achene. Name of genus *Morus* means 'delay', referring to the formation of winter buds late in the season after the weather has turned cold and the species name, *alba*, means 'white', referring to the whitish color of the buds.

Origin and Distribution

White mulberry is native to China and has long been cultivated in Europe. British then introduced plant into North America. This plant grows wild in India, China, Japan, Pakistan, Philippines, Sri Lanka, Viet Nam like countries.

Cultivation

This plant is cultivated by seed and vegetative propagation method. Mulberry seeds are treated before sowing in nursery bed with cold water for week hastens germination. Kerosene treatment protects seeds from germination. Seeds are sown during May-June in lines of 20 cm apart. Germination in mulberry from seeds take place in 6-7 days and completed within 10-12 days under favourable condition.

Vegetative propagation is also a successful method. Plants are raised by vegetative methods, are of true type and grow uniformly. Stem cuttings are prepared from mature shoots with 20-25 cm length having 3 to 5 buds. Time for planting of cutting is December-January. First cutting are sown in the nursery beds. After successful regeneration of cuttings in nursery it is transplanted in the field. Time for planting of cutting is December-January. On well cultivated soil, pits of 60 cm^3 are dug during October-November. For winter planting and May-June for July planting at the distance of 6 x 6m. When plantation is grown for sericulture, planting distance is reduced to 1 x 1m from to 2 x 1 m. Pits are prepared with a mixture of firm Yard Manure and soil (in the ratio of 1:3)

Harvesting

During December-January, when all shoots of tree are at a height of 1 to 1.5m from the ground level, prooning is carried out. Prooned trees have large sized fruits. First picking of leaves is carried out 8-10 weeks after planting under irrigated condition whereas it is done 12-15 weeks after planting under rainfed conditions. About 10 pickings of leaves are available under irrigated condition and 6-7 picking under raifed condition. Regular pruning followed by regular application of manure helps the plants to produce good amount of leaves for 15 years.

Yield

Average yield of leaves is 10-15 tonnes per hectare. When mulberry is raised as a crop, mature fruits are harvested. Harvesting is done in 3-4 pickings. A full grown tree can yield 10 to 15 kg of fruits.

Medicinal and Economic Importance

- Plant is used in cough, dyspnaea, facial dropsy, oedema.
- The leaves are antibacterial, astringent, diaphoretic, hypoglycaemic, odontalgic and ophthalmic, they are taken internally in the treatment of colds, influenza, eye infections and nosebleeds.
- A tincture of the bark is used to relieve toothache.
- Extracts of the plant have antibacterial and fungicidal activity.
- A fibre is obtained from the bark of one-year old stems.
- A brown dye is obtained from the trunk.

119

Mucuna pruriens Bak. (Kauwanch)

Botanical Name: *Mucuna pruriens* Bak.

Family: Papilionaceae, Fabaceae

Local Names

- ☆ **Gujarati:** Kaucha
- ☆ **Hindi:** Kaman
- ☆ **English:** Common kowich, Kownch
- ☆ **Sanskrit:** Atmagupta
- ☆ **Tamil:** Punali gali
- ☆ **Telugu:** Neela

Introduction

It is slender and climbing annual plant which climbs on the bushes every year. Leaves are ovate, rhomboid membranous and the petiole is 6-11 cm long, leaflets are 7-12 cm long and petiole of the leaf is 6-11 cm long. Flowers are arranged in the raceme inflorescence. Flowers are purple in color. Pods are 5-7 cm long, bristly red curved or golden colored. Glands are present on bristles. The seeds are bean shaped and 1 cm in diameter, they are white in color. Flowering time of this plant is September to October.

Distribution

Plant is grown in whole tropical and subtropical regions of Asia. It is reported from Himalayan foothills and from Punjab to Srilanka. In most of the states like Punjab, Bihar, Gujarat, Maharastra etc. kaucha plant is observed wild.

Cultivation

It is widely observed in the forest areas. Because it is very useful its cultivation is essential. Generally seed propagation is common for better growth of the plant. But direct plantation is also useful. After first rainy season during June to July, one to two seeds are sown under tree because when it grows it gets support from the tree. No irrigation is required in the case of continuous rain. But if rain is not continuous then irrigation is required. Seeds are produced during November to December. If good support is provided for plant then it can grow very quickly. It is better to plant the seed in field directly. The distance between plant to plant should be 1 meter and row to row should be 2 meters. Let the tree should be allowed to grow for one year. In the second year we can sow seeds of mucuna by side of each of the tree. For one hectare of land six kg of seeds are required. In the shade of these trees other medicinal plants like *Rauvofia serpentina* and *Gloriosa superba* can also be cultivated.

Harvesting

A specific machine is prepared from bamboo and wire on it for plucking the ripen beans. Dry beans are plucked and stored. The collected beans are beaten or crushed by sticks to separate the seeds. They are filled in bags then they are taken to market. Roots are also uprooted in the season of January; the roots are dried and stored in addition to the seeds.

Yield

The yield of seeds is 20-25 quintals per hectare. Seeds are sold at the rate of Rs. 15.00 to 25.00 per kg. Supporting plants like Agasthi, Sahjan and Jatropha are also useful as they provide extra income of Rs. 50.00 thousands to Rs.1 lakhs per hectare.

Medicinal and Economic Importance

- ☆ Seeds, roots and fruits of the plant are used for medicinal value. Various uses of seeds as promoters of virility as also Bhavamisra.
- ☆ Seeds contain L-Dopa (L-3-4-dihydroxyphenylalanine).
- ☆ Biotransformation of L-tyrosine to L-Dopa has been shown in suspension cultures. The spicular hairs contain 5-hydroxytryptamine and a proteolytic enzyme-mucanain.
- ☆ The seeds have anti-inflammatory activity. The indol bases have an unspecified spasmolytic action on spasms induced in smooth muscles by acerylcholine, histamine or oxyocin.
- ☆ In ayurveda, the seeds are used for male virility. The pods are anthelmintic and the root powder is laxative. Powder of seeds is used for curing of Parkinson diseases. The plant also decreases prolaction rise induced by chlorpromazine.

120

Murraya exotica L. (Orange Jasmine)

Botanical Name: *Murraya exotica* L.

Family: Rutaceae

Local Names

- ☆ **Gujarati:** Pilo champo
- ☆ **Hindi:** Marchula,
- ☆ **English:** Orangae jasmine
- ☆ **Tamil:** Kari vempu, Tatta veppilai, Katta nartagam
- ☆ **Telugu:** Kare paku, Naga golunga Tel, Yerra munukkudsu.

Introduction

It is evergreen shrub or a small tree, with spreading crown, often crooked, short trunk, bark is yellowish brown, corky, fragrant. Leaves are imparipinnate; leaflets are 3-9, ovate or elliptic- lanceolate or rhomboid, gland-gotted. Flowers are arranged in corymbose cymes, they are white in colour. Flowers are fragrant. Fruits are berries type, each fruit is oblong or ovoid, they are red or deep orange when they ripe. Fruits are 1-2 seeded.

Origin and Distribution

Murraya is the genus of shrubs or small trees distributed from South and East Asia to Australia. Murraya exotica is commonly grown in gardens for its glossy green foliage and large clusters of fragrant flowers. It is a popular hedge plant and is well adapted for topiary work.

Cultivation

It is widely observed throughout India upto an elevation of 1500 m. It is seen growing on all types of soils. It can tolerate shade to some extent. This plant is propagated by seeds, cuttings or layering. It can be classified as a slow growing species. One year old container plants are recommended for transplanting in the field. It can be planted at 3m x 2m x 2m spacing. Normal care like weeding and soil working are carried out in the first 2 years. Watering and application of fertilizers are also necessary to boost up the growth. After 5 years of the planting, the bark can be harvested from the trees.

Harvesting

Harvesting is done after 5 years of planting. Bark is also harvested from the trees from 6th years and onwards. The bark can be removed in small patches only. The whole length and circumference of the bole is divided into small squares of 10cm x 10cm. These squares are again grouped into 5 square groups and only one square is harvested every year from that group. It is essential to see that the squares exploited are not in continuous vertical line.

Yield

It is estimated that each tree can produce about 1 kg. bark every year. Thus a production of 1000 kg. of bark can be expected from 1 ha. The market rate of the bark could be around Rs. 20/- per kg. Total expenditure is Rs. 5,000 per ha. while grow return reaches upto Rs. 20,000/-. That means net return goes upto Rs.15, 000/- per hectare.

Diseases and Prevention

It is vulnerable for the attack of citrus stem borer. It is suggested to cut and remove the infected branches and affected portions with creosote and chloroform mixture.

Medicinal and Economic Importance

- Leaves are stimulant and astringent. They are reported to be used for diarrhea and dysentery.
- Powdered leaves are applied to cuts. Leaves and root bark are some times used against rheumatism, cough and hysteria.
- The twings are used for cleaning teeth. The leaves possess antibacterial activity against certain bacteria.
- Fresh leaves on stem distillation yield a dark coloured volatile oil with pleasant odour.
- Flowers and powdered bark are said to be important ingredients of cosmetics in South East Asian countries.
- The women in this part of the world maintain the freshness and glow of their facial skin even in their old age and this is attributed to the bark of this plant that has gone into the cosmetic items.

121

Murraya koenigii L. (Mitho Limbado)

Botanical Name: *Murraya koenigii* L.

Family: Rutaceae

Local Names

- ☆ **Gujarati:** Mitho limbado, Khadi limbado
- ☆ **Hindi:** Harri, Gandhela
- ☆ **English:** Curry leaf plant
- ☆ **Sanskrit:** Giri nimba, Kadarya
- ☆ **Tamil:** Kari vepa
- ☆ **Telugu:** Charangi, Karepaku

Introduction

Curry leaf is very common in India. It is used in the preparation of making foods of different kinds. It is small and shrub or evergreen tree which has smooth, dark brown bark. Leaves are alternate, imparipinnate, 15-30 cm long, raches terete, pubescent, leaflets are 10-25, ovate or lanceotale, 2.5 cm long, acuminate or retuse, usually irregularly crenate, unequal sided, pubescent beneath, petiole is very short. Flowers are white in color they are 1-2 cm long in terminal corymbose cymes. Fruits are berry type, globose to subglobose, pink when unripe, black when ripe. Each fruit is 2 seeded.

Distribution

This plant is distributed throughout India and other Asiatic countries like

Pakistan, Sri Lanka, Burma, Japan *etc.* In India this plant is found in Gujarat, Madya Pradesh, Rajasthan, West Bangal and all the other states.

Cultivation

Curry leaf grows well in tropical climate under all types of soils. Heavy water logging can not be tolerated. It is drought hardy and withstands fire to an extent. Numerous root suckers are produced. Curry leaf tree coppices well. Pits are prepared of 60-70 cm depth and diameter and it is filled with compost and firm yard manure. Propagation is done by two methods. (1). By root suckers and (2) By seeds

1. **By root suckers:** Root sucker method is the best method of propagation. In this method roots are broken off in the soils by a spade.
2. **By seeds:** The plant is propagated from seeds. Seeds are sown before monsoon season and they are retained on nursery bed for one month. Seeds are sown directly in the pit or seedlings transplanted.

Seedlings with one and half year old are used for planting. The spacing is kept of 3-4 meter. In raising a plantation, it is spaced 2 x 2 m and pits of 45 cm^3 are taken and half filled with topsoil and cow dung. It is planted singly in homesteads. Planting of uprooted seedlings is done in June-July.

Irrigation and Fertilization

During early growth stages, shade and daily irrigation is required. During summer months irrigation is to be given. Seeds yield of one tree is reported to be 20-25 kg. Pest and disease seldom affect the plant. Every year 5 kg of FYM and 1 kg castor cake may be applied to each plant. Leaves may be picked in small quantity after 5 to 6 months. Tree goes on sprouting even if leaves are picked up. The plant lives upto 30 to 35 years and continuously produces aromatic leaves throughout year.

Harvesting

During the summer season leaves are collected from one and half year onwards. Irrigation and manuring can improve the leaf yield. The height is always maintained to make collection easy. First harvest of leaves is done in the March to May after 8-9 months later of planting and second harvesting is carried out in the second and subsequent years. Plants are cut above ground level with a sickle, and the small branches with leaves, or the leaves separated from branches can be marketed.

Yield

Per year 100 kg of leaves are obtained from a tree so in each harvest 50,000 twings with leaves are obtained.

Medicinal and Economic Importance

- ☆ Curry leaf contains 63.8 per cent water, 6.1 per cent protein, 1 per cent fat, 16 per cent carbohydrate, and 6.4 per cent fibers, 4.2 per cent minerals, Vitamin A-12,600 (international unit/100 g), Nicolinic acid 2.3 mg/100g and Vitamin C-4 mg/100g.
- ☆ Leaves contain oil, glucoside and resins. From the leaves of this plant volatile oil and koenigin, a crystalline glucoside are extracted.
- ☆ The leaves and roots are used against Piles, skin diseases and bacterial infection. It increases digestion and and controls dysentery. It promots hair growth.
- ☆ Fresh leaves on stem distillation yield 2.5 per cent volatile oil. The oil is light yellow in colour, aromatic with spicy note.
- ☆ It is mainly used for seasoning curry and other food preparation. Tender leaves are consumed to cure dysentery. It increases digestive power and makes dishes tasty.

122

Nelumbo nucifera Gaertn. (Kamal)

Botanical Name: *Nelumbo nucifera* Gaertn.

Family: Nymphaeaceae

Local Names

- ☆ **Gujarati:** Kamal
- ☆ **Sanskrit:** Ambuja, Padma, Pankaja
- ☆ **Hindi:** Kamal
- ☆ **Telugu:** Kalung, erra-tamara
- ☆ **Tamil:** Ambal, Thamarai

Introduction

This plant is perennial aquatic herb. Root stock is stout, cylindrical, embedded in the mud. Leaves are peltate, radiately nerved. Margines are wavy. Petiole is long, inserted in the middle of the leaf. Flowers are large, solitary, handsome and fragrant, rosy or white, carpels numerous, ovoid, fleshy, sunk separately in cavities of receptacle, maturing into nut-like achenes.

Origin and Distribution

N. nucifera is a native of China, Japan and possibly India. This plant is very common throughout the world. Mostly cultivated in the ponds and swamps. Although mainly grown in tropical and subtropical regions, lotus root can withstand

a considerable degree of frost, and in India it is found from sea level up to 1,800 m. It is grows well in lakes, ponds and rivers.

Cultivation

This plant is propagated by rhizome. This plant requires large amount of moisture. From the water surface this plant can grow upto 2.5 metres deep. However, it is better to grow the plant in shallow water. The optimum depth is more than 30cm. It helps the plant in warming up quickly, results in better growth and flowering.

Planting Procedure

Lotus can be planted by using following methods.

1. Pieces of rhizome are planted with the eyes just above the soil surface and the water level is maintained at about one meter of water throughout its growing period.
2. Pieces of rhizome are placed horizontally about 15 cm below the soil surface, and water allowed to cover the soil, but with the crown of the developing plant just breaking the surface of the water. The water level is raised as the plant develops.
3. A method of planting in a filled pond is to put sprouting pieces in a basket, pot, tub or other suitable container filled with a mixture of soil and compost or FYM, and then place the container in the pond in such a way that the crown of the plant is just above the water surface. The container should be on bricks or stones, and as new growth appears the container

is lowered by removal of bricks to maintain the crowns just on or above the water surface.

4. When grown from seed, the seedlings are raised in nursery beds and planted out in the ponds after about two months in the manner indicated in (2).

Seed rate - approximately 45 kg of rhizome pieces are used to plant one hectare or 10-12 kg of seed. Planting density is low because of the very rapid growth of the rhizomes, reported as up to 15 m^2 per year.

Harvesting and Yield

The roots are normally dug by hand after the ponds are drained just before harvesting, but a mechanical harvesting system is being developed in Japan. In India the crop is reported to yield 3.5-4.5 t/ha.

Chemical Composition

The leaves contain alkaloids like nuciferinem roemerine, nor-nuciferine and the flabonoid-quercetin. The plumules yield proteins, sugars and vitamins. The receptacles contain quercetin.

Medicinal and Economic Importance

- ☆ The ripe seeds produce a wholesome effect in case of neurasthenia, spermatorrhea and metrorrhoea.
- ☆ The leaves and the seed in decoction are effective for insomnia, haemorrhage and haematemesis.
- ☆ The plumules, the filaments or the receptacles in the form of a decoction are used in treating bloody stools, heamaturia, uterine haemorrhage and haematemesis.

123

Nymphea stellata Willd. (Kanval)

Botanical Name: *Nymphea stellata* Willd.

Family: Nympheaceae

Local Names

- ☆ **Gujarati:** Kamal
- ☆ **Tamil:** Vellampal, Alittamarai
- ☆ **Hindi:** Kanval, Kokka
- ☆ **Telugu:** Allikada, Tellakaluva
- ☆ **English:** Indian water lily, East Indian lotus

Introduction

A large perennial aquatic herb with short, erect, roundish, tuberous rhizomes; Leaves are floating, peltate, sharply sinuate-toothed, flowers are large, floating, solitary, variable in colour from pure white to deep red. Fruits are spongy and many seeded. Fruits are berry type. Seeds are minute. They are grayish black when they are dried with longitudinal striations. It is an aquatic species with big white or pale blue flowers. The curry prepared out of seeds is a delicacy. The rhizomes are also picked. The flowers fetch good rate in the market. The leaves are used for packing purposes.

Origin and Distribution

The aquatic herb is native of Indian subcontinent. Plant is found throughout in

India in lake and ponds. This plant suits warm climate with humidity. It is found in ditches, tanks and ponds. Plant needs rich loam with an addition of plenty of well rotten cowdung manure. The plants are badly affected by frosts.

Cultivation

Indian water Lily is cultivated by seeds and tubers. The seeds are sown in shallow pans containing screened sand and kept in water at temperature of 22°-29°, the surface of sand being above 60 cm. deep. When the first floating leaf develops, the plants may be potted singly, allowed to grow and later lowered down in the pond in the required position.

The tubers are planted in boxes or half barrels, half filled with soil consisting fibrous loam of cow dung manure in such a way that their growing points remain upwards. More soil is put in. The upper 25 or 50 mm thick layer is of sand. Therse barrels are then lowered in the pond in required position.Tubers are put down into muck of the bottom. If they don't stay there and float on water surface, mask is wrapped around the tubers and put in the required places. As regards putting in plants, these are first placed in baskets in a soil consisting of fibrous loam and well rotten cow dung manure. These baskets are lowered down in the required position. The plants will rest properly on this soil and spread their roots at will. The best time for planting tubers is March/April.

Harvesting and Yield

The tubers start ripening after the flowering is over. There will be about 30,000 plants in one hactare. An annual production of 2,000 kg. of seeds per hectare is expected.

Chemical Constituents

This plant contains Nymphaeine, Nymphalin.

Medicinal and Economic Importance

- ☆ The rhizome is cooling, sweet, bitter and tonic and is useful in diarrhea, dysentery, dipsia and general debility.
- ☆ The flowers are astringent and cardio tonic.
- ☆ Seeds are sweet, cooling, constipating, aphrodisiac, stomachic and restorative.
- ☆ They are useful in vitiated conditions of Pitta, dipsia, diarrhea and dermatopathy.

124

Ocimum basilicum Linn. (Tulsi)

Botanical Name: *Ocimum basilicum* Linn.

Family: Lamiaceae

Local Names

- ☆ **Gujarati:** Tulsi, Maravo
- ☆ **Hindi:** Babui tulsi, Gulal tulsi, Marua
- ☆ **English:** Sweet basil, common basil
- ☆ **Sanskrit:** Munjariki, Surasa, Varavara
- ☆ **Tamil:** Tirnirupachal, karpura tulasi
- ☆ **Telugu:** Bhutulasi, rudrajada, Vepudupachha

Introduction

This is very common in India. It is erect, almost glabrous which gets height of 30-90 cm. Leaves are ovate to lanceolate, acuminate, toothed or entire, glabrous on both surfaces, glandular, flowers are white or pale purple, in simple or much branched arrangement in racemes. Flowers are 0.72-1.25 cm long,. Bracts are present they are small and staked, ovate, minute, caducuos, calyx are 5-toothed, upper tooth is rounded. Flowers are small and in purple colour. This plant is related with religious retules.

Origin and Distribution

Sweet basil is native of North West India. It is cultivated in France, Italy, Bulgaria, South Africa and USA. The genus Ocimum is an extremely versatile

group consisting of about 160 species with a geographical distribution spread over tropical, sub tropical and western temperate regions of the world ranging from sea level to 550 m in height. The species are mostly found in rain forest and tropical places receiving higher rainfall. It is found in wild form in hilly regions of India.

Cultivation

It grows on different kind of soils where pH ranges between 5 to 8.5. It grows well under varied climatic conditions. It also grows as a wild plant over hills. In the subtropical climate this plant grows in winter. In the plains of North India, it grows in summer. The areas where possibilities of falling snow or frost, the crop of *Ocimum basillicum* L. does not grow well. The most suitable temperature for its growth is near 30°C. This plant can be propagated by two methods (1) Seed propagation and (2) Vegetative propagation.

Seed Propagation

By broadcasting method seeds are sown directly in the soil. It is mixed with sand to ensure an even distribution. Before sowing, the field is marked into rows 50-60 cm apart. Seeds are sown at the depth of 2 cm in the soil. If seeds are sown deeper in the soil then they fail to germinate. The field is irrigated after 24 hours depending on soil moisture. Germination starts within 10 to 15 days. After 20-25 days when the seedlings are 10-15 cm tall, first weeding and thinning is done.

Transplanting

Seeds are sown in the nursery towards end of February to March in raised bed. Germination is profuse and is completed in about 10 days. Seedlings with

4-6 leaf stage become ready in about 6-7 weeks for transplanting. Seedlings may be transplanted at spacing of 60 x 60 cm. The fields are irrigated both prior to and after transplanting.

Vegetative Propagation

During October- December this plant can be propagated vegitatively. For this purpose cuttings with 8-10 nodes and 10-15 cm length are used. They are so prepared that except for the first 2-3 pairs of leaves the rest are trimmed off. Later, they are planted in the well prepared nursery beds or polythene bags. In about 4-6 weeks time the rooting is complete and they are ready for transplanting into the main field. Manure is useful for better growth of plant. FYM and nitrogen, entire of phosphorus and potassium are given as a basal dose, and remaining nitrogen is applied in two splits after first and second cuttings.

Irrigation

In the plains of India *Ocimum basilimum* L. sown mainly in the months of July-August. The crop sown at this time does not need irrigation, becomes its life cycle is nearly 65 to 70 days. As such it requires only 2-3 irrigation.

Harvesting

The crop needs 65 to 70 days to mature. When on nearly 80 percent of the plants, inflorescence appears and its colour changes from green to golden that is the correct time for harvesting. The crop is harvested at full bloom stage by cutting the plants at 15 cm from ground level to ensure good regeneration for further harvests. First harvesting is done at 90 days. Later on after per 75 days harvesting is carried out.

Distillation of Oil

Harvested plant is spread on the ground before distillation and allowed to dry for 6-8 hours. This helps in moisture evaporation. The oil is distilled by water vapour method. Sweet basil contains a clove like scent with an aromatic somewhat saline taste. It yields a volatile oil (oil of Basil) used as a flavouring agent and also as perfume. Characteristic and composition of oil from different regions vary. Four types of oils are recognized: (1). Methyl cinnamate type (2). European type (3). Reunion type and (4). Eugenol type.

Yield

Oil is extracted from the plant. From one acre of plant we can obtain 20 to 25 kg of oil.

Medicinal and Economic Importance

- ☆ The leaves are used mainly in medicinal uses. The oil of basil is used as flavoring agent in confectionery, baked goods, sauces, ketchups, tomato pastes, pickles, fancy vinegars, spiced meats, sausages, and beverages.
- ☆ A tincture of the herb has use in liqueurs, as flavour modifier.

- ☆ The major component of the essential oil is methyl cinnamate, linalool, or methyl chaviocol in different samples.
- ☆ Other major components include ocimene, linalyl acetate, eugenol and transanethole.

125

Ocimum gratissimum L. (Ram Tulsi)

Botanical Name: *Ocimum gratissium* L.

Family: Lamiaceae

Local Names

- ☆ **Gujarati:** Ram tulsi
- ☆ **Tamil:** Elucicham tulasi
- ☆ **Hindi:** Van tulsi, Ram tulsi
- ☆ **Telugu:** Nimma tulasi,Rama tulasi
- ☆ **English:** Lemon basil, large basil, shrubby basil

Introduction

This perennial herb is much branched, It gets height of 1-1.5 m. Stems are quadrangular, pubescent, woody at the base. Flowers are white greenish to pale yellow, small, simple, or branched racems, moderately close whorled. Inflorescence is axillary or terminal in simple or branched whorled raceme. Leaves are opposite, apiculate, pubescent on both surfaces.Margins are coarsely toothed.

Distribution

This plant is grown wild in tropic and sub tropical regions of the world. It is found throughout India. Plant grows well under warm and humid conditions.

Cultivation

This plant is cultivated by two methods (1) By seeds and (2). Vegetative method. Long days and high temperatures are beneficial for its growth. It comes up on varied types of soil, rainging from rich loam to poor laterite, saline/alkaline to moderate acidic. The soils must have good drainage. It can tolerate drought and frost, but not water-logging.

The crop can be raised either by direct sowing of seeds or by transplanting nursery raised seedlings in the field. However, the latter method is better and recommended. The raised seeds beds are prepared for raising seedlings. Sowing of seeds in the beds is carried out. For one hectare 300-400 grams of seeds are required. After sowing of seeds in the beds, a thin layer of soil is sprinkled over the seeds. The seeds take 8-12 days to germinate and the seedlings are ready for transplanting in about 6 weeks. A spray of 2 percent urea solution on the nursery raised plants, 15-20 days before transplanting, helps in getting very healthy plants for transplanting. The land is ploughed thoroughly and parallel continuous ridges and furrows are made at 60 cm apart. The transplanting of seedlings at 4-6 leaf stage is carried out on ridges at a spacing of 60 cm in third week of June The field is irrigated immediately after the transplanting. The seedlings establish well within 10 days time. Gaps are filled as early as possible to get uniform crop. In dry summer months, the field is irrigated once in 10 days and during other period; it may be irrigated as and when necessary. About 10-15 irrigations are enough during the year, and during the successive year, four hoeing,s one after each harvest are necessary. Fertilizers of 80 kg Nitrogen, 80 kg phosphorus and 30 kg of potash per ha should be applied in two three split doses. The crop is renewed after every 3 years.

Harvesting and Yield

The plantation becomes ready for first harvest after 90-95 days of transplanting. Then it may be harvested after every 65-75 days interval. In the first year, 2-3 and in the subsequent years 4-5 harvests are taken. Harvesting is usually done on bright sunny days for good oil yield and its quality. The crop should be cut at 15-20 cm above ground level in first year, 25 30 cm high in the second year, and 35-45cm high in third year. The oil is mainly available in leaves. It is therefore necessary to harvest the leaves only. The crop is harvested just before the flowering stage, when the eugenol content is maximum. On an average, 400 quintals of fresh herbage in the first year and 700 quintals in the subsequent years per ha can be obtained.

The herbage should be distilled in the fresh stage. However, oil quality and its yield do not diminish up to 6-8 hours after harvest, but further delay causes considerable loss in the yield and quality of oil. The yield of oil is 200 kg, 350 kg, 350kg per ha in the 1st, 2nd and 3rd years respectively. The present market rate of oil is around Rs.100 per kg. In the 1st year expenditure per ha. goes upto Rs.10,000 and upto 3rd year total expenditure goes upto Rs.40,000. While gross returns per ha. reaches upto Rs.20,000 in first year and Rs.90, 000 upto 3rd year and onwards. This means net returns reaches upto Rs.16, 600 per ha/year. Net returns goes upto Rs. 50,000 in 3rd year and onwards.

Chemical Composition

This plant contains an essential oil which is made up of eugenol (45-70 per cent), methyl eugenol (20 per cent), carbacol, ocimene, p-cymene, camphene, limonene, α-pinene and β-pinene.

Medicinal and Economic Importance

- ☆ Whole plant is medicinally useful.
- ☆ Plant is used in treating sunstroke, headache and influenza.
- ☆ It is considered to be diaphoretic.
- ☆ It serves also as material for the extraction of essential oil and qugenol.
- ☆ Eugenol is used widely in odontoloty and for the synthesis of vanillin.

126

Opuntia ficus-indica L. (Thor)

Botanical Name: *Opuntia ficus indica*

Family: Cactaceae

Local Names

- ☆ **Gujarati:** Thor
- ☆ **English:** Prickly pear, Bele
- ☆ **Hindi:** Thor

Introduction

This plant is perennial which grows upto three meter. Large spatula shaped stem is covered in clusters of spines, brilliant yellow flowers and roundish purple fruit.

Distribution

This plant is wildly found anywhere in India. It is native of Mexico and naturalized in sub tropical regions around the world.

Cultivation

Opuntia ficus-indica is a plant easy to propagate. However, for commercial plantations following a few rules ensures high quality and vigorous plants. Suitable cuttings also reduce the vegetative phase of the plant and extend its productive life. Prickly pear can be propagated from seeds or from cuttings. Seed propagation is not used commercially, but is used in the breeding of new varieties. When using

individual cladodes it is preferable to use complete pads. However, if the planting material is scarce, it is possible to use cladode fractions. Dividing the cladode into pieces smaller than half will reduce the initial size and number of cladodes of the new plant and it will take longer to reach full size, and also take longer to reach the reproductive stage. The best season for collecting plant is December to February. Cladodes should be allowed to heal and dehydrate for better conservation and establishment. To protect them from rotting they can be treated with Bordeaux mixture. Following are some planting methods for this plant.

Planting Methods

Upright Position

Used for early planting, during the dry season. The cladode is buried in the ground down to half of the cladode.

Flat Position

Suitable for planting during the rainy season. The cladode is placed on the ground and maintained in place with a small stone or a handful of soil. This planting method reduces the risk of rotting associated with high moisture content of the soil.

Planting Date

The shoots are highly susceptible to frost damage, and they start emerging two to three weeks after planting. Planting should therefore be after the risk of frost is over. In the highlands, prickly pear is planted during February to April.

The planting distances for fruit production depend on the possibilities for crop management, especially weed control. Orchards intended for manual weeding can be planted using 4 to -5- m between rows. In those orchards intended for mechanized weed control the rows should be at least 5 m apart. The distance between plants for both systems is 2.5 to 3 -m. Using a 4 x 2.5 meter planting layout results in a planting density of 1000 plants per hectare. In the case of 5 x 2.5 the number of cuttings needed to plant a hectare is 800. The cladodes are separated from the plant by means of a sharp knife a machete or *derho*. They should be cut at the joints to reduce the risk of infection and to accelerate healing.

Harvesting

Harvesting is carried out when fruits are metured. The fruits contain the lowest water content and, therefore have the highest nutritional value. Under rainfed conditions, a bele plant can produce at least one layer of pads. They grow and reach the maximum size at the end of the rainy season. It is advisable to collect pads not older than two years, that means collecting no more than two or three layers of cladodes.

Storage

Fresh pads should be stored in a dry shaded location. They can be either stacked or arranged in rows lying on their sides. Avoid spots that collect runoff in order to minimize rotting and sprouting. The pads in close contact with the ground should be flipped over every 4 to 6 weeks to avoid rooting. Some relief from direct sunrays can be obtained with a thin layer of straw or grass spread on top of the stacked pads. Direct sunrays induce deformation of the pads and chlorophyll degradation on the exposed area, thus reducing nutritional value.

Chemical Composition

The fruit of prickly pear contains mucilage, sugar, vitamin C and other fruit acids. The flowers contain a flavonoid.

Medicinal and Economic Importance

- ☆ Flowers are astringent and reduce bleeding, and are used for problems of the gastro intestinal tract particularly diarrhea, colitis and irritable bowel syndrome.
- ☆ The flowers are also taken to treat an enlarged prostate gland.
- ☆ The fruits are nutritious.

127

Origanum marjorana L. (Maruwo)

Botanical Name: *Origanum marjorana*

Family: Labiateae

Local Names

- ☆ **Gujarati:** Maruwo
- ☆ **Tamil:** Marruva
- ☆ **Hindi:** Maruwa
- ☆ **Telugu:** Maruvam
- ☆ **English:** Sweet marjorum

Introduction

This plant is aromatic, branched perennial; with heights of 30-60 cm. Leaves are oblong and ovate. Flowers are small and whitish or purplish in colour. They are arranged in terminal clusters. Fruit contain minute seeds, they are dark brown colour.

Origin and Distribution

This aromatic plant is much branched and perennial, native of Southern Europe, North Africa, Asia Minor and is commonly grown in gardens of India as savory plant.

Cultivation

Even though it is a perennial in nature, it is often considered as an annual under cultivation. It is propagated by seeds as well as cuttings. When seeds are utilized, the seeds are sown in nursery beds. When the seedlings attain sufficient height for safe handling of plants, the seedlings are transplanted in the field at 20-25 cm. apart in lines and the lines are at 30cm apart. The cuttings are also spaced at the same spacing. The crop is much benefited, when well rotten farm yard manure is added to the soil at the time of soil for preparation, and also when watered by sprinkler system. The crop is ready for harvesting within 3-4 months.

Harvesting

Harvesting is done when plants enter into bloom; the tops are cut and dried under shade. Since the oil content in the plant is highest before seed formation, it is advisable to harvest at that stage. After the first harvest, the plants are watered if possible. The plants sprout again and are ready for the second harvest in two months period. A production of 25,000 kg of fresh material of the plant can be expected from one hectare plantation.

Yield

The herbage yield about 75 kg of oil on steam distillation.

Medicinal and Economic Importance

- Fresh leaves are used as garnish and included in salads.
- It is also used for flavouring vinegar.
- The seeds which are aromatic are used in confectionery.
- The leaves and flowering tops, on steam distillation, yield a volatile oil.
- The yield is 0.3-0.4 per cent in case of fresh leaves and 0.7-3.5 per cent in case of dry herb. The oil is colourless or pale yellow to yellow green, with a tenacious odour like nutmeg and mint.
- Sweet marjoram oil is often confused in commerce with thyme oil and origanum oil. Oil is used to a small extent in high grade flavour preparation and perfumes, and in soap and liquor industries.
- Sweet marjoram is considered as carminative, expectorant and tonic.
- The leaves and seeds are astringent.
- An infusion of the plant is used as stimulant, sudorific, emmenogogue and galactagogue.
- It is reported to be useful in asthma, hysteria and paralysis.
- The leaves and flower tops are used in garlands and bouquets.

128

Oroxylum indicium L.Ven (Tetu)

Botanical Name: *Oroxylum indiucm* L.Ven

Family: Bignoniaceae

Local Names

- ✰ **Gujarati:** Tetu
- ✰ **Hindi:** Arlu, Aralu, shyonaka
- ✰ **English:** Tetu, Indian trumplet tree
- ✰ **Sanskrit:** Shyonaka
- ✰ **Tamil:** Achi, Pelarianthei
- ✰ **Telugu:** Pampini

Introduction

This is small to medium sized desiduous tree with grey-brown, lenticellular, soft spongy bark. It attains heights of 12 m. Leaves are large, upto 1.5 m long, pinnate, bipinnate or tripinnate, leaflets ovate or elliptic. Flowers are arranged in large, erect raceme inflorescence. They are purple flesy and foetid. Fruits are capsules, large; flat, sword-shaped, upto 90 x 9 cm valves, woody, seeds are many and flat, thin with broad silvery wing.

Origin and Distribution

This plant is distributed throughout the country. It grows well the elevation of 1200m. It is not selective of a particular soil. It produces root suckers in plenty. It is found in tropical and sub tropical regions of the world like India, China, Sri Lanka etc.

Cultivation

It requires moderate shade in the early stages. It does not need very special soil for its growth. It produces plenty of root suckers. The tree reproduces through seeds. It also can be propagated by artificially sowing the seeds in the nursery during March-April and transplanting the seedlings in the first of second rainy season. It can also be propagated by root suckers. This species is a fast growing one. It is planted in the field at 5m x 5m spacing. Workings are done by time to time. The root bark can be extracted after 10-11 years of age. The root bark becomes ready for use as extraction after ten-eleven years.

Harvesting

The stem bark can be extracted from the tree after 10-11 years when tree becomes matured. During the time tree must have produced some root suckers, which have become independent plants. The stem bark must be harvested in serpentine fashion. In this method of hgarvesting tree doesn't die.

Yield

Each tree can produce about 4 -5 kg of bark after 10 years of plantation.

Medicinal and Economic Importance

- ☆ Root, bark, leaves and fruits are used in medicinal purpose.
- ☆ Root bark is one of the well known drug Dashmoola in a Ayurvedic system and it is prescribed fresh. It is cream-yellow to grey in colour, soft and juicy without any characteristic odour.
- ☆ They are useful in vitiated conditions of Vata and Kapha, inflammations, dropsy, sprains, neuralgia, hiccough, cough, asthma, bronchitis, anorexia, dyspepsia, flatulence, colic, helminthiasis, diarrhea, dysentery, strangury, hout, vomiting, leucoderma, wounds rheumatoid arthritis and fever.

129

Pandanus fascicularis Lam. (Kewdo)

Botanical Name: *Pandnus fascicularis* Lam.

Family: Pandanaceae

Local Names

- ✰ **Gujarati:** Kewoda
- ✰ **Hindi:** Kewda, Ketki, Keura, Gagandhul
- ✰ **Sanskrit:** Ketaki
- ✰ **Tamil:** Talai
- ✰ **Telugu:** Mugali, Gajangi

Introduction

Kewda is a perennial tall growing shrub, which flourishes under hot and humid conditions. It is a shrub with fragrant flowers. The average height of the plant is 4-6 m and is exceptional cases goes up to 10 m. The flowers are 25-50 cm long with an average weight of 100-200 grams without leaves. Flowering begins in May-June and continues till September to October. The spadix takes a fortnight to develop to its full size. Older leaves drop whereas younger once are attached straight.

Origin and Distribution

Kewda plant is ancient and important plant of India. Ganjam district of Orissa is believed to be origin of Kewda. It is said that the Muslims of West Punjab were the first to start kewda distillation in the district followed by several others, especially from the north. It occurs in the coastal regions of India. Iran and Burma. In India

it is distributed over costal Andhra Pradesh, Tamil Nadu, Orissa and Gujarat. The Ganjam district of South Orissa comprising chatarpur, Gopalpur and Berhampur district are the major production center of different kinds of kewda product.

Cultivation

Kewda is capable of growing in wide variety of soils. However the soil having the high moisture holding capacity and rich in nutrients content are most suited for its growth. The plant prefers to grow near the source of water and tolerates water logging for a long period. It coppice well if main stem is removed. Under favourable weather conditions rooting is quick from branch cuttings which are normally used for propagation. Propagation is done by stem cuttings or branches which are 3-4 years old and have thickness of 10-15 cm. Sometimes new shoots which emerged from lower portions of main stem of old branches can also be removed and used for propagation. The best time for propagation is July to August. It grows well in high humidity, soil moisture and temperature. Total life span of kewda is 40-50 years. Flowering starts after 5 years of planting and continues until 40-50 years.

Harvesting

When kewda tree becomes fully grown it bears an average 15 to 20 flowers every season. The main season for collection of flowers starts from July to August and lasts for 2 to 3 months but some collection is also made in December to January and April to June. The flowers are plucked simply by breaking with the help of a hook attached at the top of a stick. A cutting of the flowering stick is not recommended as this may damage the floral prim or dia and arrest further flowering. Collection of flowers is completed by early morning so that these can be supplied to the distillery by 9 am. It is better to go for distillation o9n same day. The rate of flowers is fixed by the local committee every year.

Extraction of Oil

Oil extraction is carried out using hydro distillation unit.

Storage

Kewda oil is precious hence care should be taken for its proper storage. Normally the oil is stored in ½ to 1 lit. Aluminium bottles filled to its capacity and properly sealed.

Medicinal and Economic Importance

- ☆ Kewda oil is used in the preparation of high grade tobacco. It is also used in cosmetics, pan masala and in snuff. The kewda water is used for flavouring syrups, soft drinks and other food preparations.
- ☆ Kewda attar is one of the most popular perfumes, extracted and used in India since ancient times. It blends well with almost all types of fancy perfumes, and is used for scenting clothes, bouquets, lotion, cosmetics, soaps, hair oils, tobacco and agarbatti.
- ☆ The leaves are employed for covering huts, for making matts, cordage, hats, baskets and other fancy articles, they are also used for making umbrellas. They are good paper making material.

130

Passiflora edulis Sims. (Kaurav Pandav Nu Phul)

Botanical Name: *Passiflora edulis* Sins.

Family: Passifloraceae

Local Names

- ☆ **Gujarati:** Kaurav pandav nu phul
- ☆ **English:** Passion fruit, Purple granadilla.
- ☆ **Hindi:** Kaurav pandav ka Phul

Introduction

This plant is perennial woody climber which has simple, axillary, tendrils, often grown on trelies. Leaves are 5-12 x 4-10 cm broadly ovate, lamina is glabrous. Shallowly to deeply three lobed, lobes of laminate closely serrate, stipules are linear, small, not gland tipped. Flowers are axillary, solitary, they are 5.5-7 cm across, petals are light bright blue, corolla are purple, bluish-purple or lilac. Fruits are beriies type, globose, glabrous, orange to red brown in colour when they ripe. They are 3.5-9 cm across.

Origin and Distribution

The passionfruit is native from southern Brazil through Paraguay to Northern Argentina. It has been stated that the yellow form is of unknown origin, or perhaps native to the Amazon region of Brazil, or is a hybrid between P. edulis and P. ligularis. The plant is distributed in Australia, Kenya, Newzeland, USA like countries. In Asia, the plant is distributed in India, Srilanka, Myanmar like countries.

Cultivation

This plant is cultivated by seeds and vegetative method. It thrives best in warm, humid, sub tropical atmosphere. It is not specific to any soil. It comes up on all types of soil with moisture retentive capacity. It does not tolerate sunny slopes. It grows well on all types of soils. It responds well to irrigation. The heavy soil and soils of poor drainage and low fertility are not suitable. It can withstand light frosts only.

It is propagated by seed or by branch cutting. When the vegetative method is employed, semi hard cuttings of 3-4 m in length are used. Cuttings are taken from well matured wood strike root readily and can be transplanted in the field in about three months period. The seeds have poor viability. Therefore, when seeds are used for raising seedlings, the seeds are sown, soon after collection. Within 2-3 weeks seeds are germinated and become ready for transplating after 3 months. Seedlings should be 20-25 cm in height at the time of transplanting. The vines can be grown on trelies, pergolas, fence, bowers etc. They are grown at a spacing of 3-5 m along low wire trellies. The seedlings are manured once in a year with well rotten farm yard manure and little of chemical fertilizers. The chemical fertilizers recommended are ammonium sulphate, super phosphate and potassium sulphate in the ratio of 10:6:10. In the months of February/March, the dried and diseased parts of the branches are pruned out.

Harvesting

The plants start yielding fruits from the second year, but reach maximum level from 6th year. Maximum yield could be nine kg per plant in a year. Even though the fruiting is continued through out the year, the main harvesting period is May-June and September-October. Well coloured fruits are harvested before they are dead ripe. The fruits stand transport well, due to the hard rind or shell, but they have short storage life.

Yield

Total expenditure per hectare reaches upto Rs.25,000 in 1st 5 years and gross return reaches upto Rs. 30,000 while net returns per ha goes upto Rs.5,000 in 1st 5 years and Rs.20,000 in 6th year and onwards.

Chemical Constituents

Passion fruit contain specific component known as edulin I and edulin II.

Medicinal and Economic Importance

- ☆ Passion fruit juice is used for flavouring candy, ice-cream, cake-fillings, frostings, carbonated beverages and cordials.
- ☆ The pulp is highly acidic with a pronounced pleasant flavour.
- ☆ The juice is preserved by canning or freezing and used with other less acidic fruit juiced and in squashes, syrups, cordials, beverages etc.
- ☆ Juice from the purple variety is considered superior to that of yellow variety due to higher percentage of sugars, ascorbic acid and carotene.
- ☆ The peel of the fruit can be used for recovery of pectin, for feeding live stock and as a manure. The seeds yield on cold pressing, an oil which is used in paint and varnish industries.
- ☆ The leaves along with leaves of *P.incarnata* are used in medicines.

131

Phoenix sylvestris Roxb. (Khajuri)

Botanical Name: *Phoenix sylvestris* Roxb.

Family: Arecaceae

Local Names

- **Gujarati:** Khajuri, Kharak, Khakri
- **Hindi:** Khajoor, Sendhi, Shindi
- **English:** Khajoor, Wild date palm
- **Tamil:** Icham
- **Telugu:** Pedd lta

Introduction

This is very tall palm which reach the height of 10-16 meter. There is large crown seen and it is found up to 1500 m elevation. Trunk covered with persistent bases of petioles. Leaves are grayish green in colour and they are 3 to 4.5m long with few short spines at the base, pinnate, linear and numerous. Spines are 15-45 cm long which ends in short points. Flowers are arranged in spadices type of inflorescence. Flowers are small fragrant. Male flowers are white while female flowers are greenish in colour. Fruit is of 90 cm long. Fruit is oblong ellipsoid berry type which generally 2.5 x 3.2 cm long. Seeds are orange yellow in colour. They are 1.7 cm long, deeply grooved, rounded at the ends.

Origin and Distribution

This plant is endogenous to India. But this plant is widely seen in the tropical

and sub tropical regions of the world. This plant is seen in warmer parts of the world like Pakistan, Sri Lanka, Myanmar and some other countries.

Cultivation

Generally it prefers moist alluvial soils which are not too heavy and clayey. Natural regeneration takes place freely by seed and the young plants are little subjected to browsing. Artificial propagation may be done by transplanting nursery raised seedling. The seeds are sown in May and the seedlings are transplanted in the third year, during the rainy season in the field. The field is properly ploughed and hoed before planting is done. The plants are planted at 3 m x 3 m espacement. No elaborate after care is required to be taken. When plants are 5 years old the upper green leaves are tied up and lower yellow ones are cut away, this is repeated annually. Meanwhile the soil in the field is loosened every now and then by ploughing.

Harvesting

Flowering in the palm is occurred in the beginning of hot season and the fruits are ripen in May and June. After 8-9 years, plants which have grown artificially becomes ready for tapping. Tapping is done when stem gets height of 1m. Tapping is done for 25-30 or many more years. Tapping season lasts for 4-6 months, commencing from October-November. For tapping lower leaves along one side of the trunk are cut off and the bark removed to expose the inner soft layer to the extent of 45 cm vertically and half of the circumference of the tree horizontally. The exposed surface is brilliant white in the starting but soon it turns into brown. After few days, tapping is started in the evening by making V-shaped cut in the exposed surface and scooping out a triangular patch inside the V. The sap exuding from the scooped surface is run through a bamboo spout into an earthen vessel, internally coated with lime and is collected early next morning. This tapping schedule is continued by making fresh cuts to ensure optimum flow of the sap, fruit branches are usually cut off. By repeated tapping, the palm gets more and more hewed into, so that at the end of the season exuding surface is 7.5-10 cm deep. Cuts are made on alternate sides of the trunk in successive season giving it a zig-zag appearance. As per the new technique developed, the exposed surface is divided vertically into three equal zones and tapping is done by slicing only one zone daily, thus affording two days rest to each zone. The yield of the juice is said to be double in this method.

Yield

Nira and *tody* are the important juices which are obtained from the plant. Per tree yield of the plant is reported to vary considerably from 100 to 300 litres in a season, giving 10-13 per cent of its weight as jaggery. Nira is converted into jaggery and sugar. The product nearly as white as cane sugar also is obtained after two or three boilings. The palm molasses can be used for edible purposes and as a source of alcohol. Upto 7th to 8 years total cost of Rs. 16,000 is needed for the plantation work for one hectare. Tapping is done in 8th year but every year 500 trees per hectare can be tapped. If the net profit per kg of jaggery is taken as Rs. 2/- then the net revenue per ha. comes to Rs. 16,000 so in the 8th or 9 year cost of plantation will be recovered and from 9th year onwards it will give good return. Profit can be increased by selling sap as Nira and by increasing the output of jaggery from sap.

Medicinal and Economic Importance

- Nira and Toddy are the well known drinks that are available from the tree. Palm jaggery is prepared from the sap which is more nutritious than cane jaggery.
- Fruits are edible and also can be made into jellies and jams.
- Leaves of the palm are widely used for thatching and for making mats, fans, baskets, bags, brooms, etc.
- They also yield a fibre which beaches well.
- The petioles are beaten and made into ropes. The female spadix forms a good brush for white washing.
- Leaf bud and inner portion of the stem are eaten in times of scarcity.
- Roots are used in case of toothache.

132

Phyllanthus niruri L. (Bhoi Amali)

Botanical Name: *Phyllanthus niruri* L.

Synonym: *Phyllanthus amaras, Phyllanthus fraternus*

Family: Euphorbiaceae

Local Names

- **Gujarati:** Bhoy amali
- **Sanskrit:** Bahupatra
- **Hindi:** Bhoy amali, Jaremala
- **Tamil:** Keela nelli
- **Telugu:** Nela usirika

Introduction

This plant is annual herb which gets height of 30-60 cm. This is quite glabrous. Stem is often branched at the base, angular. Leaves are numerous, subsessile distichous often imbricating, elliptic oblong obtuse. Stipules present, very acute. Flowers are yellowish, very numerous, axillary, the male flowers are one to three and female flowers are solitary. Capsules are 2.5 mm in diameter which are depressed globose, smooth scarcely lobed.

Origin and Distribution

The plant is native of India and distributed throughout tropical and subtropical regions of the world. This plant is distributed in Central and Southern India extending to Sri Lanka.

Cultivation

This plant is cultivated by two methods (1). By seed propagation and (2).By vegetative propagation. It is propagated by seeds. The seeds can be directly sown in the field. A quantity of 5-6 kg. of seed is sufficient to cover one hectare area. The area is ploughed thoroughly well and fine tilth of soil is obtained. Afterwards, furrows are made at 25 cm. apart. The beginning of monsoon, the seeds are mixed with sand and sown in the furrows continuously. The seeds germinate soon. After the germination is complete, weeding is carried and gaps in the planting lines occurred due to failure of germination, are filled again. After two months of sowing, second weeding along with hand hoeing is carried out. Then, 100 kg of urea per hectare is applied by broad Cast method. At the end of January, the plants are ready for harvesting. Fruits are collected for the next years crop, while harvesting.

Harvesting

The harvesting is carried out in January/February. The plants are uprooted and dried in shade for 2-3 days before packing and dispatching for marketing.

Yield

It is estimated that a production of 1500 kg of the drug is available from one hectare.

Chemical Composition

In the aerial part, three crystalline lignans including phyllanthine and hypophyllanthine have been found. Five flavonoids have been identified, quercetin, astralgin, quercitrin, isoquercitrin and rutin. Four leucodelphinidine alkaloids were separated from the leaves and stems, one of them being an enantiomorph of securinine.

Medicinal and Economic Importance

- ☆ It is used in viral hepatitis, oedema, anorexia like diesese.
- ☆ The fresh root is used for the treatment of viral hepatitis.
- ☆ The plant is also used as a diuretic in oedema.
- ☆ It is also used to increase appetite and locally to relieve inflammations.
- ☆ Fresh roots are used to relieve digestive troubles of camels.
- ☆ A decoction of the leaves is used as a refrigerant for the scalp.
- ☆ Roots and leaves are used as poultice for swellings and ulcers. Latex is also applied to sores and ulcers. Latex when mixed with oil, is used in ophthalmia.
- ☆ Dried leaves are toxic, bitter principle phyllanthin, which is toxic to fish and frogs.

133

Piper longum Linn. (Lindi Pipar)

Botanical Name: *Piper longum* Linn.

Family: Piperaceae

Synonym: *Piper methysticum* Forster f.

Local Names

- ☆ **Gujarati:** Lindi pipar
- ☆ **Hindi:** Pipli
- ☆ **English:** Indian Long Papper
- ☆ **Sanskrit:** Pippali, Kana
- ☆ **Tamil:** Thippili
- ☆ **Telugu:** Pippallu

Introduction

It is slender sub-scandent herb, branchlets erect, straggling or sometimes climbing, hairless, with swollen nodes and those of creeping branching with roots at lower nodes. Leaves are alternate, variable in shape, usually egg-shaped heart shaped, 7-15 X 4-6 cm, base heart shaped and unequal, apex is acute to acuminate, margin of leaf is entire, hairless, lower leaves are with long stalks and upper ones without stalk. Lateral nerves are 5-7 and they are arising from the base. Male spikes are erect and 2-7 cm long, it is greenish to yellow in colour. Fleshy, cylindrical, with minute male flowers. Female spikes are erect, and 1-3 cm long, it is yellow in colour. Fruiting spikes cylindrically oblong, about 4 X 1 cm. Fruits are berries

globose type about 2 mm across, partly sunken in the rachis, compactly arranged, red are converted in black when they ripe.

Origin and Distribution

Piper is very large genus of shrubs, rarely herbs and trees, found throughout tropical and sub tropical regions of the world. About 30 species have been recorded from India of which P.nigrum, the black pepper and P.betle are widely distributed. Piper longum is slender aromatic climber with perennial woody roots occurring in the hotter parts of India, from central Himalayas to Assam, Khasi and Mikir hills, lowes hills of Bengal and evergreen forest of western ghats from konkan to travankore. It has been recorded also from car Nicobar Islands West Bengal, Uttar Pradesh, Karnataka, Kerala and in the most of the states long piper is common.

Cultivation

The plant needs a hot moist climate and elevation between 100 and 1000 m for its cultivation. It can be grown successfully even in areas which receive heavy rainfall with high relative humidity. In its natural habitat, the plant is found growing as an under shrub. It can grow well in rich, well drained loamy soil.

It is also cultivated on a large scale in limestone. Laterite soil with high organic matter content with good moisture holding capacity is also suitable. Propagation is done by seeds, suckers or cuttings or by layering of mature branches at the beginning of rainy season. Stem cutting is good method which is obtained from one year old growth and 3-5 internodes. Vine cuttings can be rooted in polythene bags filled with the common pot mixture. The nursery can be raised during March and April. The cuttings which have been planted in March-April becomes ready for planting in the main field by the end of May. Before planting land should be ploughed 2 to 3 times and leveled properly. Then the field is divided into convenient size of plots in which the pits are dug at a spacing of 60 cm X 60cm. These pits are filled with soil mixed with well decomposed FYM or compost. In heavy rainfall areas, channels are made to drain excess water. Rooted cuttings are planted in pits at the rate of two per pit. The pits are gap filled one month after planting. This plant is planted as an inter crop with Subabul, Eucalyptus and coconut in different parts of the country.

Irrigation-Weeding and Mulching

The crop should be irrigated once in a week, if it grown as a pure crop. In case when crop is grown as an intercrop with other crops, the irrigation provided to the main crop is sufficient. Generally it is done after the application of FYM to the beds, earthing up is done from the channels. During summer to prevent the moisture loss or losses from the soil surface, the beds should be mulched by dry leaves. Straws are also used for this purpose.

Harvesting

After six months of planting first harvest is available from vines. The spikes are ready for harvest two months after of formation on the plants. Spikes are picked when they are blackish green and most pungent. Harvested spikes are dried in the sun for 4 to 5 days.

Yield

During the first year, the dry spikes are stored in the moisture proof containers. During the first year, the dry spike yield is around 400 kg per hectare. It reaches upto 1000 kg per hectare in the 3 rd year. After three years the productivity of the vines decreases and should be replanted. Root and stem are also harvested from 18 months after sowing. While harvesting, the stems are cut close to ground, the roots are dug up, cleaned and heaped in shade for a day, after which they are cut in to 2.5 to 5 cm long pieces. The average yield of dried roots is 500 kg per hectare.

Medicinal and Economic Importance

- ☆ Fruits and roots are used for medicinal uses.
- ☆ Roots and fruiting spikes are used in treating diarrhea, indigestion, jaundice, urticaria, abdominal disorders, hoarseness of voice, asthma, hiccough, cough, piles, malarial fever, flatulence, vomiting, thirst, oedema, earache, wheezing,
- ☆ This is one of the ingredients in the Sidha medicine *trikadugu.* It is also considered a rejuvenating plant.

134
Piper nigrum Linn. (Mari)

Botanical Name: *Piper nigrum* Linn.

Family: Piperaceae

Local Names

- **Gujarati:** Kalomirich, Mari
- **Hindi:** Kalimirch, Kalamorich
- **Sanskrit:** Marich
- **Tamil:** Milagu
- **Telugu:** Kiryalatige

Introduction

It is large genus of shrubs, rarely herbs and trees throughout the tropical and sub tropical regions of the world. It is branching, climbing and perennial shrub. Branches are stout, trailing and rooting glaucous beneath, base is acute rounded or cordate equal or unequal. Flowers are minute in spikes usually deciduous. Often female bears two male stamens and the male a pistillode. Fruiting spikes variable in length and robustness, rachies glabrous, fruits are drupe type, ovoid to globose, testa is thin, albumin is hard. In the wild state the plant is generally deciduous and rarely bears fruits. Flowering time for this plant is July-August.

Origin and Distribution

This plant is found the throughout tropical and subtropical regions of the world. A numerous wild relatives of *P.nigrum* occurring in tropical/sub tropical

part of India. Black pepper is one of the most ancient crops cultivated in India. The species has probably originated in the hills of South Western India, of North Kanara to Kanyakumari. *Piper nigrum* is cultivated in the hot and moist parts of India, Ceylon and other tropical countries.

Cultivation

This plant grows well in rich soil or laterite soil or alluvial soils. Although it can also be cultivated in sandy loam soil along with alluvial river. This plant requires warm and humid climate. It grows in places where rainfall is well over 200 cm and not less that 125 cm. The maximum temperature seldom exceeds 40°C and the lower temperature 10°C. In India black pepper is grown with other crops as mixed crops like jack, mango, coconut bamboo or arecanut. The vines can be grown as an inter or subsidiary crop along with other plantations like coffee, cardamom, orange, arecanut, coconut *etc.*

Plant can be propagated by two methods (1) Vegetative method and (2) Seed method.

1. Vegetative Propagation

This method is widely selected because of slow regeneration from seeds. Although it varies from place to place. There are two categories of shoots preferred for planting, (a). The runners which are usually confined to bottom portion of the vines and also terminal shoot toward the top region. The runners form the most common propagatory material. Healthy shoots are selected and are coiled up on forks as they may not come in contact with soil and rooting prematurely. The shoots are then cut into pieces of four to five nodes and then they are planted in the field or are rooted in baskets. (b). In the another method terminal shoots are preferred. These cuttings can yield a satisfactory crop from the fourth year onwards. Propagation through layering is adopted on a small scale. Grafting propagation has also become useful. Planting is done in July or August.

2. Seed Propagation

Dried seeds are selected for raising plants. These plants remain productive for longer period and yield more in the later years. Seeds are taken from the well ripened selected fruits Seeds are sown in nursery first and then transplanted into the field at 4 -5 leaves at stage.

Harvesting

Harvesting varies from region to region according to climate. In coastal areas harvesting extends from the middle of December to middle of March, while in cooler regions it is done in the April. In the area of higher elevation cropping period is delayed.

The pepper vine starts to fruit in the 2nd or 3rd year of plantation. The yield increases year by year. In well cared gardens and 10th -25th years is the best period to get good yield. The yield decreases after 13th year. Generally there are two crops in a year; one is in the August-September and another in March-April.

Yield

Yield differs in wide range from place to place; the vine under intensive care remains productive upto 60-100 years. Different varieties give different kind of results. Like the type *Balameotta* yield 1-1.5 kg dried pepper per vine while *karimcotta* gives 1.0-1.25 kg dried pepper. Balameotta's maximum yield has been recorded in Mysore is 3.0-3.5 kg per vine.

Medicinal and Economic Importance

- ☆ Oil of pepper is a valuable adjunct in the flavouring of sansage, canned meats, soups, table sauces and certain beverages and liquors.
- ☆ It is used in perfumery, particularly in bouguets of the oriental type. The oil is used in making soaps.
- ☆ In Indian medicine it is employed as an aromatic stimulant in cholera, weakness following fever, vertigo, coma *etc.* as a stomachic in dyspepsia and flatulence as an antiperiodic in malarian fever and as an alternative in paraplegia and arthritic diseases.
- ☆ Oleoresin is obtained from pepper is used as antibacterial and fungistatic property. Pepper lowers the phagocylic activity of leucocytes. Extracts having some hypercoagulative effects.
- ☆ It retards the development of rancidity in fats and meat.
- ☆ Fruits of pepper can be used after drying as black pepper of after processing into white pepper. Green fresh peppers are used in preparing pickles. Peppers are used among the major condiments employed for seasoning freshly prepared foods.

135

Plantago ovata Forsk (Ishabgul)

Botanical Name: *Plantago ovata* Forsk

Family: Plantaginaceae

Local Names

- ☆ **Gujarati:** Ishabgol
- ☆ **Hindi:** Ishabgol
- ☆ **English:** Blonde psyllium, Psyillum, Ribwort
- ☆ **Sanskrit:** Isharbol
- ☆ **Telugu:** Ishabgaluvittulu

Introduction

Psyllum is an annual herb. This plant is stemless soft with hairy and annual leaves. The leaves are 7.5 to 23 cm long, 6 mm broad, narrowly linear or filiform, finely acuminate entire or distantly toothed, attenuated at the base and usually 3-nurved. The flowers are arranged in ovoid or cylindrical spikes. They are 1.3 to 3.8 cm long. Bracts are 4 mm long and broad. The seeds are 3 to 4 mm long, they are boat shaped or ovoid-oblong, smooth and yellowish brown.

Origin and Distribution

This plant is believed to be originated from Peris. *Plantago ovata* is indigenous to the Mediterranean region and Western Asia extending up to Sutlej and Sind in Western Pakistan. In India this plant is cultivated in parts of Rajasthan and Maharastra. It is mainly cultivated in Mehsana and Banaskantha districts of North

Gujarat, but grows elsewhere in India too. Indian plantago seeds are superior to that of French or Spanish in quality.

Cultivation

This plant is grown in marginal land where there is good drainage. (PH should be 7 to 8). This kind of soil is good for cultivation for *Plantago ovata* propagation is generally done by seeds. The seeds are sown in the second half of the October to first half of November. Sowing is carried out in rows or through the method of sprinkling or spraying. Improved varieties which are used for cultivation are Gujarat Isabgol 1 and Gujarat Isabgol-2, the other varieties which are used for cultivation are Trombay selection 1 to 10 and EC 124 -345.

Irrigation and Fertilizers

First irrigation is provided immediately after sowing. Second is given after three weeks and third is given after ear appears. One or two soil loosening may be undertaken after watering. Cow dung is utilized for improving productivity of soil.

Harvesting

When plant becomes red at that time crop is matured. First Harvesting is carried out after 100-130 days. The ripened seeds are separated by rubbing them between the palms. The cutting is undertaken in the morning hours to minimize seed loss during harvesting. After harvesting but before crushing little water is sprinkled over it, helps in separation of seeds. The husk is removed by winnowing and clean seeds are collected. The seeds are filled in bags or suitable containers after drying them for 4-5 days.

Yield

For this plant expenditure reaches upto Rs.6000 per hetare in year while gross return goes upto Rs.12, 000 and net return reaches upto Rs.6, 000/- in year per hectare.

Diseases and Prevention

Downy mildew is the main diseases of plantago ovata. *Powdery mildew, white grab,* and *Afeeds* also affect plants. These diseases and pests can be controlled by bio techniques.

Medicinal and Economic Importance

- ☆ Seeds and husks are used for the preparation of drugs based on ayurvedic and allopathic system of medicine.
- ☆ The seed husk has the property of swelling into jelly like mass, in presence of water; mucilage relieves constipation by mechanically stimulating intestinal peristalsis Peristalsis is the contraction of smooth muscles to propel contents through the digestive tract.
- ☆ Mucilage is not affected by digestive enzymes and bacteria and can pass unaffected through the intestines, lining the mucous membrane and exercising a soothing and protective action. Toxin present in the guts is absorbed by the gel of the mucilage and do not get absorbed into the system.
- ☆ Dried seeds and husks are used as emollients, demulcents and laxatives and can be used to treat chronic constipation and amoebic and bacillary dysenteries.
- ☆ Husks also work as antidiarrhoel drug, the seeds and husk cure the inflammation of the mucous membrane of gastro-instestinal tract, genitor-urinary tract, dueodenol ulcers, gonorrhoea and piles. They are also used as a cervical dilator for termination of pregnancy.
- ☆ Seeds contain a glycoside aucubin which is physically inactive. The endosperum yields an oil up to 8.80 per cent.
- ☆ Seed mucilage is also used in cosmetics and as a basic stabilizer in the ice cream industry. It is also used in calico-printing and in the preparation of chocolates.
- ☆ The husk acts as a good binder and disintegrant in compressed tablets. Treating husks with hot caustic soda and subsequent neutralization yield a jelly that is used as a substitute for agar-agar.

136

Plumbago zeylanica L. (Chitrak)

Botanical Name: *Plumbago zeylanica* L.

Family: Plumbaginaceae

Local Names

- ☆ **Gujarati:** Chitrak
- ☆ **Hindi:** Chitrak
- ☆ **Sanskrit:** Jyotiksha, Krishnavartama
- ☆ **English:** White lead wort, Ceylon lead wort, Chitrak
- ☆ **Tamil:** Chitramulam, Kodiveli, Sittragam
- ☆ **Telugu:** Agnimula, Chitramulama

Introduction

This species is distributed through out India. It is sub-scandent perennial shrub which gets height of 60-100 cm, with diffuse branches. Leaves are alternate ovate and narrowed into a petiole. Flowers are white, sub sessile, with bracts and bracteoles and are borne in single or branched terminal or axillary spikes. They are 7-25 cm long. Rachis of the spike is covered with short glandular hairs. Fruits are one seeded, capsule type and 0.8-1.0 cm long. It is oblong, 5-valved. Seeds are oblong glabrous and smooth. Bracts are larger than bracteoles.

Origin and Distribution

The native of this plant is Sikkim and Khasi hills of India. It is grown in Kerala and North India, East India and some areas of Gujarat, Rajasthan, Madhya Pradesh, Maharastra of India.

Cultivation

It is propagated by vegetative means. Off shoots or cuttings are used for propagation. It grows in forests and waste lands of semi arid and tropical climatic zones. It grows on all type of soil. Land is prepared by ploughing the land three to four times. Furrows are prepared continuously at 30cm distance. In the June, along with the one set of monsoon, branch cuttings are planted. This cutting should be of 10-15 cm in length. These are planted in furrows at 30 cm apart. The cutting soon sprouts. If the plants are very close they are thinned out at 30 x 30 cm. Along with second weeding in August, hand hoeing is also carried out.

Fertilizers

Application of 100 kg of urea per hectare is desirable for good growth of plant in August by broadcasting method. Farm yard maner is applied to obtain good root yield at the rate of 10 tone per hectare. NPK @ 50:50:50 kg/ha can be also applied. Entire P is given basally and nitrogen and potassium are applied in two equal splits, two months after planting and four months after planting.

Harvesting

Harvesting is done in the January to February. Plants are uprooted and dried under shade for 2-3 days. The soil is removed which is attached to the roots. Roots are separated from the shoot portion. These roots are cut into smaller pieces and packed in plastic bags before marketing.

Yield

It is estimated that a production of 2000 kg of the drug can be obtained from one hectare.

Medicinal and Economic Importance

- ☆ Roots and leaves of the plant is used for medicinal purpose.
- ☆ Roots of the plant are used for obtaining drug Chitramula. Roots are vesicant and abortifacient. Roots are diuretic and are useful in rheumatism. It is used as an irritant to the skin, in the treatment of dyspensia, piles, anasarca, diarrhea and skin diseases. They are also used in leprosy.
- ☆ Leaves are very useful because they possess abortifacient qualities. An infusion of the root is used for influenza and black water fever.
- ☆ Alcoholic extract from roots has been found to be very powerful aphicide
- ☆ Roots are used to treat certain types of leucoderma.

Insecticidal and antibacterial activities were also reported in some cases.

137

Pogostemon sp. (Patchouli)

Botanical Name: . *Pogostemin patchouli* Pellet. Var.Suavis **Synonym:** *Pogostemom cablin* (Blanko) Ben

Family: Labiateae

Local Names

- ☆ **Gujarati:** Pancha, Pachouli
- ☆ **Hindi:** Pachouli
- ☆ **English:** Patchouli
- ☆ **Sanskrit:** Pachi
- ☆ **Tamil:** Kattam

Introduction

It is an erect, branched, pubescent herb which gets 0.5 to 1 meter high. Leaves are ovate to oblong, acute to obtuse and coarsely singly or doubly serrate on both surfaces, swollen on the nodes, spikes are terminal and axillary, panicle, dense, sometimes interrupted, 2.5-15cm long. They are pink purple or white violet in colour. Flowering takes place in Luzon, India and Malaya. In java, it is never found flowering. The plant appears like a herb of 2-2.5 feet height covered with leaves which are sometime light green in colour. Lower foliage are always dark in colour. Two plant types patouchli are mainly recognized *viz.* Jahore type which yields a poor leafy growth but produces a superior quality of oil; the other is Singapore type that gives more foliage, therefore more oil.

Origin and Distribution

The world Patchouli is believed to be derived from Sanskrit word "Patchouli". In old times, Indian fabrics and particularly shawls prepared for export to European countries were permeated with strange odour which was believed as origin and proof of oriental origin. This plant was first described in 1845 by Pelletiesautelet, and named Pogostemon patchouli. For many years the distillation of patchouli oil was confined to British-Malaya, mainly on the ice land of Penang and Singapore. In the starting of 20th century, the patchouli oil production took a new turn when patchouli plantations were developed in the province and the cultivated leaf material was exported to Europe. In India Madhya Pradesh, Bangalore, Andhra Pradesh, Tamil nadu, Karnataka, Gujarat etc. are some of the states which produce this plant.

Cultivation

A moderate temperature with very high humidity favours the growth. As such it can be cultivated in all those tropical areas where temperatures are not very high with evenly distributed rainfall. Patchouli plants also require partial shade and grow well with shade of coconut or other plantation crops.

Patchouli is propagated by stem cuttings. For best results, apical cuttings 10-15cm long with 3-4 leaves are taken from well developed branches. These cuttings are either used for raising nursery, containing sufficient sand and leaf moulded mixture. Nursery is raised in the shady place with provision of good drainage. Cuttings are stripped off one or two fully developed lower leaf before planting. Cuttings in nursery or in seed beds are treated with root promoting hormones for quick establishment and growth. Nursery beds are kept moist for 8-10 days, till the rootings starts, through sprinkler irrigation 2-3 times daily. In north Indian plains nursery raising may be done successfully in August-September and October and also in month of February-March. In about 4-5 weeks, nursery plants become ready for transplanting in field. Mist chambers are best suited for propagation of Patchouli.

Harvesting

Patchouli crop is ready for first harvest 5-6 months after planting. After first cutting crop is harvested at about 4 months interval. Harvesting is carried out with the help of sharp sickle in early morning or in evening. All the young growing leaves and tender branches 30-40 cm above ground are cut. Because of this there occurs delay in the regeneration and long gap between harvest, thus affecting the total oil production. Drying is done on racks made of bamboos. In any case it should ensure that complete drying is done in 3-4 days time. This plant remains productive for three years. Dry leaves give good quality of oil.

Disease

Golden nematode seriously damages this crop. No effective measures to control the nematode are known.

Yield

On an average 125-150 quintals per hectare fresh herb is harvested from three harvests in a year, which on drying remains to 25-30 quintals. Total expenditure per hectare goes upto Rs.10,000 per year and gross return reaches upto Rs.25,000/- while net returns per hectare is Rs.15,000 in first year.

Medicinal and Economic Importance

- ✫ The dried leaves are used for scenting wardrobes. The leaves and tops are added in bath water for their antirheumatic action.
- ✫ It is also used as a masking agent for alcoholic breath.
- ✫ Tenacity of odour is one of the virtues of Patchouli oil and one of the reasons for its versatilities *etc.*
- ✫ Patchouli oil is extensively used as a flavour ingredient in major food products including alcoholic beverages, frozen dairy products, candy, baked foods, gelatin, meat and meat products.
- ✫ It blends well with sandalwood, geranium, vetiver, lonones, cedarwood derivatives, clove oil, lavender, bergamot and many others. The oil is almost perfume by itself.

- ☆ It is widely used in soap, cosmetic, tobacco and incense.
- ☆ The oil gives one of the finest attars when blended with sandalwood oil. The oil possesses antibacterial activity and it is used as an ingredient in insect repellant preparations.

138

Portulaca oleracea L. (Khursa)

Botanical Name: *Portulaca oleracea* L.

Family: Portulacaceae

Local Names

- ☆ **Gujarati:** Khursa
- ☆ **English:** Garden purslane, purple flowered purslane
- ☆ **Hindi:** Kursa
- ☆ **Sanskrit:** Brihalloni, Lonika, Lonamia
- ☆ **Tamil:** Kariikeerai
- ☆ **Telugu:** Peddapaylijura

Introduction

This plant is prostrate annual herb. Stems are succulent, green or reddish. Leaves are alternate, fleshy, shining glabrous. Base is attenuate. Apex is truncatel nerves inconspicuous. Flowers are bright yellow in terminal cluster, without stalks. Capsules are globose or ovoid. It remains open transversely. Seeds are numerous and shining black.

Origin and Distribution

The origin of the plant is unknown but possibly it can be India or Western Asia. It occure in Austrailia, America, Europe, Africa like countries. In Sudan and

Egypt it is very popular. The plant is cultivated throughout in India in wild and wet condition.

Cultivation

Portulaca oleracea L. is propagated by seeds and vegetative method. The plant requires a moist light rich, well drained soil, not like dry condition but it can be grown as half hardy annual plant. Seed is best sown under protection in early spring and can then be planted out in late spring. Seeds are very small, around 1000 seeds weigh only 0.2-0.4 g. Seed rate should be 20 kg/ha. It is recommended that seeds should be sprinkled on the field and then covered with compost. For proper management organic manure should be incorporated at the rate of 20-30 tones per hectare. Stem cuttings are also used for cultivation method. Outdoor sowings *in situ* take place from late spring to late summer, successive sowings being made every two to three weeks. It requires ample amount of water and temperature range between 18-38°C.

Harvesting

Harvesting is done after 3-4 weeks of sowing and 2-3 cut and 2-3 intervals are possible in commercial production. Cutting should be done low to stimulate new growth. Some time plants are uprooted in over harvesting method.

Handling after Harvesting

Purslane can be stored in plastic boxes at 0-1° C for 2-5 days with high relative humidity.

Yield

In the tropical countries 10-12 tones per hectare crop have been yielded.

Chemical Composition

The whole plant contains carotene, vitamins C, B, P, Ca, Mg, Na, K salts, organic acids, nicotinic and oxalic acids, noradrenalin, and the bioflavonoid liquiritin.

Medicinal and Economic Importance

- ☆ Exept for the roots, the entire plant is used as an antibacterial, anti inflammatory and anthelminthic.
- ☆ It is used in treating bacillary dysentery and dysuria.
- ☆ A combination with equal parts of *Euphorbia thymifolia* is also used.
- ☆ The juice extracted from 100g of pounded fresh plant and diluted with water serves as an anthelminthic against oxyuriasis and ascariasis.
- ☆ It is administered in the morning, for 3-5 days.
- ☆ Poultices of fresh leaves are used to treatmastitis, boils and impetigo.

139

Premna integrifolia Linn (Agnimanth)

Botanical Name: *Premna integrifolia*Linn.

Family: Verbenaceae

Local Names

- ☆ **Gujarati:** Ghiti, Agnimanth
- ☆ **Sanskrit:** Agnimanthah
- ☆ **Hindi:** Agetha, Arni, Ustabunda
- ☆ **Tamil:** Munnay, muney miray
- ☆ **English:** Headache tree
- ☆ **Telugu:** Pomanti, Pedda narva, Gaebbu nelli

Introduction

This plant is very large shrub or a small tree which gets height of nine meter with yellowish lenticellate bark, spinous large branches and yellowish brown woody aromatic root. Leaves are simple, opposite, sometimes whorled, elliptic-ovate, membranous when they become young. When they become matured they become coriaceous. Leaves are entire or irregularly toothed, primary lateral nerves are in 4-6 pairs. Flowers are small, greenish, yellow or greenish white with a strong disagreeable odour, arranged in corymb or cyme panicles inflorescences. Fruits are globose drupe type, when they ripe they become black with persistent saucer shaped calyx surrounding its base.

Distribution

This plant is grown wildly throughout in India. Perticularly in tropical forest this plant is found.

Chemical Composition

The chief active principals are the three alkaloids, premnine, ganiarine and ganikarine.

Medicinal and Economic Importance

- ✰ Roots are astringent, bitter, acrid, sweet, thermogenic, anodyne, anti inflammatory, alexetric, cardiotonic, alterant expectorant, depurative, digestive, carminative, stomachic, laxative, febrifuge, antibacterial and tonic.
- ✰ They are useful in vitiated conditions of *vata* and *kapha*, neuralgia, inflammations, cardiac disorders, hepatopathy, cough, asthma, bronchitis leprosy, skin diseases, dyspepsia, flatulence, colic, anorexia, constipation, haemorrhoids, fever diabetes and general debility. Leaves are stomachic, carminative and galactagogue, and are useful in dyspepsia, flatulence, colic, agalactia, cough and catarrh, fever, rheumatalgia, neuralgia, haemorrhoids and tumours.

140

Psoralea corylifolia Linn. (Bavchi)

Botanical Name: *Psoralea corylifolia* Linn.

Family: Papilionaceae

Local Names

- ✰ **Gujarati:** Babchi
- ✰ **Hindi:** Bakuchi
- ✰ **English:** Babchi seeds
- ✰ **Sanskrit:** Vakuci, Bakuci
- ✰ **Tamil:** karpogam, Karpokkarisi
- ✰ **Telugu:** Bapunga, Bavangalu

Introduction

The plant is an erect, annual herb. It grows 30-60 cm tall under natural conditions and up to 160 cm under cultivation. There are profuse branches found there. The stem and branches are covered with conspicuous glands and white hairs and bears simple leaves, broadly elliptic, rounded and mucronate at the apex. The axillary, solitary inflorescence is observed in which 10 to 30 flowers are arranged with a hairy pedicle. It bears a single seeded pod, which is indehiscent, and the pericarp is usually found adhering to the seed.

Distribution

The genus Psoralea is widely distributed in the tropical and sub-tropical regions of the world. Out of the four species in genus *Psoralea* occurring in India,

the seeds of *Psoralea corylifolia* Linn. alone are used in medicine. This species has a large distribution in the Southern districts of Madhya Pradesh and Uttrar Pradesh from where commercial collections are being made.

Cultivation

The crop is suitable for dry tropical regions. Warmer climate is good for this plant. It is hardy plant which grows well in areas with low to medium rainfall during the summer months. It can grow on any type of soil ranging from sandy medium loam to black cotton soils. However sandy loam soil with good organic matter is best. Before one set of monsoon, seed beds should be prepared by light ploughing and furrowing. Bauchi is propagated through seeds and *Psoralea corylifolia* is having poor germination (5-7 per cent) due to dormancy. The germination percentage can be improved by breaking the dormancy by mechanical puncturing of seed coverings or by treating the seeds with sulphuric acid for 50 minutes

A pretreatment of seeds in one percent sulphuric acid has been found efficacious, repeated washing in water should wash off the sulphuric acid before the seed is sown. The seeds are dibbled in rows, preferably 45-60 cm apart, keeping plant to plant spacing around 30-45 cm, depending upon the fertility of the land. A seed rate of 7 kg is sufficient to plant one hectare land.

Irrigation and Fertilizer

Farm Yard manure at 20 t/ha gives a good initial growth and increases the seed yield significantly. Besides a fertilizer dose of 100 kg N, 60 kg P and 50 kg K/ha is recommended. Of this, half a dose of N and a full dose each of P and K is given as a basal dose, while the remaining 50 kg/ha is applied in one dose as a top dressing after 45 days of sowing. Rainy season is the best for plant therefore the irrigation requirement of bauchi is moderate. However, after the rainy season is over in June-September, the crop may be irrigated fortnightly. It needs about 6 to 8 irrigations until the crop is finally harvested.

Harvesting

After 8-9 months of sowing crop is matured. The maturation of seed is continued process; there fore the seed picking can be done from mid December onwards to the end of February. In all, 4 to 5 pickings are usually taken. At maturity, the single seeded fruit turns brownish black and emits and mild odour.

Yield

An average yield of about 2 t/ha of dry seeds may be obtained. An yield of 4000 kg. of seeds can be expected from one ha of plantation. The market rate of the seeds revolves around Rs.10/- per kg.

Medicinal and Economic Importance

- ☆ The seeds of babchi are used in the indigenous system of medicine in the treatment of leucoderma, leprosy and psoriasis.
- ☆ The seed is surrounded by a sticky, oily pericarp, which contains coumarins, of which Psoralen and Isopsoralen are therapeutically important.
- ☆ Besides treating psoriasis, psoralin is being investigated as a cure for several diseases including AIDS.
- ☆ It is also used in the treatment of intestinal amoebiasis and the healing of wounds and ulcers.
- ☆ There are several reports in literature on the antimicrobial, antifeedant and insectidial activities of babchi, suggesting other possible uses.

141

Pterocarpus marsupium Roxb. (Biyo)

Botanical Name: *Pterocarpus marsupium* Roxb.

Family: Fabaceae

Local Names

- **Gujarati:** Biyo
- **Sanskrit:** Rakta chandana
- **Hindi:** Lal chandan,Rakta-chandan
- **Tamil:** Ratha Sandanam,Chenkunkumam, Sivappu chandana
- **English:** Red Sandal wood
- **Telugu:** Agaru gandhamu,Yerra chandanamu, Rakta chandanam.

Introduction

This is medium sized to large tree which gets height of 15-30 meter, bark is dark brown which has shallow cracks, exfoliating in thin flakes and exuding a red gummy substance (Gum kino) on injury. Leaves are compound, imparipinnate, leaflets are 5-7 in numbers and coriaceous, oblong, obtuse, emarginated or even bilobed at the apex, glabrous in both surfaces lamina nerves numerous, prominent. Flowers are yellow in terminal panicles, corolla are with crisped margins. Fruits are nearly circular, glabrous, flat winged pods, convexly curved between stipe and style. Wings are veined. Seeds are 1-2, convex bony.

Distribution

This plant is distributed throught in India particularly in deciduous and evergreen forest.

Cultivation

This plant is cultivated by two methods: (i). By seed propagation and (ii). By vegetative method.

Chemical Constituents

Pterocarpus marsupium contains l-epicatechin. Heart wood yields liquiritigenin, isoliquiritigenin, a neutral unidentified component, alkaloid an dresin. The wood also contains a yellow colouring matter and an essential oil and a semi drying fixed oil. Kino contains a non-glucosidal tannin kinotannic acid, kinoin, and kino-red in addition to small quantities of catechol protocatechuic and resin, pectin and gallic acid.

Medicinal and Economic Importance

- ☆ The heart wood is astringent, bitter, acrid, cooling, anti-inflammatory, union promoter, depurative, urinary astringent, haemostatic, revulsive, anthelmintic, constipating, anodyne, alternat and elephantiasis, inflammations, fractures, bruises, leprosy, skin diseases, leucoderma, erysipelas, urethrorrhea, diabetes, rectalgia, rectitis, ophthalmopathy.

- This plant is also used in haemorrhages, verminosis, diarrhea, dysentery, odontalgia, gout, rheumatoid arthritis, cough, asthma, bronchitis and greyness of hairs.
- Leaves are useful in boils, sores and skin diseases.
- Flowers are bitter, sweet, cooling, appetizing and febrifuge and are useful in vitiated conditions of Pitta, anorexia and fevers.
- Gum is biter, styptic, antipyretic, anthelmintic and liver tonic.
- It is useful in spasmodic gastralgia, vitiated conditions of pitta, boils, gleet, urethrorrhea, odontalgia, diarrhea, psoriasis, wounds and ulcers, helminthasis, intermittent fevers, hepatopathy and ophthalmia.

142

Rauvolfia serpentina L. (Sarpa Gandha)

Botanical Name: *Rauvolfia serpentina* L.

Family: Apocynaceae

Local Names

- ✰ **Gujarati:** Sarpagandha
- ✰ **English:** Sarpagandha
- ✰ **Hindi:** Sarpagandha, Chandrabhaga
- ✰ **Sanskrit:** Sarpagandha

Introduction

This is one of the most important medicinal plant of India which is mostly found in moist deciduous forests at elevations ranging from sea level to 1200 m. It is small, evergreen, perennial undershrub, attaining a height of 15-45 cm (rarely 90 cm). Its taproot is tuberous, soft, sometimes irregularly nodular; bark is pale brown, corky, with irregular longitudinal fissures, rarely lenticellate. Leaves are thin, they are in whorls of three, large. They are 7-17 x 4-7 cm elliptic lanceolate or obovate, acute or acuminate, dark green above, pale green obscure owing to the blade running down into the petiole.

Distribution

Sarpagandha is one of the most important native medicinal plants of India, The plant is of deciduous forest and rarely found in evergreen forest. It grows well in

[A] ***Rauvolfia serpentina* L with flowers and [B] Root**

areas where rainfall is around 2500 mm to 5000 mm with proper drainage. Tropical or subtropical zones preferably with south-west rains are considered ideal for the plant. The place where severe winter and frost are not suitable. It grows on a wide variety of soils, ranging from sandy alluvial loam to red lateritic loam or even stiff and dark loam.

Cultivation

The land is prepared in the month of May. The plant is propagated from seeds as well as vegetatively. Besides direct sowing or transplanting root cuttings, stumps and stem cuttings can be employed for propagation. As direct sowing in the field does not give good results. Seedlings are raised in the nursery before transplanting in the field. The seeds have very poor and variable germination percentage (10-50 per cent). Before planting, the seeds should be soaked in 5 per cent sodium chloride solution.

The pre-treatment of seeds with some fungicide such as Ceresan or Captan, before sowing imporves the crop. The seeds are sown in mid May and the seedlings are transplanted at the break of monsoon. The soaked seeds are dibbled about 0.6 cm deep in lines 7.5 cm apart. Thick sowing should be avoided to facilitate easy uprooting of seedlings, without injury to long tender roots for transplanting. About 5.5 kg of seeds are sown in 0.05 hectare give adequate number of seedlings to plant one ha. Seedlings are then transplanted during the monsoon. The seedlings are planted at 15-20 cm in deep holes in rows 60 cm x 30 cm apart.

Vegetative propagation is also good method. Clones with high alkaloid contents are propagated. To restore alkaloide content vegetative propagation is advaisable. The large taproots as well as lateral secondary not exceeding 2.5 mm in diameter. Horizontal planting is done in holes of 5 cm deep in the monsoon.Fields are irrigated soon after planting, till they establish.

Harvesting

After 2 to 3 years harvesting is done. The digging of roots is best done with the help of trenching hoes. Irrigation of the field before digging facilitates easy picking of main as well as fibrous roots. The fibrous roots are rich in alkaloidal content and are not discarded. The root bark is richer in alkaloids than the woody portion. The harvested roots are cleaned and thoroughly air dried, so that moisture content drops to 12 to 20 percent. The moisture content is further brought down to 8 percent artificially for increasing the keeping quality of the roots. The dried roots are broken into pieces of 10-15 cm and packed in air tight containers for storage in a cool dry place to prevent moulding.

Yield

The yield of roots per ha. is about 1,200 kg. in the case when plants are raised from seeds. An irrigated two years old plantation on sandy-clay-loamy soil is reported of yield 5,500 kg and a three year old plantation 8,800 kg of air dried roots per ha.

Pests and Diseases

Following are some of the pest and diseases causing damage to the plant

- **Caterpillars of various spp.** Cause extensive damage to the leaves and render the plants defoliated. For that Dusting 5 per cent, Diptrex of 2 per cent Folidol is recommended for effective control. Cockchafer grub and Scarabaeide grub also cause extensive damage to the roots of the plants. Mixing BHC (10 per cent) powder with the soil at the time of preparation of the land can control the menace to some extent. This plant is prone to infection by fungi which causes various diseases.
- **Leafspot** also affect development of plant. Small dark brown dots are developed on the upper surface of the leaf and yellowish brown on the lower surface. The affected leaves turn yellow and dry. In another case inconspicuous, oval shaped grayish black, small spots develop on the leaves. The disease can be controlled by spraying Dithane Z-78, or Diathane M-45.
- **Target leafspot** are dark brown spots on the upper and yellowish brown spots on the lower surface of the leaves. They are developed throughout the growing season. It can controlled by spraying Captan (0.25 per cent solution in water) in early June and subsequently at month intervals.

In addition to above, powdery mildew, die back, wilt and mosaic virus, Leaf blotch, leafblight and bud rot also affect the growth and development of plant.

Medicinal and Economic Importance

- In recent years sarpgandha and its preparations are being used as antihypertensives and sedatives.
- Root and its bark are used as raw materials for extraction of isolated alkaloids, chiefly reserpine, resinnamine, ajmaline and ajmalcine for preparation of extracts with standardized alkaloidal content, and for preparing powdered root by process.
- This drug was used for centuries by the Ayurvedic and Unani systems of medicine for various types/problems of ailments, ranging from disorders of the central nervous problems such as mania insomnia to intestinal disorders, child birth and opacity of cornea.

143

Rosa spp. (Gulab)

Botanical Name: *Rosa centifolia* L., *Rosa damascena* Mill, *Rosa galica* L.

Family: Rosaceae

Local Names

- ☆ **Gujarati:** Gulab
- ☆ **Hindi:** Gulab
- ☆ **English:** Golap, Cabbage rose, Pale rose
- ☆ **Sanskrit:** Alikulasankula, Atikesra
- ☆ **Tamil:** Eros, Isora, Paninirppa
- ☆ **Telugu:** Gulabi, Roja

Introduction

Rosa damascena is a perennial, erect, sarmentose or climbing shrub with a life span up to 50 years. The plant gets height of 2.5 meters. The stem are somewhat arching and have numerous moderately stout and hooked falcate prickles of unequal size. Leaves are pinnate; leaflets are serrate, stipules adnate to the petioles. Flowers are terminal, solitary and corymbose, fine pink, red bracts rarely they are persistent. Calyx tube is persistence globose, ovoid or pitcher shaped, mouth is contracted, lobes, leafy imbricate in bud. Petals are so many and large. They are inserted on disk. Carpels are many and they are rarely few in the bottom of calyx tube. Style is sub terminal, free or connate above, stigma are thickened, ovule is one which is pendulous. Achenes coriaceous or bony, which is enclosed in the fleshy calyx tube.

Origin and Distribution

Rose is distributed in temperate and tropical regions of the world. Rose oil perhaps the costliest and sweetest fragrant material known to the world for thousand of years in various centers of civilization which includes Roman, Greek, Middle-East and India. Oldest record of findings of distillation of rose occurs in ancient Hindu testament "Ayurveda". However recent archaeological excavation carried out in Pakistan indicates that probably rose water was distilled in prevedic period during the Indus valley civilization (5000-6000 years) ago. Rosa has 150 species distributed throughout the world. In India the genus is represented by 30 species but none has value for essential oil and otto worth for industrial uses. In many countries of the world, scented rose is cultivated for the production of rose oil, rose water. Important rose growing countries are Bulgaria, France, Italy, Turkey, USSR, China and India. *Rosa centifolia* is largely grown in France and Italy while *Rosa damscena* is grown in Bulgaria, Turkey, USSR, Persia and India. Bulgaria is the major producer and supplier of rose oil to the world. In India scented rose cultivation is concentrated to certain pickets in Rajasthan and Uttar Pradesh, Gujarat and all the states along the India.

Cultivation

The suitable soils for the crop of Rosa sp. are excess soft, alluvial or brown and full with high contents of manure and mineral nutrients. If the soil is saline then lime is mixed in little quantity. For cultivation of *Rosa sp.* dried soil should be mixed with little humus. In north India Gangatic alluvial soil is suitable. Land should be prepared in April to June, irrigate slightly and it is kept for some time about 10-15 minutes, then it is digged in one meter deep and round shallow pails in the field. Now this soil is kept in open sun. This way pest insect, wild grass, seeds and roots are removed. In the August to October soil is removed from pails and spread in the bed. Before this loosening of soil of beds is done. Manure and cow dung are utilized for making the soil fertile. Cutting the bud is useful for growing the plant. This plant is not grown by seeds. Cutting is dipped in clean water for 24 hours. Before planting the cutting in the bed, soil is wrapped on its both the ends. Now soil is digged straight deep one feet or 9 inches. Cuttings in nursery beds are planted in rows 20-30 cm apart at 15 cm distance between the cuttings. These cuttings are kept in nursery bed until July-August when these are shifted to the main field. Cuttings should be treated with any of the root growth promoting hormones for early rooting. The optimum time for taking cuttings is December when roses are pruned. Nursery raised cuttings are raised in the pits of 30 x30 x30 cm or 45 x45 x 45cm with 1 x 1 m spacing. In the temperate regions the row to row and plant to plant distance may be kept at 2 x 2 m or even 2.5 x 2.5 m, pits are filled with soil mixed with well rotten farm yard manure at the rate 15 kg per pit.

Harvesting

Rose in sub tropical plains starts flowering towards middle of the March. The flower yield increases by the end of the March.There after it gradually declines and practically ceases by middle of the April. Sporadic flowring occurs from September and continues till the end of November. In the early morning flowers are harvested because rose flowers have maximum oil content in early morning, Generally plant starts giving flower in second year of planting. But for commercial purposes, it should be harvested from the third year onwards. Flowers are picked by labour and brought to processing unit in cloth bags or woolen bags.

Yield

Depending upon the soil and management practices flower yield in sub tropical climate from third year onward which ranges from 2000-3000 kg/ha/year. In the temperate climate flower yield ranges from 4000-5000 kg/ha/year.

Grading and Processing

After collection of flowers they are then taken to the distillation unit. There are three methods for distillation of rose oil.

1. Deg and Bhapka Method

Majority of rose distillers use this method for distillation of oil from rose. Equipment is made up of round kettle (Deg.) and a receiver (Bhapka) which also serve the purpose of condenser. These are made of copper. Kettle and receiver are connected with bamboo (chonga) through which steam from kettle reaches to the receiver. Kettle has capacity of 50-100 kg flowers per batch and takes about 6-8 hours to complete the distillation. After distillation is over the receiver is removed and oil is separated. This method gives a recovery of 0.01-0.015 percent oil which is significantly lower than the other method described below.

2. Direct Fired Distillation Unit

The unit consists of a kettle (still), a column, condenser and receiver. The whole unit can be of stainless steel. The capacity of still may range from 100-250 kg flowers per batch. To increase the fuel efficiency and to generate sufficient steam flue tubes are fitted in the bottom of the still. The outlet of the receiver is connected with the column to recycle the condensate in the still, after the separation of oil takes place in the receiver. Process is known as Cohobation which has been found to improve the recovery of oil.

3. Boller Operated Unit

This is most modern method of distillation and suitable for large scale production of oil. The whole plant consists of three units, a boiler, a distillation unit and distillate after passing through a receiver fed to another still through cohobation column to ensure complete recovery of oil. Part of the oil is also collected from first receiver and the remaining from the second. It is advantageous to collect the distillate water. Complete distillation in boiler operated distillation unit takes 4-5 hours. The oil unit may cost Rs. 6-8 lacs. The other advantage with the system is that several distillation units can be connected with the boiler.

Medicinal and Economic Importance

- Flowers of Rosa damascene are used for the production of rose attar. Oil of rose is one of the oldest and most valuable perfumery raw materials.
- Rose oil imparts characteristics flowery top notes to perfumes and tends depth to blended material. It is necessarily a constituent of a high grade perfume, cosmetics and flavouring materials of tobacco, foods, soft drinks and beverages.
- It is also used in traditional medicines for various ailments.
- Rose attar prepared by rose flower on sandalwood oil is used in pan masala and tobacco industry.
- Flowers of rose are used in the preparation of rose water and glukand.
- In aromatherapy rose oil is utilized in baths and massage. One of the least toxic of all essences, it is particularly good for older, drier skins.

144

Salvadora persica L. (Pilu)

Botanical Name: *Salvadora persica* L.

Family: Salvadoraceae

Local Names

- ☆ **Gujarati:** Pilu, Mithi jar, Mitha pilu
- ☆ **Hindi:** Pilu
- ☆ **English:** Khakan, Pilua, Pilu
- ☆ **Sanskrit:** Pilua
- ☆ **Tamil:** Cittua, Kodumaavali, Opa,
- ☆ **Telugu:** Waragu, Venki, Gogu, Gunica

Introduction

Genus Salvadora has two well known species in Gujarat. (1) *Salvadora persica* and (2) *Salvadora oleoides* Decne.

1. ***Salvadora persica* L:** In this species elliptic lanceolate and ovate type of leaves are seen. They are 3-7 X 2-4 cm. Some what fleshy or coriaceous, tip is obtuse type and base is narrowed, they are occasionally rounded, glabrous. Flowers are axillary and terminal. Inflorescence is open panicles type; it is 5-13 cm long, they are many in upper axils, shortly stalked a greenish yellow in colour. Calyx is cup shaped and corolla are very thin and they are also cup shaped. Stamens are 4 in the flowers. Fruits are 0.6-0.7 cm across globose, smooth berry type.

2. ***Salvadora oleoides*** **Decne**: In this species leaves are linear lanceolate or elliptic-lanceolate, they are 3-10 x 3-1.2 cm whitish green, coriaceous or fleshy when mature, acute or sub-obtuse, often mucronate. Flowers are panicles types; they are reduced of axillary fascicles of short spikes, greenish white. Calyx is cup shaped and 0.15-2 cm long, while corolla is 0.15 -0.25 cm long, stamens are 4. Fruits are 0.4-0.5 cm across, ovoid and globose berry subsessile type.

Origin and Distribution

The species of Salvadora are found in the arid and semi arid areas and in saline tracts and salty marshes. Both the species are large evergreen shrubs or small trees with numberous shining drooping branches. Salvadora is mainly found in the Gujarat, Maharastra, Madya Pradesh and Bihar.

Cultivation

Salvadora comes up well in the dry arid zones and in salt marshy areas; they are frost and drough resistant species. They can withstand in severe heat, thanks to the construction of the leaf. The areas along the sea coast and around the mouth or rivers and also low lying pockets of desert areas are most suited for these species. These species are propagated best by seeds. The seeds do not retain viability beyond 2 months period. So they should be sown after their collection soon. Fresh seeds are sown in polypots. They are watered regularly. The germination starts quickly and is complete within 60 days. By the end of next May-June, the one year old seedlings will attain a height of 30-45 cm. When they will be ready for transplanting in the field. Pits of 30 cm are dug in the field by the end of April, in strips of two lines, each.

The pits are dug at two meter distance in the lines and pits in the adjoining lines are staggered and the strips are placed at four meters distance from each other. Within a hectare, there will be about 1600 can be planted. If the area is highly saline and low lying, then continous trenches of trapezoidal shape, 50cm at top, 30 cm at bottom and 50 cm depth are dug in between two lines. This is done to drain away salts in monsoon season. The dug up soil may be used to pile up at the planting spot, so that the pits are dug in the piled up soil. Transplanting of polythene bag seedlings in the pits is carried out with the onset of monsoon. Weeding and soil working around the plants are also done to keep soil fertile. Fertilizers are also applied at the rate of 50 grams per plant in two split doses in the first three years. From 4th year onwards, deep ploughing is done twice in the area once at the beginning of monsoon second at the end of monsoon when the moisture in the plot dried up. This operation is carried out every year to get good result.

Harvesting

Plant starts yielding fruits from the 3rd year onward. The yield slowly increases up to 15-20 years of age and remains constant later on for a period of 25 to 30 years. Harvesting is done by hand. Fruits are plucked by hands or they are allowed to drop on ground.

Yield

In the third year the yield may be as low as 100 grams per plant, but by 20 years of age, it increases to about 20 kg and more of fruits.

Medicinal and Economic Importance

- ✰ The seeds form 44-46 percent of whole fruits and yield a green, semi hard to soft oil, which is used for soap manufacturing and in medicines.
- ✰ The yield of oil varies from 30-50 percent of the seeds. Purified seed fat can be used for making soaps, candles and is a potential industrial substitute of coconut oil for this purpose.
- ✰ The fat is used in the treatment of rheumatism and is used as the base for ointments.
- ✰ The leaves and branches are lopped for fodder. The root bark, leaves fruits are used for various diseases in native medicine.

145

Santalum album Linn. (Chandan)

Botanical Name: *Santalum album* Linn.

Family: Santalaceae

Local Names

- ☆ **Gujarati:** Chandan, Sukhada
- ☆ **Hindi:** Safed Chandan
- ☆ **English:** Sandal wood
- ☆ **Sanskrit:** Chandanah, Srikhand
- ☆ **Tamil:** Chandanam, Selegam
- ☆ **Telugu:** Candanama

Introduction

Sandal wood tree grow wild, in evergreen forests, sometimes reaching a height of 18m and 2.4 m in girth. It is small fragrant tree with slender drooping branches, soft wood while heart wood yellowish brown and strongly scented. Leaves are opposite, simple, glabrous, dark green above and paler below, ovate, acute, entire up to 5 x 3 cm. Inflorescence is auxiliary and terminal branched, paniculate cymes. Flowers are greenish yellow, turning into reddish purple, bracts are minute, perianth is tubular and tube is oval in shape. They are connate at base with ovary inferior, one celled, ovules are 3 and style is short, stigma is bilobed. Fruit of the tree is drupe type, subglabose and green when young. It becomes pink at the maturity. Flowering time of this plant is February to April. Fruit development takes place during July to September.

Origin and Distribution

Sandal wood oil, often referred to as East Indian Sandalwood oil, which is obtained by steam distillation of heartwood of *Santalum album*. Sandal wood oil is perhaps one of the earliest essential oils known to human being for almost 3000 years in Indian Subcontinent. It is mentioned in the two main epics, Ramayana and Mahabharata in India. It has been associated with medicine, cosmetics and religious rituals in India for a very long period.

The tree is native to India and Indonesia, which are the major producers of sandalwood oil. The genus Santalum is distributed from Indonesia in the west to Jaun Fernander Island in the east and from the Hawai in Archiplago to the North New Zealand in the South. In India, Sandalwood tree is distributed over 480 km from Dharwar in the north to the Nilgiris in the south and 400 km horizontally from Corg in the west to Kyppam in East. Some trees are also found in Rajasthan, Uttar Pradesh, Madhya Pradesh, Assam, Gujarat, Orrissa and West Bangal.

Cultivation

This tree can be grown in wide varieties of soils which grows well in clay to sandy loam soil. However, most of the sandalwood is found growing wild in laterite soil. For planting trees, soil should be well fertiled. The plant is propagated through seeds. Vegetative propagation by cuttings has not been successful. Seeds are obtained from plants, over 20 years old. Fresh seed obtained from October fruiting are depuled, dried and sown in seed beds. Gibberlic acid has been found to be effective in brining down the dormancy period and inducing quick and uniform germination. After germination seeds are put in polypots of the size of 15 x 25 cm. It is parasite plant whose roots can grow on so many of hosts. *Cassia siamea* is sown in the polypot when the seedlings are 25-50 cm high and the basal portion becomes darker.

The area for planting is cleared off all vegetation, including removal of all roots. Sandal seedlings along with rest of the seedlings are planted from late June to October at a spacement of 2.5 to 4 m. In case the seedlings of sandal wood are overtapped by the host plant, the host is tapped to provide enough light to the seedlings. Climbers if any are cut and removed. Sandal wood is also propagated by micro propagation in tissue culture.

Harvesting

Harvesting of the sandal wood is done in the period when tree becomes 30-60 years old. Generally at that time girth of the tree reaches up to 40-60 cm. Harvesting of sandal wood for the production of oil take place when the tree reaches a minimum age of 30 years. Sandal tree is harvested by uprooting and not by cutting. This is to ensure that no part of the roots system which is richest in oil content escapes collection.

Yield and Storage

Yield of the oil varies from part of plants used for extraction. Generally it is highest in the roots (10 per cent) and it is lowest in the chips (1.5-2 per cent) which constitute a mixture of heartwood and sapwood. The yield from the heart wood varies with maturity and localities. On an average yields vary from 4 -10 percentage. Storage of the Oil should be stored in small alluminium or S.S.containers, preferably at low temperature.

Grading and Processing

The oil is obtained by steam distillation of heartwood. The soft wood is first removed; the heartwood is chipped and then converted into powder in a mill. The powder is then soaked in water for 48 hours and then distilled in a boiler operated still at a steam pressure of 10-20 PSI. Distillation takes place in 48-72 hours. In large factories capacity of the still is one tonne but in small factories capacity of the still is very small. The use of higher pressure increases the oil yield and reduces the time of distillation; the oil quality is not appreciated by the perfumer as compared to the oil obtained at low pressure.

Medicinal and Economic Importance

- ☆ Sandalwood oil is used primarily in perfumery because of its fixative outstanding properties. It is used in preparing all types of perfume compositions, especially Indian attars like Rose, Kewda, Jasmine and Hina in which the natural essential oils from distillate of floral distillation is absorbed in sandalwood oil.
- ☆ Sandal wood oil with neem oil can be used as contraceptive. In medicine it is used as healing wounds. In the blisters caused by the small pox vaccination, sandal wood oil proves to be the best and most effective treatment.
- ☆ This plant gives finest woods for carving. The wood is smooth with uniform fiberes. Several items like boxes, jewel cases, combs, bookwork, picture frames, walking sticks, table lamps, animal replicas *etc.* are made out of heartwood of sandal.
- ☆ Saw dust from heartwood is mostly used in incense for scenting clothes and cupboards. Also, the oil is used as a base for co distillation of other essential oils, particularly the very delicate floral essential from *Mumusops elengi, Anthocephalus cedamba, Pandanus* spp. *etc.*
- ☆ The wood has been used for carving or powdered and mixed with coconut oil as perfume. The oil obtained by distillation for more than 2,000 years in India, has been used for cosmetics and medicinal purposes.
- ☆ This plant is used in skin diseases, rib inflammation and sexual diseases like diseases.

146

Sapindus emarginatus Vahl. (Aritha)

Botanical Name: *Sapindus emarginatus* Vahl.

Family: Sapindaceae

Local Names

- ☆ **Gujarati:** Aritha
- ☆ **Hindi:** Ritha
- ☆ **English:** Soapnut
- ☆ **Tamil;-** Pounanga, Ponam
- ☆ **Telugu:** Kunduku, Kukudu

Introduction

It is medium sized deciduous tree upto 20m in height with grey smooth bark, peeling off in scales, leaves are pinnate type. Leaflets are 2-3 pairs, terminal pair being the largest of them. Flowers are white in colour. They are polygamous. Male flowers are many and bisexual flowers are few. They are all arranged in pubescent panicle. Fruits are fleshy drupes; the pulp becomes saponaceous wrinkled rind on drying. Seeds are black in colour.

Distribution

This plant is distributed throughout in India. It is although found in sub tropical and tropical regions of the world. This plant is found at low elevations also. It is distributed in the western and southern India. Plant is seen in the Gujarat, Maharastra, Madhya Pradesh, Tamil Nadu etc. like states.

Cultivation

It comes up in areas with rainfall varying from 250mm to 500mm and temperature varying from 15°C to 47°C. The tree comes up on variety of soils. But thrives best on loamy, clayey and black cotton soils.

Thus it will be seen that this tree has a wide adaptability to soil and climate. Seed propagation is good method. Seeds remain viable for a period of one year. The germinative capacity is 67 per cent. Container seedlings are best for propagation. Pre treatment of seeds is done by soaking the seeds in cold water for 24 to 48 hours. The seeds when swollen are readyfor dibbiling them in containers. The germination commences within 7 days and is suitable to complete by 20 days. In the month of April, pits are dug at an espacement of 7m x 7m. Planting is carried out at the onset of monsoon. Later on the soil around the plants is kept loose. Fertilizers are useful to the plant growth; they are given at the rate of 100 grams per plant in two doses of 50 grams. Each along with the first two soil workings. In the subsequent years, casuality replacement is done and soil workings are also done. Open space in the field is ploughed and kept loose immediately after the rains and just before the rain. Space between the lines of plants can profitably be utilized for growing vegetables, cotton, pulses, etc. as per the suitability of the local condition during the initial four periods.

Harvesting

From the 6th year tree starts flowering and fruiting. However all the plants start bearing fruits from 8th year onwards and continue to do so upto a period of 60-70 years. The flowering is seen in November to December and the fruits ripen on the tree in the month of March to April. The ripe fruits start falling on the ground and the juicy berries start drying. Generally, fruits are plucked up by hands from the tree when major portion of the fruits ripe. They can even be picked up from the ground without any deterioration in the quality.

Yield

Average production of the trees and the care taken to maintain their vigour. However an average production of fruits per tree of 10 years age may be taken as 10 kg and this goes up with the age of 30 years of age. An average tree of 25 years will yield about 50 kg of fruits. There are instances, where average tree of 25 years age produced more than 100 kg of soapnuts. Thus a most conservative estimate of soapnut production per ha. is estimated as 1000 kg in 5th year, 2000kg in 10th year, 4000 kg in 13th year and 6000 kg. in 16th year and 8000 kg in 20th years and onwards. Total expenditure on plantation goes upto the end of 10th year comes to Rs. 15,000. and total production of soapnuts goes to 9000kg which will fetch Rs.45,000. Thus net return will be Rs. 30,000 per ha. per year upto 10th year. From 11th year onwards, the net revenue works out to Rs. 10,000 to Rs. 25,000.

Medicinal and Economic Importance

- Fruit is most useful part of the tree which has alkaline principle known as saponins. This is good for cleaning hair, woolens, fine fabrics, gold, etc.
- Semi solid oil is obtained from the kernels of the fruit, which is used medicinally.
- The root is used as expectorant.
- The fruits are used in the diseases like asthma, hysteria, epilepsy, etc. A solution of the fruits is used in curing skin diseases and the powder of the seed is used as an insecticide and fungicide.
- The flowers are sweet scented and appear on the tree in between November and December.
- The flowers are excellent source of nectar for honey bees. The entire indigenous production of soapnut is consumed in the country itself.

147

Saraca indica Linn. (Ashoka)

Botanical Name: *Saraca indica* Linn.

Family: Caesalpinioideae

Local Names

- ☆ **Gujarati:** Ashoka
- ☆ **Hindi:** Sita
- ☆ **English:** Asoka
- ☆ **Sanskrit:** Ashok
- ☆ **Tamil:** Asogam
- ☆ **Telugu:** Ashokamu

Introduction

Asoka is very handsome tree which is medium sized and evergreen. It has erect habit. The branches spread in all directions and form dense canopy and shapely crown, which obstruct sunrays to reach the ground. The stem is covered with smooth, dark brown bark. Bark becomes grayish brown in colour when it becomes old. The leaves are shiny and dark brown bark. Each is leaf is a 25 cm long. There are 3-7 pairs of leaflets. These leaflets have wavy edges. Young leaves are droopy, coppery red and flaccid. They grow alternately on the branches. The flowers are large and compact. They are orange red in colour which forms into clusters. The fowers with a deep green foliage background appears stunning beauty to the viewers. It is propagated through seed. It flowers profusely in February to March.

Origin and Distribution

This tree is native to India and it is extensively found in Malayan, peninsula, Myanmar, Sri Lanka and Bangladesh. In India, it is commonly found in Khasi hills of Assam, hilly areas of West Bangal, Western Ghats of Maharashtra and northern circars. The tree was found wild in evergreen forests of Northeastern area. This plant is related with Hindu religion. Hindus regard this tree as highly sacred. It is considered as a symbol of love and is dedicated to *Kama* and the Indian God of love.

Cultivation

The plant prefers moist well drained soil. Water logged soils are not suitable for the growth of the plant. Red laterite alluvial soil is suitable for the growth of plant. It thrives well in the area receiving annual rainfall ranginig from 2000-4000 mm. Seed propagation is good method. Mature seeds develop on plant in the month of February to March, which is collected from the ground on falling. Seeds are soaked in water for 12 hours and sown in the elevated beds of required size. Seeds take about 20 days for germination, thereafter; two months old seedlings should be transplanted to polybags. When plant becomes half to one year old, they are used for field planting. Pits of 30-45 cm are taken in monsoon season and filled with topsoil sand and dried cow dung.

Harvesting

Tree is cut at 20 years to remove the bark. It must be done in rainy season to promote sprouting and again after 4-5 years the coppice shoots can be harvested.

Medicinal and Economic Importance

- ☆ Ashoka bark has been very widely used in Indian medicine from time immemorial for the treatment of uterine and menstrunal troubles, particularly in uterine haemorrhages, dysmenorrhoea and menorrhageia.
- ☆ The drug derived from plant acts on central nervous system.
- ☆ Flowers are used in the treatment of bleeding piles, scabies in children and other skin diseases.

148

Schleichera oleosa (Lour) Oken (Kusum)

Botanical Name: *Schleichera oleosa* (Lour) Oken

Family: Sapindaceae

Local Names

- ✰ **Gujarati:** Kusum, Koshymb
- ✰ **Hindi:** Kusumb
- ✰ **English:** Lac tree, Kusum, Koshymb
- ✰ **Sanskrit:** Mukulakah, Raktamrah
- ✰ **Tamil:** Kumbadiri, pava pumarata
- ✰ **Telugu:** Botanga, mavidavitiki, Posuku

Introduction

It is medium sized tree. Bark of the tree is grey or brown, reddish inside, exfoliating in small, round, irregular flakes of 8 mm thick, leaves are paripinnate, 20-40 cm long, leaflets are elliptic or elliptic oblong, coriaceous, 2-4 pairs. Flowers are minute; they are yellowish green in colour. They are either male or bisexual, in axillary racemes. Berries are smooth or slightly prickly, globose or ovoid, hard skinned. Seeds are brown, irregularly ellipsoidal, 1 cm in size, slightly compressed, oily, enclosed in a succulent aril, which dries upon this seed.

Origin and Distribution

This tree is found in deciduous and sub deciduous forests. In India this tree is found in the most of the states like Madya Pradesh, Gujarat, Bengal, Tamil Nadu and Karnataka.

Cultivation

The natural habitat of this tree receives a rainfall ranging from 750 mm to 2500 mm and its temperature fluctuates between 2°C to 47°C. It is found up to an altitude of 900 m only. It comes up on all types of soil. It is a shade bearer, frost and drought hardy. It produces root suckers and withstands moderate pollarding. Seed propagation is good method. Seeds are collected from the trees in the month of May-June and used immediately after collection. Seeds of 750 g to 1 kg are required to raise 500 seedlings. The seeds should be immersed in hot water for 24 hours before sowing. The germination starts within a period of 5-6 days and is complete by 4 weeks. The seedlings are ready for planting in the field in the next June, by the time they attain 30-40 cm height. Pits of 30 cm cube are dug at a spacing of 7m x7m by the end of April. The nursery raised polypot seedlings are planted in the pits with the onset of monsoon. Fertilizers are added at the rate of 50 g per plant into two split doses every year. The irrigation should be applied once in the month.

Harvesting and Yield

Kusum starts flowering and giving fruits around 8-10 years. The kernels constitute 60-65 percent of the seeds. A middle aged tree is expected to yield about 18-37 kg. of seed per year. The average annual yield of seeds may be taken as 5 kg per plant in 10th year, 15 kg per plant in 15th year and 25 kg per plant in 20th year and onwards. The kernels yield a fatty oil of about 59-72 per cent, which is equivalent to 35-45 per cent of the entire seed. Kusum oil is yellowish brown semi-solid, with a faint odour of bitter almonds. When allowed to stand, a lighter coloured, solid fat separates and settles down. Oil cake has good manurial value.

Medicinal and Economic Importance

- This tree is known as lac tree because it yields good quality lac.
- Seeds are useful for medicinal purpose. Fatty oili often obtained from the seed which is known as Madagascar oil of commerce.
- Kusum oil is quite used for hair dressing; it is used for culinary and lighting purposes. It is also used for skin troubles, for external massage in rheumatism, for soap manufacturing and as lubricant.

149

Semecarpus anacardium Linn. (Bhilamo)

Botanical Name: *Semecarpus anacardium* Linn.

Family: Anacardiaceae

Local Names

- ☆ **Gujarati:** Bhilamo
- ☆ **Hindi:** Bhela, Bhilva
- ☆ **English:** Marking nut, Bhilaus, Bibba
- ☆ **Sanskrit:** Aruskarah, Agnikah
- ☆ **Tamil:** Shencottel, Erumugi
- ☆ **Telugu:** Bhallataki, jidi

Introduction

It is a moderate sized deciduous tree which attains height of 12-15 m and girth of 1.23 m with dark brown, rough bark. Leaves are large and simple. They are 17-60 x 10-30 cm, obovate-oblong. Flowers are small, dull greenish yellow, dioecious in terminal panicles. Fruits are 2.5 cm long, obliquely ovoid, drupe type of fruits are found. When they ripe, they become shining black, they are suited on a fleshy orange coloured receptacle.

Origin and Distribution

Bhilamo is medium sized tree which is found in all hotter parts of India. This tree is found in Maharashtra, Rajasthan, Punjab, Gujarat and some other states of the country. There is no certainty of the origin of this plant.

Cultivation

This species is found naturally growing in all dry deciduous and moist deciduous and sub tropical forests. It is an associate of Sal and Teak wood tree. It comes up on all types of soil, but grows well in moist patches in forest areas. It tolerates moderate shade. Seed propagation is adopted in most of the places. In fact it is not grown on a large scale any where in India. Seeds are collected fresh and used soon after collection. Seeds are sown in polypots which is filled with soil. One or two year old seedlings are transplanted in the field. Pits of 45 x 45 X 45 cm are dug at 6m x 6m or 7 x 7 m. Spacing. Seedlings are transplanted in the pits at the beginning of monsoon. Though this plant does not require artificial watering, few watering during non rainy season will help the plants. Addition of chemical fertilizers or farm yard manure to the plants is recommended. Tree starts flowering and fruiting at an young age of 6 to 7 years.

Harvesting

Crop is harvested in 6th to 7 years. Fruits are obtained from the tree. They are collected by hands of picked from the ground. The fruits and nuts are separated and dried separately.

Yield

Each tree produces about 10 kg of fuits. Thus, it is expected that a production of 900 kg of fruits would be available from one ha.

Medicinal and Economic Importance

- The fleshy orange cup of the fruits is eaten when fully ripe. It is slightly astringent.
- The pericarp abounds in a black, oily, bitter and highly vesicant juice, which has been traditionally used for marking linen. Of late, it is used extensively in paints, varnish, plastic and allied industries.
- The vesicant juice is known as Bhilawan shell liquid (BSL) in commercial circles. It is obtained from Bhilawan nuts by extraction with a hydrocarbon or other solvent, by hot expression in hydraulic prees and by other methods.
- The BSL is converted in to non-vesicant, semi solid resins, which are used as bases for the manufacture of varnishes, lacquers, enamels, paints, moulding compositions, water proofing and insulating materials.
- BSL is found to be very useful in deriving various products like metal surface coating materials, adhesives, insecticides, antiseptics, termite repellants, mildew and moth proofing agents, synthetic, herbicides, fire-proof plastics, rubber goods etc.

150

Sesbania grandiflora (Jacq.) (Sasi Ikkad)

Botanical Name: *Sesbania bispinosa* (Jacq.)

Family: Laguminoceae- Papillionaceae

Local Names

- ☆ **Gujarati:** Ikad, Sasi-Ikad
- ☆ **Hindi:** Ikad, Dadon, Dhuicha, Dhaincha
- ☆ **Sanskrit:** Jayanti, Itakata
- ☆ **English:** Dhaincha, Ikad, Prickly Seban.
- ☆ **Marathi:** Ran Shevari
- ☆ **Kannad:** Dhaincha

Introduction

A suffruticose, shrubby annual, native of Australia and cultivated and found wild throughout in marshes and swamps. It is found upto altitudes of 1220m. The stems are green, sparingly prickly, branched from the base. Leaves are abruptly pinnate, with linear-oblong leaflets, which are glabrous. Flowers are in 3-4 flowered racemes, 1.25cm long. They are pale yellow in colour. Sometimes spotted or unspotted red to black. Fruits are pods type, 15-25 cm long, straight or slightly curved with slightly indeted margins.

Origin and Distribution

This plant is believed to be originated from Australia. This is cultivated throughout in marshes and swamps. This plant is distributed in India, Sri-lanka, Pakistan, and Burma like countries.

Cultivation

It comes up in areas of arid, semi arid and tropical climatic zones. It adopts itself to various soil types. It grows well in loamy, clayey, black and sandy soils. It is drought resistant and withstands water-logging and salinity. Daincha is propagated by seeds easily. Seeds are usually sown in the field at the beginning of monsoon. Seeds are sown directly drilled into soil or broadcast method. They are drilled in lines at 30 cm apart, for which 20-60 kg. of seeds are required. When broadcast 90-100 kg of seeds are required to cover one ha. Daincha is generally grown as rainfed crop, but only occasionally as irrigated crop. When irrigated the crop should get 3-4 watering. Addition of farm yard manure to the field is helpful.

Harvesting

When crop is grown for green manure, the irrigated crop yields 3 cuttings and the rainfed crop only one crop. One can expect a production of 11-12 tones of green material from irrigated crop. Daincha seeds are collected generally from the plants grown in wind breaks. The pods are hand picked and seeds are easily separated.

Yield

The average yield of seeds is reported to be 1000-1600kg/ha. Seeds are sold around rate of Rs.8/-per kg in the market. Total expenditure per hectare is Rs. 3,000/- and gross return will be around Rs.10,000.While net return reaches upto Rs.7,000/-.

Diseases and Prevention

Stems and roots are bored into borer attack, then the plant becomes weak and breaks off. Leaves are affected by larvae twists. The young larvae twists the terminal leaflets into small galls and live inside. BHC and DDT are effective against these pests. No disease of considerable magnitude is reported.

Medicinal and Economic Importance

- Daincha has gained lot of importance in the recent past as green manure crop for rice, cotton, sugarcane and coconut crops. It is also cultivated mainly for its fibre which is used for fishing nets and lines and sails.
- Alkaline soils when treated with 1 per cent of dried increasing exchangeable calcium and available phosphorus.
- Daincha is valued as a fodder for cattle, sheep and goats.
- Daincha is also grown as a cover crop in plantation crops.
- The plant is also grown as poultry food in South Africa.
- Daincha seeds are extensively used in cattle feeds. They contain about 53 per cent protein on dry weight. Colloidal substance, similar to that of marine algae, guar gum, gum tragacanth, also found in seeds.
- The kernel yields (220-25 per cent) a reddish brown, semi drying oil with a pleasant taste.
- The kernel oil is used as a wood preservative against white ants and as a lubricant for wooden axies of carts.
- The fruits are acrid, hot and anthelmintic. It is considered beficial in ascites, tumours, warts, acute rheumatism, asthama, neuralgia, epilepsy and psoriasis.
- The juice of the pericarp and of trunk is a powerful vesicant and is used for tattooing and by mahouts for chobing elephant's feet.
- The juice is found to be possessing anti bacterial properties. The fruit is found useful against certain cancers and several other dangerous diseases.

151

Sida cordifolia Linn.
(Bala)

Botanical Name: *Sida cordifolia* Linn.

Family: Malvaceae

Local Names

- ☆ **Gujarati:** Bala, Khopat, Kharenti
- ☆ **Hindi:** Kungavi, Baldana, Kharenti
- ☆ **English:** Country mallow
- ☆ **Sanskrit:** Badi yalaka, Bala, Baladhhya, Balini
- ☆ **Tamil:** Paniar tutti, Nilla turti
- ☆ **Telugu:** Tella antisa, Tella gorra, Chirubenda

Introduction

Bala is a small downy erect shrub which gets height of 1.5 m. Branches are very long, sometimes rooting occurs at the nodes, bark is yellowish to white. Leaves are cordate-oblong, ovate or ovate-oblong; they are very downy on the both sides. Flowers are tawny-yellow or white. Fruits are with a pair of awns on each carpel. There are two other varieties of Bala. *Sida cordata* and *Sida rhombifolia* are also very common in India.

S.cordata: It is hairy herb which has lot of branches. Main stem is short and long branches are there that occasionally root at places of contact with the soil. Leaves are long petioled and seeds are brownish

S.rhombifolia: It is an erect, minutely hairy, branched undershrub with affirm woody stem and intricate branches. Leaves are short petioled, obovate, truncate of more often retuse and serrate. Flower is yellow in colour. Seeds are black in colour and smooth.

Origin and Distribution

Sida cordifolia is very common erect herb which is very common in the India. It is plant which belongs to tropical and sub tropical climatic zones. It comes up on all types of soil, but it thrives best in moist loams. It grows well upto 1000 m altitudes.

Cultivation

For this plant seed propagation is best method. Fresh seeds are collected and they are kept ready before monsoon. Area is ploughed thoroughly well and furrows are made at 1 m. apart in parallel lines. Seeds are sown directly on the ridges continuously at the beginning of monsoon. The seeds germinate soon. After the germination is complete, when the seedlings get height of 8-10 cm, plants are thinned out by the removal of excess plants and these excess seedlings are transplanted in the patches, where germination does not take place. Generally 20,000 plants can be sown in one hectare.

Harvesting

Harvesting is done in the April, plants are uprooted and the roots, stems and leaves are separated. They are cut into small pieces of proper size. The leaves are dried under shade. The roots are more useful and are marketed at higher rate than the stem and leaves.

Yield

A production of 1500 kg of roots and 500 kg. of stem with bark and leaves are expected from one hactare.

Medicinal and Economic Importance

- Root leaves and seeds are slightly bitter in taste and are used in medicines.
- Root of the plant is one of the ingredients of Dashamula, an Ayurvedic drug. The root is considered to possess astringent, diuretic and tonic properties. An infusion of it is given in urinary diseases, bilious disorders and haematuria.
- The root is administered internally in combination with asafetida and rock salt, for curing nervous disorders like haemiplegla, sciatica and facial

paralysis. The root bark is powdered and is given with milk and sugar to relieve frequent micturition and leucorrhoea.

- ☆ Seeds are known for their demulcent and laxative properties and are given in bowel complaints like piles, colic and tenesmus. The seeds, leaves, roots and bark contain different alkaloids (main being ephedrine).
- ☆ *S. rhombifolia* Linn. Is good source of fibre and leaves contain ephedrine. Roots contain alkaloids. Plant is useful as demulcent, emollient, diuretic and febrifuge and in skin diseases. Roots are used in the treatment of rheumatism and leucorrhoea
- ☆ *S. acuta* Burnm. is another plant of this genus, similar to that of Bala. Leaves are used in rheumatic affections. They are used on ulcer and sores. Juice of leaves is applied in testicular swellings and in elephantiasis. Decoction of leaves and roots is considered emollient and tonic and is used in the treatment of hemorrhoids and impotence. Leaf juice is used to get relief in chest pains and as an anthelmintic.

152

Solanum surattense Burm f. (Kantkari)

Botanical Name: *Solanum surattense* Burm. f.

Synonym: *Solanum xanthocarpum* schrad and wendl

Family: Solanaceae

Local Names

- ☆ **Gujarati:** Kantkari, Bhoi ringani
- ☆ **English:** Indian solanum
- ☆ **Hindi:** Kateli

Introduction

This is much branched, prickly, perennial herb with desely covered star shaped hairs. It has yellow shining prickles of about 1.5 cm in length. Leaves are very prickly, sparsely hairy and ovate. Flowers are purple in colour. Fruits are yellow, rounded with green veins. Seeds are smooth and numerous in numbers.

Origin and Distribution

This plant is indigenous to India and it is found often in wastelands and roadsides. Fruit of the plant constitutes the drug. The fruits yield capesteral, glucoside alkaloids and solanocarpine. Plant is of tropical climate, occurs every where in arid and semi arid regions, especially on the soils with moisture retentive capacity. It comes up on all type of soil. It is found throughout in India.

Cultivation

This plant is grown by seeds. After ploughinhg the area, the seeds are sown in the field at the beginning of monsoon in parallel lines, which are spaced at 50 cm apart. In the lines the sowing is done at 50 cm apart. After the germination is complete, the gaps formed due to failure of germination are filled again by seed sowing or transplanting the seedlings. Seven to eight kg of seed is required to cover one ha. (40,000 plants/ha.) The plants start flowering in October. The fruits can be harvested in December/January. Even though it is a perennial plant, it is advisable to maintain it as an annual in rainfed areas.

Harvesting

When the fruits start drying on one branch other branches may be bearing flowers. Therefore, when the plants are bearing maximum number of mature fruits, the plants may be uprooted and dried under sun for 4-5 days. Afterwords, the dried fruits are separated and kept for selling in the market. It is also possible to pick up ripe fruits from the plants in October/November and finally uprooting the plants in the final harvesting.

Yield

It is estimated that about 3,000 kg.of dried fruits would be available from one ha. The drug will fetch a market of Rs.10/- per kg.

Medicinal and Economic Importance

- Root is used in the treatment of several ailments.
- Root is mild purgative. Drug is useful in clearing catarrh and phlegm, therefore it is used in respiratory disorders like asthma, bronchitis and cough.
- The herb is used to treat constipation and flatulence.
- Drug possesses anthelmintic properties and is found useful in expelling intestinal worms.

153

Spilanthes paniculata Wall. (Marethi)

Botanical Name: *Spilanthus paniculata* Wall.

Family: Asteraceae

Local Name

- ☆ **Gujarati:** Marethi

Introduction

This plant is erect or sub erect herb which contains hairy stem. Leaves are opposite, ovate, acute, irregularly crenate-serrate, glabrous. Heads have 6-12 mm diameter. It is ovoid, becomes elongated in fruit.

Florets are yellow in colour. Heads some times consist of tubular florets only. Some other heads have ray florets on the outer row. All the florets have yellow colour. Flowering time of the plant is October to November.

Distribution

The plant is distributed in Taiwan, Bangladesh, Sri-lanka like countries in Asia. This plant is mainly found in the tropical and sub tropical forest.

Medicinal and Economic Uses

- ☆ Whole plant is useful but mainly floral head of the plant is used for medicinal value.

- ☆ Chewing the heads gives benefit in toothache.
- ☆ Tincture is made from the flower heads is applied in some lint to the teeth and gums.

154

Sterculia urens Roxb. (Kadayo)

Botanical Name: *Sterculia urens* Roxb.

Family: Sterculiaceae

Local Names

- ☆ **Gujarati:** Kadayo
- ☆ **Hindi:** Kadayo
- ☆ **English:** Kulu, Kadaya
- ☆ **Sanskrit:**
- ☆ **Tamil:** Kuvalam, Puttali, Sendalai,
- ☆ **Telugu:** Ettapanaku, Kavilu, Ponaku, Tapasi, Tanuka

Introduction

It is moderate sized tree, stout trunk and smooth, greenish white, shining, thick bark peeling off in thin papery flakes. Leaves are simple and alternate, they are 20-30 cm in diameter, 5-lobed, more or less glabrous above, velvety-tomentose beneath, petiole terete, 13-23 cm long. Flowers are yellow in colour and they are numerous. Fruits are follicle and 4-6 in numbers; they are bright purple in colour, ovoid to oblong. Each fruit is 2-5 cm long, densely pubescent. Seeds are 3 to 6 and black in colour. They are oblong.

Origin and Distribution

There is no certainity about the origin of the *Sterculia urens* but it is found throughout India and other tropical and sub tropical regions of the world. It is mainly

seen in the deciduous forests of the India. It is observed in Gujarat, Maharashtra and Bihar.

Cultivation

It is propagated from seed collected in the months of February to April. About 6000 seeds weigh to a kg. and 300 grams of seed are required to get 1000 seedlings. No pretreatments of seed are required. The seeds are sown in containers filled with soil, sand and farm yard manure, in the month of June. Germination starts within 7 days and is completed in 45-50 days. Seeds have a germination capacity of 40-65 percent. The seedlings are protected from severe sunlight and hot winds in the initial stages. The seedlings should be protected from severe sun light and hot winds in the initial stages. The seedlings attain a height of 20-25 cm by next June, when they are ready for transplanting in the field.

Pits are prepared of 30 cm cube in April at a spacing of 7 X 5m, the distance between lines should be 7m and distance from plant to plant in the lines is kept 5 m. The one year old seedlings raised in containers are transplanted in the pits with the onset of monsoon. Fertilizers are applied to the plants at the rate of 40 grams in the first year in two split doses. 50 grams in the second year and 3rd year in two split doses and 50 grams in one dose in the 4th and subsequent years. The plant is given fertilizers in monsoon season. When trees obtain age of 15 years they become capable of yielding gum.

Harvesting

Harvesting is done in the 15 th year after cultivation. Tapping is done by taking some precautions. Tapping is done to trees to 90 cm and above height from the ground.Tapping must be done in October to June, depending upon the locality.

Yield

According to time and depth of the girth yield of the gum varies. Annual yield of the gum per tree of 0.9 to 1.35 m girth with two blazes varies from 2.5 to 5 kg. However, a safe estimate may be taken as under. Gum exudes from the blazes all the year round. The flow is copious in hot weather, the best quality gum being collected between April and June before the onset of monsoon. The gum collected during rains being darker and poor in quality. Up to 120 cm girth – 1 kg per tree per annum is obtained, wheras up to 120-125 cm and above, 1.5 kg gum per tree per annum is obtained. Up to 150 cm and above 2 kg per tree per annum gum is obtained.

Medicinal and Economic Importance

- ☆ The stem of the plant provides specific gum which is known as Katira, Karaya gum, Indian gum. It is substitute for gum *Tragacanth.*
- ☆ Karaya gum is important raw material in the textile, cosmetic, food, pharmaceutical and in other industries.
- ☆ It is also used in throat affection and in dental fixture powders, lozenges and pasters.

155

Stereopsermum suaveolens DC. (Patala)

Botanical Name: *Stereopsermum suaveolens* DC.

Family: Bignoniaceae

Local Names

- ☆ **Gujarati:** Patala, Patla
- ☆ **Hindi:** Pal, Paral, Parur, Padiala, Padarina
- ☆ **English:** Patala
- ☆ **Sanskrit:** Abhipriya, Amova, Kalavrinti
- ☆ **Tamil:** Padri
- ☆ **Telugu:** Ambhuvasini, Kala goru, Patla

Introduction

Patla is a large, desiduous tree which attains height of 18 m, 1.8 m girth and a clear bole of 9m. The bark is grey or dark brown with horizontal furrows, exfoliating in large, flat scales. Leaves are impairpinnate, 30-60 cm long, leaflets are 5-9, broadly elliptic. Flowers are dull purple in colour, yellow within, fragrant, in large panicles. Fruits are capsule type, straight, cylindric, 30-60 cm x 1.7 cm slightly ribbed, dark-grey, some what rough with elevated whitish specks. Seeds are pale yellow, brown, 3.2 x 1.3 cm large membranous wings.

Origin and Distribution

This plant is found in the mixed desiduous forest and Sal forests and it is common in sub Himalayan tract upto 1500 m elevation. It is found in most of the states like Madhya Predesh, Bihar, Tamil Nadu, Karnataka and Gujarat *etc.*

Cultivation

This plant is moderate demander and it is found as frost resistant. It can be propagated by direct line sowing and by transplanting the nursery raised seedlings. For nursery, fresh seeds should be used during April/May. When seedlings attain 75 cm in height by next rainy season, they are ready for transplanting in the field at the beginning of rainy season at 2m X 1m spacing. The root (22cm) and shoot (2.5cm) cuttings planted at the onset of monsoon also give good results. There will be 5000 plants/ha. and they are looked after carefully by regular weeding and soil working and application of fertilizers. The plants can be harvested at the end of 5 years.

Harvesting

In the 6th year, plants are uprooted carefully so that no root portion remains in the soil. The root bark is separated out, cleaned of dirt and dried. Then the root bark is cut into smaller pieces, packed and marketed.

Yield

It is estimated that 1250 kg of root bark can be obtained from one hectare.

Medicinal and Economic Importance

- ☆ Root bark is very useful in the Ayurveda as one of the ingredients in Dashmula.
- ☆ A decoction of root is used in intermittent fevers, inflammatory affections of chest and brain.

- The bark is diuretic and possesses tonic properties.
- The flowers are given with honey to control hiccups. The gum exuded from the trees is also used in several diseases as antiviral and anticancer drug.
- The seeds yeld fatty oil. The seeds yield a dark green, non drying fatty oil (5 percent).

156

Strychnos nux-vomica Linn. (Zer Kochlu)

Botanical Name: *Strychnos nuxvomica* Linn.

Family: Strychanaceae (Loganiaceae)

Local Names

- ☆ **Gujarati:** Kuchala, Zer Kochlu
- ☆ **Hindi:** Kuchala
- ☆ **Sanskrit:** Kuchala
- ☆ **English:** Nux vomica
- ☆ **Tamil:** Etti, Kansirai
- ☆ **Telugu:** Mushadi

Introduction

This is deciduous tree grows well throughout tropical India. It has a thin grey bark and shiny elliptical leaves. The flowers are arranged in terminal cymes, and are greenish white. The fruits are bitter red berries. Seeds are collected from southern states, mainly Andhra Pradesh, Tamil Nadu and Kerala.

Origin and Distribution

This plant grows well throughout tropical India. it has a thin grey bark and shiny elliptical leaves. It is distributed in southern and western states of the India. it is distributed in Andhra Pradesh, Tamil Nadu, Kerala, Gujarat, M.P., U.P and some other states.

Cultivation

Seed propagation is good method for cultivation. The seeds are soaked in water for 12 hours before sowing them into polybags. This process helps to sprout seeds easily. Nursery should be developed in scattered shade and watering should be done by sprinkler method. Seed should be sown in the month of October. The plants will be ready in July for transplantion. Distance between two trees should be kept 15 feet. In the first week of July the ditches should be filled with 20 kg of cow dung and 5 kgs of neem or madar leaves. During Second or third week of July when 1-2 rain showers take place transplanting should be carried out. Further watering should be undertaken at the intervals of 5-7 days in little amount.

Harvesting

During July to August flowering starts in the third year and in October fruits get ripened. On rippinig the fruits fall on the ground, otherwise they can be plucked by hands. Fruits are broken into two pieces and allowed to be dried up. Later seeds are separated from the fruits and allowed to dry fully.

Yield

From 6th years and onwards up to 50-60 years of age 8000 kgs of seed per 1000 trees can be obtained. Seeds are sold at rate of Rs. 20 to 40 per kg. It can be safely stored for 2 years.

Medicinal and Economic Importance

- ☆ Seeds bark and fruits are of medicinal importance. Seeds are source of drug nux vomica which is used as tonic, stimulant and in the treatment of paralysis and nervous disorders.

- ☆ The toxic potential of the Strychnos tree has been recognized and exploited by all the ancient cultures of the world. Alkaloids are obtained from all the parts of the plant tree. Strychnine is famous which is widely studied.
- ☆ Leaves and bark contain 0.99 per cent alkaloids of which strychnine is 0.7 per cent and brucine is 0.276 per cent. Various method of extraction is being explored to enhance the commercial value of the alkaloids.
- ☆ Ayurvedic physicians utilized the medicinal properties of this plant by processing its products through *Shuddhi* which is method of detoxification. The compound thus prepared is used as a remedy for diarrhea and nervous disorders.

157

Strychnos potatorum L. (Nirmali)

Botanical Name: *Strychnos potatorum L.*

Family: Strychnaceae

Local Names

- ☆ **Gujarati:** Nirmali
- ☆ **Hindi:** Kaya, Nirmali, Neimal
- ☆ **English:** Clearing nut, Clearing nut tree
- ☆ **Sanskrit:** Ambuprasadah, Katakatphalam
- ☆ **Tamil:** Tattan-koteei
- ☆ **Telugu:** Chilla, Indupa cettu

Introduction

It is medium sized tree which is found in the deciduous forests of India, up to elevation of 1200 m. Bark of the tree is black in colour. There are crackes in the bark. Leaves of the plant are sub sessile, elliptic or ovate, generally acute, base is rounded or acute, petiole is minute. They are 4-8 x 2-4 cm in size. Flowers are white, fragrant, in axillary cymes, peduncles and pedicels very short, calyx is glabrous, segments ovate, corollas are 5 lobed and oblong. Fruits are black, globose and berry when ripe with hard rind, seeds are 1 to 2 circular and yellow and lenticular.

Cultivation

It grows on all type of soils. But it prefers the areas receiving annual rain fall of 1000-2000 mm and semi arid in nature. Serious efforts are not made so far to cultivate

this plant on a big scale. The production of seeds is obtained from trees standing in forests areas. It can be raised easily by seeds. One year old polypot raised seedlings can be tried to raise the plantiation at 5m x 5m spacing. The planting is done in the beginging of monsoon season. Watering is not necessary, but if irrigation facility is available than it benefits the plants. Fertilizers are given to the plants in first 3 to 4 years. The plants start flowering and fruiting from the 10th year.

Harvesting

When fruits are fully ripen, they are collected. They are washed to remove pulp of the seeds. The clean seeds thus obtained are dried under sun for 2-3 days and then packed in base. The seeds are then marketed.

Medicinal and Economic Importance

- Bark, seeds, leaves and roots are mostly used in medicinal treatment.
- Ripe seeds are used in clearing muddy water. Therefore, they are often used to remove impurities from coal-washing waste. In Bihar labours of coal mines use seeds for clearing the water before consumption.
- Seeds are said to be stomachic, tonic, demulcent and emetic. They are used in acute diarrhea, diabetes, gonorrhoea, eye disease (conjunctivitis).
- Roots are also used in leprosy.

158

Symplocos racemosa Roxb. (Lodhra)

Botanical Name: *Symplocos racemosa* Roxb.

Family: Symplocaceae

Local Names

- ☆ **Gujarati:** Lodhra, Lodar
- ☆ **Hindi:** Hara, Lodh, Lodhra
- ☆ **English:** Lodhra
- ☆ **Sanskrit:** Lodhra, Marjana, Tilaka
- ☆ **Tamil:** Lodhra, Velli-lethi
- ☆ **Telugu:** Erralodduga, Lodduge, Lodh

Introduction

It is an evergreen tree which is distributed throughout India. It is tree or shrub, which gets length of 6-8.5 m tall, often found gregariously. Leaves are dark green above, orbicular, elliptic-oblong, 12.5 x 5 cm coriaceous, glabrous above. Flowers are white, turning to yellow, they are fragrant in axillary simple or compound in raceme inflorescence. Drupe type fruits are seen. They are purplish-black in colour. Some time sub cylindrical and smooth. Fruit is 1-2 seeded.

Cultivation

It grows in warmer parts of India, ascending in hills up to 1400 m of elevation. It is not exacting to any particular soil. However, well drained loamy soils are preferred. The soil should be moist, but no water logging. Generally this plant is

propagated by seed. The seeds are used to raise seedlings in polypots in nursery. One year old seedlings are transplanted in the field at 2m x 1m spacing, at the beginning of monsoon. There will be 5,000 plants per ha. and they are kept clean of weeds. Support watering and application of fertilizers will help the plants to a great extent. The plants can be harvested after 5 years.

Harvesting

Harvesting is done in the 6th year and onwards. The aerial portion of the plants is cut and the bark of the stems is peeled off from the stems and branches. Then it is dried under sun and dispatched for marketing.

Yield

It is estimated that each plant would contribute 500g of dry bark 2500 kg can be the expected output from one hactare. The market rate of the drug is around Rs.10 per kg. Total expenditure per hectare is Rs. 10000 for 5 years and onwards. Gross return will be Rs. 25,000/- per ha. and net return will be Rs. 15,000/- per hectare.

Medicinal and Economic Importance

- The bark of the tree yields the commercial drug. The leaves and bark yield and yellow dye.
- The bark is used as mordant with other dyes, for dyeing silk yellow.
- It is used in combination with turmeric and gum bergfong.
- A decoction of the bark is used to stop bleeding of gums.
- The bark is used in the treatment of monorrhagia and other uterine disorders.
- The bark is reported to contain three alkaloids namely loturine (identical with harman) and colloturine.
- Of late, the drug has attracted lot of attention in allopathic research.

159

Syzygium cumini (L.) Skeel. (Jambu)

Botanical Name: *Syzygium cumini* (L.) Skeel.

Family: Myrtaceae

Local Names

- ☆ **Gujarati:** Jambu, Jamun
- ☆ **Hindi:** Jamun, Jambhal, Jaman
- ☆ **English:** Jamun, Jambolan
- ☆ **Sanskrit:** Jambuh
- ☆ **Tamil:** Naval, Kottainaval
- ☆ **Telugu:** Neredu

Introduction

It is medium sized tree, which gets height of 15-30 m with smooth light grey bark having dark patches. Leaves are simple, opposite, variable in shape and they are about 2.5 cm broad and 7.5-15 cm long, acuminate, nernes joining in a distinct intramagrinal nerve, gland-dotted, acuminate, margins are wavy, flowers are white, sweetly fragrant in 1-8 flowered cymes at the bifurcations of the branches, lobes of corolla overlapping to right in the bud. Fruits are follicles, 2.5 to 7.5 cm long, ribbed and curved, orange or bright red within, irregular, enclosed in a red puply aril arrangement. There is another species of this plant which is famous as *Jal Jambu*. (*Syzygium heyaneanum*) Wall Ex. H and A. It is smaller to S. cumini in appearance in all respect, expect that it has linear leaves (1-4.5 broad) while the leaves of *S.cumini* are 3.5-9.5cm broad.

Origin and Distribution

Jamun is distributed throughout India and in tropical and sub tropical region of the world. Jamun is seen in the India, Sri-Lanka, Burma, Austrialia etc. It is found in most of the states like Gujarat, Maharastra, Madhya Pradesh, Bihar *etc.*

Cultivation

This plant is very common in India. It is semi-shade bearer. It comes up on all types of soil and tolerates salinity to some extent. It is propagated very easily by seeds. Nearly 1200 seeds make one kg. They have poor viability. They must be sown within in 10 days after their collection. The germinative capacity is 90 percent. It can also tolerates salinity to some extent. It can also be propagated by stumps and transplants. Transplanting of one or two year old seedlings raised in polypots in the field at the beginning of monsoon is the best method. They are planted at 8 m-10 m distance. The plants take minimum 15 years before flowering and fruiting.

Harvesting

Harvesting is done in the beginning of monsoon because fuits are ripen at the same time. The fruits may be collected directly from the trees or from the ground when they drop down.

Yield

Good sized tree yields about 100 kg per year.

Medicinal and Economic Importance

- ☆ Fruits, seeds and bark of the stem are useful.
- ☆ Fruits are edible. The Juice of the unripe fruits is used to prepare vinegar, which is carminative and diuretic and possesses cooling and digestive properties. From juiceof the ripe fruit and alcoholic drink and a liquid sauce are prepared.
- ☆ Fruits are aromatic and they are also edible. Mixed with salt, the juice is dried to form a powder and is used to augiment digestion.
- ☆ Seeds are rich in protein and carbohydrates and have a little amount of calcium, hence it is used as concentrated cattle feed.
- ☆ Seeds are considered to be a cure for diabetes and are a good antidote against Nux-vomica poisoning.
- ☆ Bark is good for asthma, sore-throats, ulcers, bronchitis, dysentery.
- ☆ It is also given to purify blood and as a gargle.
- ☆ The fresh juice of the bark is used to strengthen the fishing nets.It is also used in gout, haemorrhages, syphilis, leprosy, dermatopathy, diarrhea, colic helminthiasis, wounds, ulcers, stomatitis and tyrosine.
- ☆ Tribal people believe that the fruits will digest the hairs which have gone into digestive system by mistake.
- ☆ Leaves are used in streanthening the teeth and gums. The tender leaves are used for vomiting.
- ☆ Flowers are also good source of high quality honey.

160

Tamarindus indica L. (Ambali)

Botanical Name: *Tamarindus India* L.

Family: Cesalpinaceae

Local Names

- ☆ **Gujarati:** Amli, Ambala
- ☆ **Hindi:** Imli, Tetur, Tental
- ☆ **English:** Imali, Tamarind
- ☆ **Sanskrit:** Abdika, Amala, Amali
- ☆ **Tamil:** Tindruni, Tindiram
- ☆ **Telugu:** Tetali, Tintrini

Introduction

This is large sized, very common tree in India. This is evergreen with spreading crown and dark grey, thick, rough, shallow fissured bark. Leaves are compound, paripinnate. They are 5-12 cm long, leaflets and 18-20 pairs are there, 0.7- 2.5 x 0.4 X 0.8 cm oblong, rounded at both ends. Flowers are few and are arranged in racemes at the end of branchlets, creamy white with pink lines. Fruits are pod type, brown in colour. They are 8-15 x 1-1.5 cm, thick with brittle epicarp, brown fibrous acid pulp, somewhat flattened, scurty, curved, constricted at intervals, seeds are 3-12, smooth, compressed, shining and dark brown.

Origin and Distribution

It is found in throughout India. It is believed to be originated in Abyssinia and

Central Africa, but naturalized in India. It is often cultivated in garden. In India particularly in South India, it is grown mostly. It is seen in most of the states like Gujarat, Madhya Pradesh, Maharashtra, Tamil Nadu, Karnataka *etc.*

Cultivation

It can be propagated in any kind of soil but good rich in organic compound soil is suitable for the growth of the plant. It grows well in tropical region but found best on deep alluvium. It does well on black cotton soils also. It is drought resistant and frost tender species. It can tolerate salinity in the soil in some extent. Seed propagation is good method for cultivation of tamarind plant.

One year old nursery raised container seedlings are transplanted in the field for better growth of this plant. Some grafting techniques are also used to be yielding encouraging results. They are planted at 10 x 10 m spacing. For that pits are prepared of 60 x 60 x60 cm. These pits are filled with dug and farm yard manure. Seedlings raised in containers in nursery are planted in the pits at the beginning of monsoon. Two soil workings and weeding around the plants are carried out in the 1st and 2nd

years, in July and August-September. Each plant is given 30 grams of Urea in two equal doses along with soil workings. In 3rd, 4th and 5th years, one soil working is carried out in July August and one dose of Urea (20 grams) is also given to each plant. From 6th year and onwards, inter cultivation is done once or twice to keep the soil loose. The tree starts flowering in 10th year, but the real commercial production can be expected from 15th year only and this continue for 50-60 years.

Harvesting

Fruits of tamarind plant are ripen by February March. They are collected from the trees in March-April. The ecpicarp is removed; the seeds are separated out by cut-opening the fibrous pulp. The cleaned pulp, pulp with seed and whole fruits are marketed as per the local needs

Yield

An average adult tree yields about 150-200 kg of fruits. Thus one may expect, a minimum production of 10,000 kg per ha/annum. The fruits can be sold in the market at the rate of Rs.5/kg. Total expenditure per hectare reaches to Rs. 10,000 in 15th year and net return per hectare goes to Rs. 40,000 per hectare.

Medicinal and Economic Importance

- ☆ The fruit is used as astringent, for making condiments and in general cooking. Even the tender leaves, flowers and young pods are used as vegetables.
- ☆ The tamarind is popularly considered to be cure for dysentery.
- ☆ A polish for silver, brass ware is made out of the pulp of the fruit. In Java a beer like drink is prepared by mixing tamarind, sugar and lemon juice in an iron container.
- ☆ The mixture ferments and the liquid is consumed. Fresh seeds are powdered, mixed with gum and used as an effective cementing material for sticking wood.
- ☆ An extract from the seeds is used for sizing certain cotton and jute fabrics and woolens. The amber-coloured oil extracted from the seeds is an ingredient of a varnish used for planting dolls and idols and also for food.
- ☆ Dried seeds are used in local medicines, and also fried and eaten. The seeds may be ground to make a palatable live stock feed. The leaves are considered as good fodder.
- ☆ A poultice for boils and an yellow dye for silk are prepared from leaves.

161

Terminalia arjuna Roxb. (Arjun Sadad)

Botanical Name: *Terminalia arjuna* Roxb.

Family: Combretaceae

Synonyms: *Pentaptera arjuna* Roxb.

Local Names

- ☆ **Gujarati:** Sadado, Sadad
- ☆ **Hindi:** Arjuna
- ☆ **English:** Arjun
- ☆ **Sanskrit:** Arjunah, Gandivi, Indradru
- ☆ **Tamil:** Attumaruthu
- ☆ **Telugu:** Yerramaddi

Introduction

This plant is large evergreen tree with buttressed trunk and spreading crown. It has drooping branches with dark green leaves. The bark is smooth, grayish outside and flesh colored inside, peeling off in pieces. Leaves are simple, sub opposite, oblong or elliptic, coriaceous, crenulate, dark green above, pale brown beneath, often unequal sided, reticulate. The flowers are white in panicles or spikes. Fruits are ovoid or oblong with 5 to 7 short hard angle winged. Germination is epigynous and germination power is less. Flowering season is April to June. In October to November seeds get matured. If collection is made before ripening the seeds will not germinate. Seed is a drupe and each Kg of seed has 200-235 seeds. Size varies with tree and within a tree.

Origin and Distribution

This tree is found throughout Indian subcontinent, Myanmar and Sri Lanka. It is observed in tropical moist deciduous and dry deciduous forests. Generally this tree likes to grow in riverbanks and the sides of water bodies. It is generally found in moist deciduous forests. It is also grown at the 1500 m altitudes. Generally drier part of India which includes Gujarat, Madhya Pradesh, Uttar Pradesh, Bengal, Karnataka, Tamil Nadu *etc.* Mainly it is wildly grown in the forests.

Cultivation

The plant is a light demander, even though it tolerates shade to some extent in seedling stages. It is highly susceptible to drought and frost as its habitat is an evergreen or moist evergreen tract. It can be propagated through seeds. Under favourable conditions like good soil, sufficient moisture and no competition from weeds and grasses, it comes up naturally. Transplanting of one or two year old seedlings raised in polypots, in the field at the beginning of monsoon, is desirable for its successful germination. The germination percentage of seed is 50-60 percent.

Plants are planted at 2 m x 2 m distance. Plants are looked after by regular removal of weeds, application of fertilizers and watering if necessary. Plants may be allowed to grow into bigger trees and the bark is taken from a small portion of the bole of the tree, without causing mortal damage to the trees.

Harvesting

After 10 years of planting, the bark is harvested. Then bark is cut into convenient smaller lengths and dried under sun. After drying, the bark is sent for sale.

Yield

It is estimated that each plant of 10 years age yields about 2 kg of dry bark.

Medicinal and Economic Importance

- ☆ Bark is used for tanning the hides. Bark is also employed as astringent, tonic and remedy for ulcers. Powder is employed as febrifuge, styptic and antidysenteric.
- ☆ Tender twings are used for mouth washing in case of tongue blisters.
- ☆ Pulverized bark gives relief in hypertension and acts as a diuretic, incihosis of liver.

162

Terminalia bellerica (Gaertn.) Roxb. (Baheda)

Botanical Name: *Terminalia bellerica* (Gaertn.) Roxb.

Family: Combretaceae

Local Names

- ☆ **Gujarati:** Baheda
- ☆ **Hindi:** Bahera, Vibhitaka
- ☆ **English:** Belleric myrobalan, Bastard
- ☆ **Sanskrit:** Aksha, Taliphala
- ☆ **Tamil:** Akkam, Ambalatti, Sadanga
- ☆ **Telugu:** Ahera, Jhera, Tani

Introduction

The plant is found in dry forest tracts of the country. It is common tree occurring throughout deciduous forests of India. It is also grown as avenue tree. Plant grows up to 35 m height and up to 3 m in girth. The tree has long straight bole. The bark is ashy gray with fine longitudinal crack. The leaves are having long petioles, obovate, quite entire and glabrous. Spikes are axillary, solitary. Flowers are sessile and bisexual. Male flowers are greenish white and shortly pedicellate. Flowering time of this plant is March to May. The fruit is obovate, obscurely angled with five edges, fleshy, covered with grayish silky pericarp. The dried fruit is having an irregular

shape. The ripe fruits are available from November to February. Germination of the seed takes place about 14-30 days and seed remain viable for one year.

Origin and Distribution

The native of this plant is India. In dry forest this plant is very common; Except western Rajasthan, Punjab, Chathisgarh, Uttaranchal, and Haryana and temperate Himalayan tracts, it is found in greater part of India, upto altitude of 900 meter. Tree is seen in Burma, Bangladesh, Nepal, Thailand, Indo-china, Malaysia and Sri Lanka also.

Cultivation

This is large desiduous tree which is seen with other trees like teakwood, shal etc. It is susceptible to frost and drought. Baheda plant is propagated through seeds and coppice or stumps.

1. **By Seeds:** Seed propagation is good method for propagation of the plant. Germination percentage is up to 80 percent. One kg contains about 1,200 seeds. Germination commences from 40-45 days of sowing. Considerable amount of water is needed for good germination. If sufficient shade and moisture is available, most of the mature seeds start to germinate. Generally there are many seedlings naturally regenerating under a *T.bellerica* tree.

   Nursery is prepared in March to April. The dried fresh seeds are taken and soaked in water for a night to enhance germination. The alternate soaking and drying of seeds also help to enhance germination percentage. The treated seeds are sprinkled over the bed, with a layer of sand or soil and forced into soil, using thumb. Thrashing using a wooden plank to force the seeds deep into the soil is recommended. Seedbeds should be irrigated well daily. In the wet weather transplanting of seedlings to polybags is carried out. The tap root may be too long to pull out, if age of the seedlings crosses one year. Seedlings are planted out into the field when they are one year old. In the monsoon season, pits are prepared of 30 cm^3 are taken, and then they are filled with cow dung and soil before planting. Studies have shown that germination is more in medium sized seeds (69 percent).

2. **By Stumps:** In some areas stump planting is as successful as entire planting. Sumps are prepared from 12-15 months old plants and planted in July-August after the rains have set in pits or crowbar holes. For stump planting pits are prepared of 30 cm^3.

Harvesting

Fruiting starts around the age of 15 years. After the fruits are ripen, they fall on the ground below, after then they are collected and dried for 2 to 3 days before dispatching them for marketing.

Yield

The tree have longevity of about 100 years, therefore the trees will give fruits from 15th years to 100 years. Each tree is expected to give 20 kg of fruits in the starting which reaches to 100 kg in 30th years and onwards. On an average, each tree yields about 30 kg between 15th to 30th years and 100 kg from 31st to 50th year. The production of fruits from one hectare works out to 3000kg to 10,000 kg and 15,000 kg from 51st year and onwards.

Medicinal and Economic Importance

- The berberis roots and bark are important for the medicinal value. Berberis is employed as bitter tonic, alterative, astringent, stomachic, diaphoretic, gentle aperient and as a curative of piles.
- A thick extract (gum) is obtained from this tree which is known as *Rasaut* or *Rasanjan.* Root bark, root, lower stem wood are boiled with water to prepare this extract. It is bitter and astringent and fairly soluble in water. *Rasaut,* mixed with butter and alum or with opium and lime juice and painted over the eyelids, is useful remedy in case of conjunctivitis and chronic opthalmia. This gum is similar to that of Gum Arabic.
- *Rasaut* is said to give a colling effect when mixed with drinking water. *Rasaut* is found to be effective in curing oriental sore.
- The tincture and decoction are highly reputed as affective antipyretics and antiperiodics.
- The drug is used as a general febrifuge.Dried fruit is good astringent, the astringeny renders them valuable in the acts as well as a substitute for falls

for lotions, injections and so on. It is also used in dry prolonged coughs, dropsy, diarrhoea and leprosy.

- ☆ However overdoses can act a narcotic poison. Fruits are reported to have antibiotic activity against a wide variety of microorganisms.
- ☆ Wood pulp is suitable for wrapping paper. It is used to extract a yellow dye from the seed coat. As said earlier, the tree forms important part of Myrobelans.
- ☆ Fruits are used for dyeing and tanning, and are also used to increase the potency of spirits. Kernels yield oil, which is reported for preparing soaps.

163

Terminalia chebula Retz. (Harde)

Botanical Name: *Terminalia chebula* Retz.

Family: Combretaceae

Local Names

- ☆ **Gujarati:** Harde, Hardo
- ☆ **Hindi:** Harrir, Harana
- ☆ **English:** Yellow myrobalan, chebulic myrobalan
- ☆ **Sanskrit:** Balya, Chetaki, Bhishakpriya, Girija
- ☆ **Tamil:** Kadukkai
- ☆ **Telugu:** Haritaki, Karata, Karakkai

Introduction

This tree is large sized to medium sized tree which is generally found in deciduous forest. It has short trunk and round expanding crown. The bark is dark grown, longitudinally fissured. Leaves are dark green. They are arranged in opposite maner when they become old. But when they are young they look like clothed with glossy, silky hairs. The spikes are terminal. Flowers are small and they are white in colour. The fruit is a drupe type which is oval in size, glabrous and irregularly and darkly grooved with five edges. Falling of leaves takes place during February to March. Tree flowers during March to May. Fruits become ripe in November and become hard walled after repining. Seeds remain viable up to 12 months. The tree attains full size in 30 years.

Origin and Distribution

This tree is believed to be of Indian origin. It is distributed throuout in dry deciduous and sometimes moist deciduous forests of India. Harde is found in Uttar Pradesh, Chattisgarh, Uttarakhand, Jarkhand, Himachal Pradesh, Punjab, Haryana, Maharastra, Gujarat, Madhya Pradesh, Karnataka etc like states. It also is grown up to 1500 meter elevation.

Cultivation

Harde plant is observed in wild range of habitats along the length and breadth of the country. Rainfall requirements are in the range of 750-3,200mm. The tree is resistant frost and drought. Propagation is done by two methods.

1. Natural Regeneration

When seeds fall on the ground, natural regeneration take place. Although seeds have poor germination. But when plant is regenerated it has good quality. The natural regeneration is poor in forests due to damage of animals.

2. Artificial Regeneration by Coppices

The plant can be regenerated artificially. When seeds get matured, they start falling on the ground. They must be collected. Hard pericarp of drupe is split open and then seeds are dried in sun light. Seeds are thrashed in a gunny bag using poles to break the seed coat. Then they are soaked in winter for 48 hours and sown in nursery beds. Soil is prepared for nursery plant material before rains and spray seeds on beds, then a layer of sand or soil is sprinkled and the exposed seeds are pushed into the soil. Thrashing or pressing with a wooden plank is done to force the seeds deep into topsoil. Germination commences with in 16 to 17 days. Seedlings are transplanted to polybags with care for roots and watered well. They must be kept in shade for few days and can be exposed later. The soil is irrigated, if rains are absent. One to two month old seedlings should be transplanted to poly bags. One year old seedlings can be transplanted in the field. Sometimes seedlings are picked from forest areas are utilized for plantation. In the monsoon season pits are prepared of 30-40 cm^3 at a distance of 7-8 meter. They are filled with a mixture of cow dung and soil before planting.

Harvesting

When fruits become ripe harvesting is done. After 10-15 years when fruits are ripen, harvesting is done for getting fruits mainly. The fruits fallen on the ground are collected and also plucked from the tree.

Disease and Prevention

Insects feed on leaves of tree and damage is not serious. Seeds need protection from insects and rodents during storage. The seeds must be either well dried or the pulp must be removed for long term storage.

Medicinal and Economic Importance

- Fruits are mainly used for medicinal purpose. They are rich source of tannin compositions.
- Fruits are also used in medicines as laxative, somachic, tonic and purgative.
- The outer coat of fruit is the component used. It is the important constituent of *Triphala.*
- Unripe dried fruits are recommended as purgative. They are called as Indian or Black myrobalan.
- Well rubed with an equal proportion of *Acacia catechu-* Harra is used in aphthous problems.

164

Theobroma cacao L. (Cacao)

Botanical Name: *Theobroma cacaa* L.

Family: Sterculiaceae

Local Names

- ✰ **Gujarati:** Cocoa
- ✰ **Hindi:** Cocoa
- ✰ **English:** Cocoa, Chocolate tree
- ✰ **Sanskrit:** Cocoah

Introduction

This is the plant of humid tropic region. This is shrub or a small tree. It requires well distributed rainfall of 90-100 mm a month. It gets height of 10m. Leaves are borne on the erect branches in spiral system. On the lateral branches they are in two rows and in one plane. Petiole is short, pulvinous at each end, 15-30 cm long, obovate- oblong, elliptic-oblong, abruptly acuminate, dark green, thin and firm. Flowers are cauliflorous, on the older leafless parts, yellowish pink and small. Fruits are pod like berry, 30 cm long, and 10cm in diameter. They are elliptic-ovoid, red, yellow, purplish or brown, outer wall is thick, leathery. Seeds are 20-40 numbers in each fruit flat or rounded white pink or brownish, pale-violet or purple, mucilaginous and aromatic pulp.

Origin and Distribution

This plant is grown throughout tropics and semi tropical region of the world. This is grown in the India, Sri Lanka, Pakistan, Burma *etc.* In India this plant is

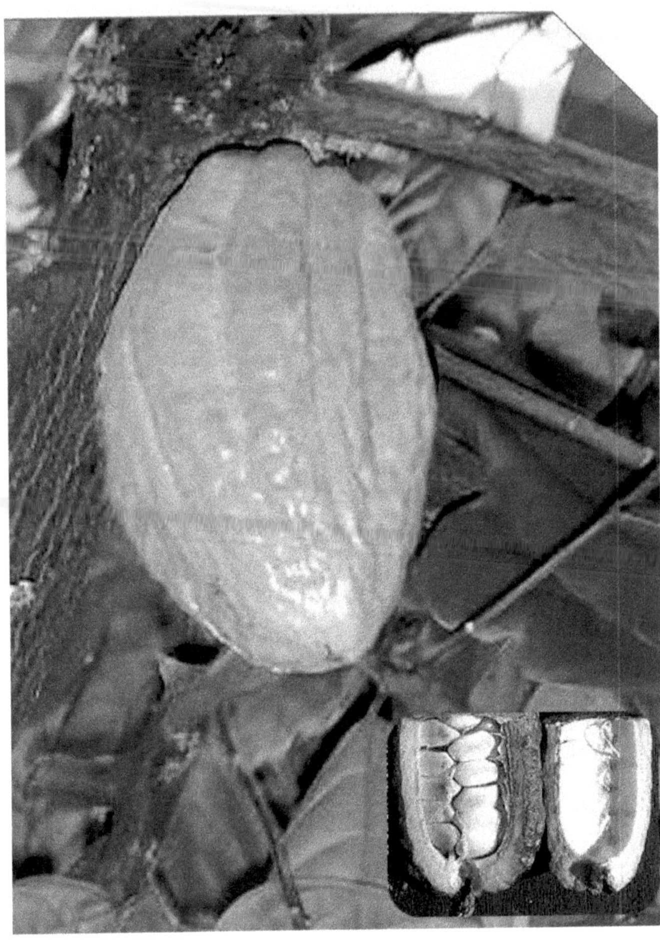

mostly seen in the western and central India. It is found in the Gujarat, Maharashtra, Punjab, Rajasthan *etc.* like states. Native of this plant is Mexico and Central America.

Cultivation

This plant is seen in arid and semi arid areas of the world. It is quite suitable to the area where there is very less rain fall. Artificial watering is required during the hot season. It can be grown in most of the soils. But soil should be capable into have capacity to retain moisture through dry season. There fore, deep and fairly heavy soils are the best. Seed propagation is good method. Seeds are sown directly or by nursery raised container seedlings. The seedlings are raised in poly bags under shade. Fresh seeds are sown in containers filled with top soil and compost. The germination is complete in 20 days. The seedlings will attain 45 cm height in 4 months period, when they are ready for planting. They are planted at a spacing of 2.5 x 3.0 mt. It requires partial over head shade. It is generally raised as under crop with coconut, etc. The plants are watered regularly once in a week. Weeding is very useful to the plant. Soil working is also done from time to time. 50 grams of nitrogen, 20 grams of phosphorus and 20 grams of potash are added to each plant in two split doses per year in the first three years. In the subsequent years, the quantity of fertilizers is doubled.

Harvesting

The plant starts flowering in the second year and keeps continue to flower till the whole year. Generally main season for flowering is between September to January and April to June. Flowers are young fruits and ripe fruits can be seen in one time in the tree. Ripe fruits are collected from the trees by cutting the stalk of the fruit by knife. The seeds are then taken out by cutting the pods cross-wise after2-3 days. They are dried and fermented. It is generally done in sweating boxes made of wood and having holes at the bottom to allow the sweating from the pulp to drain down and to air to enter are used. Beans are removed from the ripe fruits are placed in the box and they are covered with few layers of banana leaves. For uniform fermentation, the seeds are moved up and down at interval of 48 hours. After 5-6 days, the temperature begins to fall and ammonical smell is developed. When the fermentation is considered complete. Then seeds are dried.

Yield

From 1 hactare of plantation average annual production is 1000kg. Dried beans are sold at the rate of Rs. 10/- per kg.

Disease and Prevention

This plant is affected by many pests and disease. They are

1. **Mealy bugs:** They suck the sap from the parts of the trees particularly tender parts, flowers and fruits. It can be prevented by spraying 0.05 per cent Dimethoate (1.6 ml/lit. of water).
2. **Leaf eating caterpillar:** They eat tender parts of the tree. It can be controlled by spraying of carbaryl 0.1 per cent.
3. **Cockchaffr beeties:** nder leaves are eaten, BHC suspension controls the menace.
4. **Stem borer:** It bores into stem and branches, with the result the aerial portion above the point of attack dries up. It can be controlled by manual picking of the caterpillars and plugging the holes with BHC paste.
5. **Black pod disease:** In monsoon, fruits are affected in which infected tissues shrink and become dark brown and corky in texture.
6. **Charcoal rot:** The affected portion shrivels and hangs as mummies in black cololur. Spraying with 1 per cent Boprdeaus mixture controls the disease.

Medicinal and Economic Importance

- ☆ Seeds are principal source of coco or cocoa powders which is highly prized as a nutritious beverage and the chocolate used as food the world over.

- The powder is used in bakery and confectionery. The cocoa butter is used for chocolates, pharmaceutical preparations and cosmetics.
- There are two main varieties of this plant, namely *forastera* and *criollo*, of which the former is more often cultivated as commercial crop.
- The cotyledons of *forastera* are mauve coloured when fresh and dark choclate brown after fermentation, while that of criollo are white when fresh and cinnamon coloured after fermentation.
- The raw seeds are rich in vitamins, especially B and D vitamins.
- Cocoa is nervous system stimulator. In central America and the Caribbean seeds are taken as a heart and kidney tonic.
- The plant may be used to treat angina, and as a diuretic. Cocao butter makes a good lip salve and is often used as a base for suppositories and pessaries.

165

Thespesia populnea (L.) Ex.Correa (Paras Piplo)

Botanical Name: *Thespesia populnea* (L.) ex.correa

Family: Malvaceae

Local Names

- ☆ **Gujarati:** Paras piplo, Paras
- ☆ **Hindi:** Paras pipu, Porush, Paras-pipal
- ☆ **English:** Thespesia, Indian tulip tree, Portia tree
- ☆ **Sanskrit:** Gardha-banda
- ☆ **Tamil:** Poovarasam, Kallal, Cheelanthi,
- ☆ **Telugu:** Gangaraavi, Muligangaraavi, Gangareenu

Introduction

This is evergreen, bushy tree when young but thins out with age. It grows to 13 m (40 ft) or more with a spread of 3-6 m (10-20 ft). It grows rapidly under favorable conditions. Bark is corrugated with scaly twings. It is corrugated. Leaves are heart-shaped, shiny green, usually ranging in size from 5 cm to 20 cm long. Flowers are showy, yellow with a maroon to purple center, borne singly in the leaf axils. They are produced intermittently throught the year in warm climates. Fruits are capsule type and flattened. Seeds are brown in colour. They are 0.7 to 1.2 cm long. Both the capsules and the hard seeds are buoyant and can be dispersed to very long distances by sea water.

Origin and Distribution

This tree is believed to be originated in India. But it is very common plant of coastal strands across old world tropics. It has naturalized in Florida and West Indies. It is also cultivated in Central and South America and has probably naturalized there. It has taken over beaches used by nesting sea turtles on St.John, U.S. Virgin Islands. It grows in Tahiti and Austrailia. It India it is grown in most of the states. The main feature of this plant is that it is mangrove assoiate, Like other mangrove associates, these plants provide shelter and food to many creatures of the mangroves.

Cultivation

Thespesia is cultivated by. (1). Seed propagation (2). vegetative method (through cutting)

1. **Seed propagation:** Seed propagation is very easy method. The seed pods are indehiscent, that is the seed pods do not open when mature. The capsule can be opened by hand and the seeds are removed. The seeds should be scarified (the seed coat penetrated). This can be done using an emery board, sand paper, or nail clippers. Care must be taken to avoid damaging the inner part of the seed. Seeds should be planted in sterile potting mix at a depth of about twice the diameter of the seed. Germination takes 14 to 28 days.
2. **Vegetative method (Through cutting):** Cutting method is also good way to propagate this plant. Small cuttings of about 30 cm long will root easily although larger cuttings can also be used. Top three or four leaves are kept, some rooting hormone is applied, if available plant in small container. The

cutting should be protected from direct sun until it is well established. Rooting may take place within a month if conditions are favourable.

In the garden situation in a sub tropical or tropical area, this plant grows quite quickly into a tree. Its blooms are attractive both when they are pale yellow and also when they age to a deep pink.

Medicinal and Economic Importance:

- ☆ Indigestion, Pelvic infection, appetite loss, ulcers and worms are treated with the bark.
- ☆ A decoction of the bark and fruit is mixed with oil and used to treat scabies.
- ☆ An infusion of the bark is used to treat intestinal diseases.
- ☆ The inner bark is used to treat constipation and typhoid.
- ☆ The crushed fruit is used in treatment of urinary tract problems and abdominal swellings.
- ☆ Extracts of fruits and flowers are used for antibacterial activity.

166

Tinospora cordifolia (Wild.) Miers ex Hook. (Gado)

Botanical Name: *Tinosproa cordifolia* (Wild.) Miers ex Hook.

Family: Menispermaceae

Local Names

- ☆ **Gujarati:** Galo, Gado
- ☆ **Hindi:** Gulancha, Giloe
- ☆ **English:** Giloya, Gulvel
- ☆ **Sanskrit:** Guduchi, Amrita
- ☆ **Tamil:** Amridavala, Amudam, Asasi
- ☆ **Telugu:** Amrata, Duyutige, Jivantika, Guduchi

Introduction

This is climbing shrub which grows throughout India, from Kumaon and Kanyakumari. It is commonly found in dry deciduous forests. It is a shrub which is distributed throughout tropical India and the Andamans. Plant is glabrous, succulent, climbing shrub, often growing very tall. It spends down long aerial roots which have nodal swellings. The bark is creamy white to grey, deeply cleft spirally, the space in between being spotted with large rosette like lenticels.

Origin and Distribution

This plant grows throught India, from Kumao and Kanyakumari. It is commonly found in deciduous and dry forests. Origin of the Galo plant is still not clear. It is seen in most of the states like Gujarat, Bihar, Madhya Pradesh, Maharashtra, Tamil Nadu *etc.*

Cultivation

The sandy alluvial soil with good water drainage and hot climate is most suitable. Field is ploughed for 2-3 times and it is mixed 10-12 tones per hectare in the soil. The plant is propagated by cuttings method. The creeper needs some support to climb upon. For this purpose, the trees of Agasthi, Jatropha and *Moringa oleifera* should be planted one year in advance and the cuttings of *Tinospora cordifolia* are planted by their sides. Per hectare, 20,000 cuttings are required. Plant should be irrigated at the intervals of 15 days.

Harvesting

When leaves of *Tinospora cordifolia* start falling on the ground, the creepers should be cut from one feet, above the ground. The twings should be collected. The stem cuttings are cut in to small pieces and dried under the shade. They are stored in the gunny bags. After the first cutting the mother plant will start re-growing in rainy season.

Yield

Per hectare the yield is 10-12 quintals. This is generally in the form of dried stem. The market value of the product is Rs. 15.00 to 20.00 per kg. Cultivation of *Tinospora cordifolia* gives double beneficial because one can get benefit from the product of *Tinospora cordifolia* and supporting trees. Such mix farming can generate Rs. 1 to 2 lakhs income annually to farmers.

Medicinal and Economic Importance

- ☆ Root, stem, leaves and *satwa* or starch are used for medicinal value.
- ☆ Dry stems constitute drug, which is used as tonic and in diarrhea and chronic dysentery.

- ☆ It is used as a febrifuge and diuretic.
- ☆ The drug prepared from it is used to cure induced oedema and human arthritis.
- ☆ Plant contains glucoside, alkaloids, bitter principles, crystalline components etc. Gilenin and gilosterol are also present.

167

Tribulus terrestris L. (Gokhru)

Botanical Name: *Tribulus terrestris* L.

Family: Zygophyllaceae

Local Names

- ☆ **Gujarati:** Gokhru
- ☆ **Hindi:** Gokhru
- ☆ **English:** Land caltrops, Puncture-vine
- ☆ **Sanskrit:** Goksurah, Svadamstra
- ☆ **Tamil:** Nerinci
- ☆ **Telugu:** Palleru, Cinnapalleru

Introduction

This plant is distributed through out the height upto 5400 m, as weed along road sides and waste places. It is annual or perennial, prostrate herb with many slender spreading branches and silky villous young parts. Leaves are abruptly simple, pinnate, opposite, leaflets almost sessile, rounded or oblique at the base, mucronate at the apex. Flowers are bright yellow, solitary, pseudo axillary and leaf opposed. Fruits are five angled or winged spinous tuberculate woody schizocarp, separating into five cocci, each coccus having two long, stif, sharp divaricate spines towards the distal half and two shorter ones nearer the base. Seeds are one or more in each coccus.

Distribution

This plant is distributed in sub tropical and tropical regions. It is found from Kumao to Kanyakumari. It is wildly observed in any places near the mountainous area, on both sides avenue of road side, on wastelands *etc.*

Cultivation

It is wild plant so there is no need of specific technique for cultivation of the plant. This plant is propagated by seed only. This plant grows plenty in the wild. The agrotechnology for its cultivation is not developed.

Harvesting

Plant is uprooted with the help of small knife or falket.

Medicinal and Economic Importance

- ☆ Whole plant is useful. Roots and fruits are sweet, cooling, diuretic, aphrodisiac, emollient, appetizer, digestive, anthelmintic, expectorant, anodyne, anti-infalmmatory, alternate, laxative cardiotonic, styptic, lithontriptic and tonic.
- ☆ The seeds are astringent, strengthening and are useful in epistaxis, haemorrhages and ulcerative stomatitis.
- ☆ The ash of the whole plant is good for external application in rheumatic arthritis.

168

Trigonella foenum-graecum Linn. (Methi)

Botanical Name: *Trigonella fornum-graecum* Linn.

Family: Papillionaceae

Local Names

- ☆ **Gujarati:** Methi, Methi ni bhaji
- ☆ **Hindi:** Methi, Kalanusari,
- ☆ **English:** Fenugreek, Greek Hayes, Goat horn,
- ☆ **Sanskrit:** Methika, Kalausari
- ☆ **Tamil:** Ventayan
- ☆ **Telugu:** Mentulu, Mentikura

Introduction

It is an annual erect herb, growing to a height of about 45-60 cm with green leaves and small white flowers. Leaves are alternate, pinnately trifoliate, stipulate, leaflets are about 2.5 cms long, obovate to oblanceolate. Flowers are 1-2, axillary, sessile and they are arranged in raceme inflorescence. They are whitish or lemon yellow in colour. Pods are 5.7 cms long with a persistent beak, hairy with 10-20 seeds. Fenugreek is also called as goat horn or cow horn because the seed pods resemble like horn. Its seeds are used as spice and the foliage is fed to the cattle. The green plants are used as potherbs. It is said to be native to India. But it is grown throughout tropical countries. Two types of methi are found in India. One is common methi and other is Champa or Kasturi methi. These two differ in their growth habits also. The methi is quick growing, produces upright shoots and produces two or

three white flowers at the base of each leaf. Kasturi methi is slow growing initially and remains in a rosette condition during most of its vegetative growth period. Kasturi methi produces two or three white flowers in heads formed on long stalks. There are two types of varieties. (1) Varieties without smell or desi varieties and (2). Scented methi varieties.

1. **Varieties without smell or deshi varieties:** It is quick growing and an upright crop and produces white flowers. Pods are straight, 3 to 10 cm long and each pod contains 10-15 seeds.
2. **Scented methi varieties:** These are slow growers. For *e.g.* Kasturi methi, Marwari, Champa methi. The pods are smaller than those of deshi varieties and are sickle shaped. The improved varieties are: Hissar sonali, Hm-103, lam selection-1, Rajendra kanti, Rmt-1, Rmt-143, C.S.381, C.S.690 (from TNAU), Pusa Early, Bunching (IARI), Lam Sel. I, Guntar local (AP), UM.5, UM 17, UM 32, UM-34, UM-35, NLM and IC 9955

Origin and Distribution

It is an herb which is native to Southern Europe and it is now grown mainly in North India. *Trigonella polycerata* L. is found to be distributed in North India. India is one of the main producers of fenugreek. In India Rajasthan, Gujarat, Uttar Pradesh and Madhya Pradesh are important states in India growing fenugreek on a large scale.

Cultivation

Fenugreek performs well in the area of low to moderate rainfall. But it cannot withstand heavy rainfall. It is grown mostly as an irrigated crop. The plant needs loamy well-drained soils. It is also grown as a dry crop in the black cotton soils with moderate rainfall. The land is brought to fine tilth by giving two or three ploughing. Propagation is generally done by seeds. Seeds are sown by broadcast method in beds and surface raked to cover it. Sowing in rows 23 to 25 cm facilitates weeding and inter cultural operations at the initial stages. The common methi and 20 kg of Kasturi are required to cover one ha.

The seeds are sown in the middle of June or early July through four tined seed-drills in a spacing of 22.5 cm apart if sown pure @ 20 kg/ha in slightly moist seed bed. The seeds sprout quickly and show above ground in three days. In some areas, for example, India and Pakistan, two sowings are done in October and December. The temperature conditions in areas where Fenugreek is grown are described as hot summers with either mild or cool winters.

The seeds may also be soaked in water or GA (at 25, 50 or 100 ppm) for better germination before sowing. For Deshi methi 30-36 kg/ha seeds are required while for Kasturi methi 23-27 kg/ha seeds are required. They should be sown with a spacing 22-30cm apart.

Fertilizers and Manure

After ploughing a basal dressing of 10 tones of farm yard manure and 25 kg of P_2O_5 is given. A manorial schedule of 25 kg N, 25 kg P_2O_5 and 50 kg K_2O per ha is recommended. In the garden land cultivation, it requires frequent irrigation. If we apply molybdenum as ammonium molybdate increases the yield of leaves and grains.

Harvesting

Harvesting of this plant is done after 2.5 to 3 months of sowing. In the black cotton soils. During harvest, the plants are pulled out, then threshed and dried in the sun. After drying, they are beaten with sticks. Seeds are separated, winnowed and cleaned. Then they are stored and taken to market.

Yield

An average yield of 1000 to 1500 kg of seeds and 9000 to 10000 kg/ha of leaves are obtained.

Medicinal and Economic Importance

- This plant contains a number of steroidal sapogenins, especially diosgenin found in the oily embryo. Two furastanol glycosides, F-ring opened precursors of diosgenin have been reported, as also hederagin glycosides.
- The alkaloid trigonelline, trigocoumarin, trimethyl coumarin and nicotinic acid are also present. Mucilage is a prominent constitute of the seeds.
- Seeds are rich in protein, minerals and vitamins. They are mucilaginous, demulcent, diuretic, tonic, carminative, astringent, smelliest and aphrodisiac.
- They are used by the Indian women for its alleged power to promote lactation. Ground fine powder of fenugreek seeds mixed with cottonseeds and is fed to the cattle to increase the flow of milk.
- In Greece, the boiled seeds or raw are eaten with honey. The teste is slightly bitterf and is used in maple and rum flavours. It is used as a conditioning powder. Seeds are used in the diabetes.
- Fenugreek seeds produce improvement in pancreatic function. It reduces blood level of glucose, total cholesterol and triglycerides.
- Fenugreek seeds contain substances like volatile oil, fixed oil, protein, cellulose, starch, sugar, munerals, alkaloids and enzymes.
- Recent studies reveal that fenugreek seed contains the steroidal substances diosgenin, which is used as an oral contraceptive. The eather extracts of the seeds had an effect on inflammation induced in rats by cotton pellet insertion, or formalin or carrageenin exposure, comparable to that of salicylates.
- Leaves are used as fodder and vegetable, and seeds are used as spice and condiment. Fenugreek leaves are very rich in protein cotaining about 18.6 to 40.9 percent at different stages of growth. Dried plant is mixed with stored grain as an insect repellent. The leaves of *Trigonella polycerata* L. are used as vegetable. Leaves and seeds are used in obesity.
- Fenugreek is a popular ingredient of bred in Egypt and Ethiopia. This plant has been mentioned in early literature as a hypoglycemic andante inflammatory agent.
- Antiarthritic property of *methi* seed powder is widely known in many parts of India. It forms an important medicinal use of the plant in rheumatic disorders and spondylosis. This drug is also used for chronic bronchitis and hepato and splenomegaly in the unani system of medicine.

169

Tylophora indica (Burm.f) Merrill (Dam Vel)

Botanical Name: *Tylophora indica* (Burm.f) Merril

Synonym: *Telophora asthmatica* L.

Family: Asclepiadaceae

Local Names

- ☆ **Gujarati:** Dam-vel
- ☆ **Sanskrit:** Lataksiri
- ☆ **Hindi:** Dam vel, Jangali pikvam
- ☆ **Tamil:** Naippalai, Nancaruppan
- ☆ **English:** Emeticswallow-wory,Indian or Country ipeacacuanha.
- ☆ **Telugu:** Vettipala, VErripala, Tellayadala,Kakapala.

Introduction

This plant is slender laticiferous climber with much branching. Roots are long fleshy and knotty. Flowers are arranged in umbel inflorescence. They are greenish yellow outside and purplish within. Pedicels filiform with a number of filiform hairy bracts at their base. Leaves are simple, opposite and ovate to orbicular, cordate sometimes apiculate, glabrous or acuminate, more or less pubescent beneath. Fruits are fusiform, divaricate, glabrous, follicles tapering to a fine point at the apex. Seeds are with coma, they are ovate type.

Origin and Distribution

This plant is distributed throughout in India. It is wildly grown in open forest in India upto 900 meters. This plant is seen growing naturally in the moist deciduous and dry deciduous forests of India. It comes up on variety of soils, prefers moist loamy soil and tolerates partial shade.

Cultivation

This plant is propagated by seeds or by vegetative propagation method. Seeds are collected and sown in the polybags. Seeds germinate fast. The plants will be ready for planting in the field within 5 to 6 months period. It can be raised with other tree crops well. These plants are allowed to climb on old trees. Manuring and watering is good for plant. Harvesting can be done in 2 years. Before harvesting the roots of the plants, another batch of plants can be raised and planted near the supporting tree on the opposite side of the old plants. By doing this second rotation crop is on its way before first crop is harvested. The plants can be raised in similar manner as that of betel vine plantation. Spacing between plant to plant is kept at 1m x 1m. The vines may be allowed to climb Agathio (*Sesbania grandiflora*) trees raised for this purpose.

Harvesting and Yield

The leaves are harvested once in 3 months, and roots and stem are harvested after 2 years. The leaves are plucked carefully without damaging the main vine. An yield of 100 kg. of leaves per plucking can be obtained. Similarly, an yield of 1000 kg of roots can be expected from one hectare plantation. Thus a total yield of 500 kg. of leaves (5 pluckings) and 1000 kg. of roots can be expected from one hectare plantation.

Chemical Composition

Dam-vel contains three alkloid. Two alkaloids are desmethyl tylophorine and desmetyhyl tylophorinine. Third alkaloid is not identified. It contains a substance with emetic properties and essential oil. Leaves yield α-amyrin, tylophorine, kaempterol and quercetin.

Medicinal and Economic Importance

- ☆ Leaves and roots are sweet, acrid and aromatic, purgative, expectorant, vulnerary, diaphoretic, stomachic and antiviral.
- ☆ They are useful in asthma, bronchitis, whooping cough, dysentyery, diarrhea, hydrophobia, wounds, ulcers, dyspepsia, flatulence, haemorrhoids, gout, vitiated conditions of vata, cancerous tumours and murine leukaemia.

170

Uraria picta Desv. (Prisna Parni)

Botanical Name: *Uraria pricta* Desv.

Family: Fabaceae

Local Names

- **Gujarati:** Prisna parni, Pilvan
- **English:** Prisnaparni
- **Hindi:** Dabra, Prisniparni, Prishtaparni, Chitraparni
- **Tamil:** Sittirappaladai
- **Sanskrit:** Prasniparni

Introduction

It is erect, little branched, perennial herb which gets height of 90-180 cm. Leaves are variable and 1-3 foliate, they are 30 cm long and blotched white. Leaflets are 4-6 and linear, oblong or lanceolate, obtuse and mucronate with subulate stipules. Flowers are purple in colour and are in dense, long cylindrical, terminal racemes. Pods are 3-6 jointed and white in colour.

Distribution

Mostly plant belongs to dry grass lands, waste places and opens forest in the sub Himalayan tract from Kashmir to West Bengal and Assam, upto altitude of 1800m and all over the plains of India. In the world plant is distributed in the Australia, Africa and Asia regions.

Cultivation

This plant is propagated by seeds only. It comes up throughout the tropical India. It is freely propagating in the grass lands of India upto an elevation of 1900 m and all plains. It grows on variety of soils. It tolerates shade to some extent. This plant is propagated easily by seed. The area is ploughed thoroughly after adding farm yard manure, at the rate of 50 cartloads per hectare. Having got a fine tilth of the soil, ridges and furrows are made in the field at a distance of 50 cm. The seeds collected from the older plantation are sown on the ridges continuously at the beginning of the monsoon. After germination is over, excess plants from the lines are thinned out, so that the plants are spaced at 30 cm distance. The area is regularly weeded. After the plants attain 15-20cm height, super phosphate is applied at the rate of 100 kg/ha during the rainy season only. In the next May. *i.e.*, after 11 months, the plants are uprooted and the roots are obtained.

Harvesting and Yield

The area is ploughed to take out the roots easily. The roots are cleanced and cut into smaller pieces of 10 cm, long dried and bundled. A production of 1000 kg of roots per ha can be expected.

Medicinal and Economic Importance

- ✰ The plant is one of the components of Dashmoola and is credited with fracture healing properties.
- ✰ Its total extract has been found to effect better and quicker healing of fractures in the experimental animals due to early accumulation of phosphorus and more deposition of calcium.
- ✰ Roots contain aphrodisiac activity and they are given to the patients of coughts, chills and fevers.

171

Urtica dioica L. (Bichhu)

Botanical Name: *Urtica dioca* L.

Family: Urticaceae

Local Names

- ☆ **Gujarati:** Bichhu, Bichhudo
- ☆ **Hindi:** Bichuu, Bhuti, Bichua, Chichery
- ☆ **English:** Stinging nettle, Nettle

Introduction

This is perennial shrub which gets height of 1.5m with lance-shaped leaves green flowers with yellow stamens in axillary cymes. Leaves are ovate or lanceolate; usually they are cordate and serrate. Flowers are greenish and they are arranged in axillary cymes.

Distribution

This plant is mostly found in deciduous and semi evergreen forests of Himalayas, Western ghats and Peninsula. It comes up in cool and sheltered pockets. Although it comes up on variety of soils. It avoids saline soils. Nettle is seen in the temperate regions throughout the northern hemisphere, southern Africa, the Andes and Australia. In India this plant is found in the most of the areas where there are hills and mountains.

Cultivation

This plant likes to grow in cool climate. It does not like saline soil although it can be grown in any kind of soils. It is propagated by seeds. As an inter crop, it can be raised in old tree orchards. The area is ploughed thoroughly and ridges and furrows are formed at 1m. distance in parallel lines. The seeds are dibbed on the ridges at a distance of 1 m. at the beginning of monsoon. The seeds germinate soon. When seedlings obtain height of 15 cm, weeding is carried out. The plants are given fertilizers. Generally urea is utilized at the rate of 10 grams per plant during August.

Harvesting

In the month of April/May plants are uprooted. The leaves, roots, stems *etc.* are separated and dried in shade.

Yield

On an average 1000 kg of leaves, 1000 kg of roots and 100 kg of seeds are expected from one hectare of plantation.

Medicinal and Economic Importance

☆ The irritant property of the nettle has long been used externally to excite activity of paralyses limbs, and internally for treatment of haematoptysis and other haemorrhagea.

- ☆ It is used in the treatment of vomiting of blood, uterine haemorrhage, bleeding from the nose, sciatica, rheumatism and pasly.
- ☆ In the treatment of paralysis, fresh bundle of twings are used for slapping or pricking the patient once or more in a day.

172

Vernonia anthelmintica Linn. (Kali Jiri)

Botanical Name: *Vernonia anthelmintica* Linn.

Family: Asteraceae

Local Names

- ☆ **Gujarati:** Kali jiri
- ☆ **Hindi:** Kali jiri

Introduction

This plant is erect herb which gets height of 0.75 -2 meter. Leaves are lanceolate, acute, serrate, tapering into long petiole. They have 5-12 X 2.5-5 cm size. Outer involucral bracts are herbaceous. Corolla are regular, tubular, purple and green in colour.

Papppus is reddish or whitish, the outer row is very short and inner row is of flattened hairs. They are generally shorter than corollas. Achenes gets height upto 6 mm, they are 10 ribbed and pubescent. Test is bitter. Flowering time for plant is October to December.

Distribution

This is plant of desert and semi desert area. It is found in the Western part of India particularly in Gujrat, Rajasthan and some areas of Madhya Pradesh. In the Gujrat this plant is seen in Saurastra areas like Porbandar and Junagadh. It is found in the Barada hills and Giranar forest.

Cultivation

Cultivation of this plant is done through seed propagation.

Medicinal and Economic Importance

- ☆ This plant is used wildly for medicinal value. Seeds are acrid, astringent to bowels anthelmiontic. They are applied in ulcers, skin diseases, leucoderma and fevers.
- ☆ Seeds are purgative also so they are used in asthma, kidney troubles, and hiccup like problems.
- ☆ They are also applied in inflammatory swelling. They are used to remove blood from liver, good for sores and itching of eyes.

173

Vitex negundo Linn.
(Nagod)

Botanical Name: *Vitex negundo* Linn.

Family: Verbinaceae

Local Names

- ☆ **Gujarati:** Nagod,
- ☆ **Hindi:** Nirgundi, Samhalu, Saubhalu
- ☆ **English:** Five leaved chaste tree
- ☆ **Sanskrit:** Nirgundi
- ☆ **Tamil:** Nirkundi, Nallanocci
- ☆ **Telugu:** Nallavavili, Vavili, Tellavacili

Introduction

This plant is available throughout India and waste lands upto 1500m height. It is an aromatic large shrub or small tree of about 3m in height with quadrangular branches. Bark is fissured and they are found in river bends. Leaves are opposite, exstipulate, long petioled and digitately 3-5 foliate, all leaflets with middle one longer, flowers bluish purple in panicles upto 30 cm long, fruits are globose or ovoid or obovoid. They are four seeded drupe. It becomes black when ripe.

Origin and Distribution

This plant is grown throughout India. It is most common plant found in river beds of dry deciduous and moist deciduous forests and wastelands of India. It is

also grown upto elevation of 1500 m. It grows well on all types of soil but prefers alluvial flats.

Cultivation

It is most common in the river beds of dry deciduous and moist deciduous forests and wastelands of India. It can be grown on all types of soils. But it is best grown in rich alluvial soils. This plant can be propagated easily by branch cuttings of slightly matured branch. It should be of '1-2 cm in diameter and 15-20 cm in length. These cuttings are planted in polypots during month of November-December. Water is given regularly. Cuttings sprouts very soon and it will be ready for transplanting in the field by the end of May. The soil should be ploughed thoroughly and furrows are made at 1 m. distance. At the onset of monsoon generally in the first week of June, the nursery plants are transplanted in the field in furrows at 1m. distance. The plants establish soon and start growing. In summer, the plants are watered two to three times. By the end of first year, the plants get height of 1.5 to 2m and will be ready for exploitation. Since this is a perennial crop, there is no need to replace the crop for several years. After every picking of leaves, one watering is recommended.

Harvesting

The leaves can be harvested from the 2nd year. Leaves are plucked 3-4 times in a year. The leaves may be plucked from the aerial shoots of the plants or aerial shoots can be cut, dried and leaves are separated. After the leaves are collected, they are dried under shade for 3-4 days and packed in plastic bags.

Yield

A production of 3000 kg leaves can be expected from one hactare of plantation.

Medicinal and Economic Importance

- ☆ The leaves are useful in vitiated conditions of *vata*, kaphajajvara, cephalagia, sprains, orchitis, gout, splenohepatomegaly, ottorrhoea, inflammations and ulcers. The bark is useful in diarrhea, cholera, fever, haemorrhages, hepatopathy and cardiac disorders.
- ☆ Leaves contain two alkaloid nishindine and hydrocotylene. Fresh leaves yield pale greenish yellow oil.
- ☆ A decoction of leaves and vapours are employed in bath for the treatment of febrile, catarrhal and rheumatic affections. Decoction of the leaves arrests the development of swelling of joints and sprains. The juice of leaves is said to be useful for the treatment of foetid discharges. The drug possesses tranquilising effect. An ointment made from the juice is applied as a hair tonic.
- ☆ Leaves possess insecticidal properties and are used to ward off insets in grains. The freshly collected leaves on steam distillation, yield a pale greenish yellow oil
- ☆ Powdered root is pale yellow. The root possesses tonic, febrifugal, expectorant and diuretic properties. They are used in dyspepsia and rheumatism and also for boils. The powdered root is prescribed as an anthelmintic and as demulcent in dysentery and also for piles.

174

Withania somniferum Dunal. (Ashwagandha)

Botanical Name: *Withania somnifera* Dunal.

Family: Solanaceae

Local Names

- ✰ **Gujarati:** Ashwanagandha
- ✰ **Hindi:** Askandhatilli, Asgandh, Punir
- ✰ **English:** Winter cherry
- ✰ **Sanskrit:** Ashwagandha, Varahkarni
- ✰ **Tamil:** Amukkira, Amukkiran, Kizhangu
- ✰ **Telugu:** Paneru, Pulivendram

Introduction

This plant is erect, herbaceous, evergreen, tomentose shrub and 130-150 cm high. All its parts are clothed with whitish, stellate hairs. Branching is extensive, the leaf is ovate, entire, and thin, its base is cuneate and is densely hairy beneath. Flowers are bisexual, greenish or lurid yellow, axillary, in clusters of about 25 forming umbellate cymes, sessile or sub-sessile. The fruits are angled, pubescent with persistent calyx. The fruits turn orange red in colour when they mature. The seeds are yellow in colour and reniform in shape. The flowering season is from July to September and the ripe fruits are available in December.

A variety *Jawahar Asgandh-20* has been released from a single plant selection from the Jawaharlal Nehru Krishi Vishwa Vidhyalaya, Regional Agricultral Research

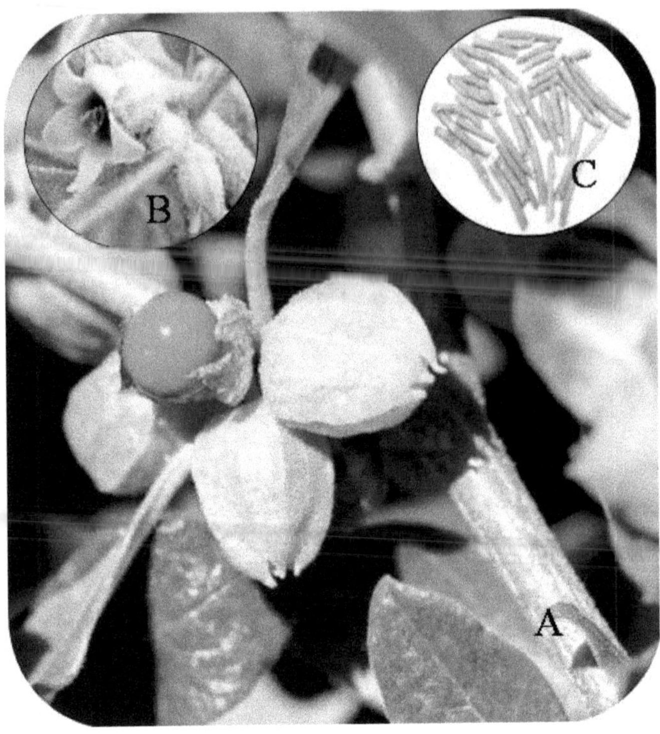

Station, Mandsaur (M.P.) This variety has recorded the highest dry-root yield, consistently over the others.

Origin and Distribution

This plant suits sub tropical and tropical regions. It is planted in late rainy season and prefers dry weather for its successful growth. It is found in Rajasthan, Madhya Pradesh, Maharastra, Uttar Pradesh, Haryana, Punjab and Gujarat. Ashwagandha is found wild in grazing ground in Mandsaur and the forestlands in Bastar district of Madya Pradesh, all over the foot hills of the Punjab and Himachal Pradesh and Western Uttar Pradesh, in Himalayas. It is also found in the wild in the Mediterranean region in North Africa. Another species *Withania congulans* Dunal, a rigid gray under shrub, 60-120 cm high is found growing wild in Punjab, Sindh and neighbouring regions. In India *W. somnifera* is cultivated in 4,000 ha. in India, mainly in the drier parts.

Cultivation

This plant suits dry climate and sub tropical regions. It is planted during the late rainy season and prefers dry weather for its successful growth. In these conditions, 1-2 late winter rains are enough for its roots to develop fully. Places, which receive 660-750 mm of rain fall, are suitable for its cultivation. It grows well in sandy loam or light red soils with good organic matter and drainage. Under such soil condition, it is also easy to dig the roots out without damaging them. The soil should not have sepage problem and should not be saline. Seed propagation is good method. The crop can be grown directly by sowing seeds into the field, as well as by raising seedlings and transplanting them.

Seed Propagation

In the monsoon season seeds are sown by broadcasting method. After receiving one or two showers, the field is thoroughly prepared, divided into plots of convenient sizes and the seeds are sown during the second week of July. When the seedlings are to be raised for transplanting, they should be sown in well prepared, raised nursery beds. About 5 kg of seeds are required to provide enough seedlings for sowing one hectare land. To avoid nursery disease, treatment of Dithane M-45 at the rate of 3g/kg of seeds before sowing. The seeds in the nursery beds are sown in lines spaced at 5 cm and covered with light soil. The germination commences within 6-7 days of sowing and in about 10 days after sowing complete germination take place. When the seedlings are 6 weeks old and sufficiently tall they are transplanted in 60 cm spaced rows, 60 cm apart in good land.

Irrigation and Weeding

The crop needs no more fertilizers and manure. No frequent irrigation is required for good growth but weeding is done by time to time. The directly sown crop is thinned at 25-30 days to maintain a plant population of 20,000 to 25,000 per hectare.

Harvesting

Harvesting is carried out in the month of January and continuous till March. For harvesting first dig the soil using tractor and collect the roots. Roots are separated from the stem. The crop matures within 150-170 days after sowing. The maturity of the crop is judged by the drying out of the leaves and the berries turning red. The entire plant is uprooted and the roots are separated from the aerial parts by cutting stem 1-2 cm above the crown. They are then transversely cut into smaller pieces of 7-10 cm for drying. Occasionally the roots are dried as a whole. The berries are plucked from the dried plants are threshed to obtain the seeds.

Yield

An average yield of 300 to 500 kg of dried roots and 50 to 75 kg of seeds can be obtained from one hactare. The roots are being sold in the market at the rate of 50-60 per kg. Good crop of Ashwagandha gives 10-15 quintals of dry root per hectare. The market rate of these roots is Rs. 40-60/- per kg. It is used in large quantities by Ayurvedic and Unani Pharmacies. They should be contacted for the sale of the produce.

Diseases and Prevention

There are no serious pests infesting this crop. But following are some of the diseases which are known to disturb the growth of plant.

(1) Damping Off and Seed Lig Rot

Casual organism: *Alternaria alternate* (Fr.) keissler.

Symptoms: damping off and seedling rot mortality caused serious damdage to the crop. The plant population is drastically reduced which resulted in reduced root yield. The infected plants turned yellowish and finally dried up. In case of damping off, the leaves of affected plants showed mottling followed by necrotic lesions, curling and burnt margins. In severe cases of infection whole plant was burnt and died. This can be prevented by use of fungicidal treatment of seeds for the control of seedling diseases. Deltan, Dithane M45, Brassicol and Bavistin could protect the seedlings.

(2) Leaf Blight and Dieback

Casual organism: *Alternaria alternate* (Fr.) Keissler.

Symptoms: Characteristic symptoms of the disease are the appearance of small light brown spots on leaves and flowers, which are enlarged, coalesced resulting in blight symptoms. In case of dieback, the disease caused necrosis of tender twings from the tip backwards. The entire branch or the top of the plant withered. The dead twings were grayish brown or straw coloured in the advanced stage of infection. This can be prevented by use of seeds from disease free plant, spraying of the crop with fungicides, such as Dithane M-45 (0.3 per cent) or Rovaral (0. 2 per cent) have been found to check spread of the disease. Seed dressings by suitable fungicides have also been found to control the disease to a large extent during the early stage.

Medicinal and Economic Importance

- ☆ Whole plant is useful medicinally. Roots contain starch, reducing sugar, hentriacontane, glycosides, dulcitol, withanicil (0.08 per cent), an acid and neutral compound. Withaniol was later found to be a mixture of two withanolides and the minor component remained unidentified.
- ☆ The free amino acids identified in the roots include aspartic acid, glycine, tyrosine, alanine, praline, tryptophan, glutamic acid and cysteine.
- ☆ The roots of South African variety yield small amount of light brown, pungent volatile oil. Root is given in the form of a decoction as well to cure scrofula and constipation.
- ☆ Roots are mostly used for curing rheumatism and dyspepsia. In Punjab they are used to relieve joint pain.
- ☆ Green fruits contain high proportion of the amino acids. The fruits and seeds are diuretic in nature.
- ☆ The commercial drug consists of the dired roots of *W.somnofere,* which occur in small pieces 10 to 17.5 cm long and 6.12 mm in diameter. Base of the stem is also used.
- ☆ Ashwagandha is mentioned as an important drug in ancient Ayurvedic literature. Several types of alkaloids are found in this plant, especially in roots.
- ☆ The total alkaloid content in the roots of the Indian types has been reported to vary between 0.13 to 0.31 per cent though higher yield have been recorded elsewhere.

175

Zingiber officinale Rose. (Aadu)

Botanical Name: *Zingiber officinale* Rose.

Family: Zingiberaceae

Local Names

- ☆ **Gujarati:** Aadu, Adarakh
- ☆ **Hindi:** Adarakh, Ada
- ☆ **Sanskrit:** Viswabhesaja
- ☆ **English:** Ginger, Zingiber
- ☆ **Tamil:** Inchi
- ☆ **Telugu:** Allamu, Ardrakamu

Introduction

Ginger is herbaceous rhizomatous perennial plant with underground modified stem. It gets height of 90 cm when it becomes fully grown. The herb develops several lateral shoots in clumps. Leaves are 15-30 cm long and 2-3 cm broad, with sheathing bases, the blade gradually tapering to a point. The rhizomes are aromatic and thick which are pale yellow in colour. Leaves are simple and alternate distinchous narrow, oblong lanceolate. The ginger of commerce is the dry product of the rhizome. Flowers are borne on a spike, condensed, oblong and cylindrical with numerous scar bracts. They are numerous, bisexual, epigynous, yellow in colour with dark purplish spots. Stamens are only one, ovary is inferior, three carpel led. Fruits are oblong and capsule type. Seeds are glabrous and fairly large. India is largest producer of ginger, accounting for 50 per cent of the total world production. Other

major ginger producing countries are Jamaica, Niegeria, Sierra Leone, Thailand, Taiwan, China etc.

Origin and Distribution

This plant is believed to be originated from the tropical South-East Asia. Later it was introduced into West Indies, Africa and other warmer parts of the world. Ginger was described as amedicinal plant in India and China long ago. It was known as Zanzabil. The plant was used by Greek and Romans as a spice. Jamaican ginger, which is widely used now, propably was introduced into the islands by the Spanish travelers. The plant is distributed in the Bangla Desh, Taiwan, Jamaica, Nigefia, Sierra Leone, Sri Lanka and South East Asian countries. In India, it grows well in the warmer and moist areas. It is grown into Andhra Pradesh, Bihar, Gujarat, Himachal Pradesh, Karnataka, Kerala, Madhya Pradesh, Manipur, Meghalaya, Nagaland, Orissa, Rajasthan, Tamil Nadu, Tripura, Uttar Pradesh, West Bangal, Arunachal Pradesh, Mizoram and Sikkim. Kerala is the major ginger producing State in India particularly the Wynad area.

Cultivation

Generally it is cultivated in tropical and sub tropical regions. The crop needs good rainfall and high temperature during the growing period. In South West India it is grown as rainfed crop in the regions having 250 cm annual rainfall. In India, it is grown in the plains of Kerala and upto an altitude of 1300-1600 m in the Himalayas. Light sandy loam soil are best suited for growing ginger. Stiff clays or coarse sandy soils are unsuitable. It should be rich in fertility, since ginger is an exhaustive crop. Water logged condition is not so good. Propagation is done by rhizomes. Each rhizome bit contains an eye or a bud 2.5 cm long. Large sized rhizomed (45-75 cm gives better yield than planting small sized rhizomes. The selected rhizomes from the previous years crops are planted during March-April. In the case of irrigated crop, planting is done from December to January in kerala, where field is formed into beds and a spacing of 30cm is given between the row and 15-20cm in the raw. Land is ploughed thoroughly and thrown into ridges. Planting materials are planted a few cm below the surface at 20-30 cm apart. They are given treat ment of agallol (100 g) before planting. Leaves are spread over as mulch to keep the soil moist to a depth of 1.5 to 2 cm.

Harvesting and Yield

Harvesting is done when stalks begin to wither and dry. It matures early. Care is taken not to injure the rhizome. The hand *i.e.* complete rhizome and adherent roots are piled in heaps. The fiberous roots are broken off and the soil and dirt are removed immediately as otherwise it is difficult to get the finished ginger white. The crop is harvested in December to January which is generally 8 months of planting. When there is great demand for ginger in the market early harvest is done. Damage is less when the crop is harvested 15-20 days earlier than at full maturity. The average yield of green ginger is about 20-25 tones per hecter.

Fertilizers and Mulching

Farm yard manure is very useful for the growth of ginger plant. 75 kg N, 50 kg P_2O_5 and 50 kg K_2O/ha are useful for the better growth. Nitrogen and potassium are supplied in two equal splits at 60 days and 120 days after planting. The whole quantity of P is given as basal application. Application of leaf mulch during planting and six weeks leter using a total of 20 tonnes of green leaves per ha. resulted in 200 per cent increase in yield over the non mulched crop. Mulching with 15 tonnes of green leaves was sufficient under Wynad conditions.

Preparation of Dry Ginger

Green ginger is washed thoroughly in water twice or thrice and then is soaked in water for a day. Next day the skin of the rhizome is peeled off carefully against the coir or small cot and or with sharp bamboo knives. Care is taken to remove only the thin outer skin. Otherwise the cells below will get damaged and loss of essential oil will result. Peeling can be done effectively and safely by keeping the rhizomes inside the rotating wire mesh drums. After peeling, the ginger is washed and dried in the sun. The bleached ginger is obtained by soaking in 3 per cent lime water for a few minutes and then fumigated with sulfer and then dried. This process is repeated once or twice to get a fully bleached white product. The recovery of dry ginger is 18 to 20 hours.

Diesease and Prevention

It is affected by following insects and diseases.

1. **Soft Rot Disease:** During storage rhizomes rots are produced due to Pythium aphanidermatum. It is serious disease which causes 50-80 per cent loot. It is prevented by 0.25 per cent wettable cerasan.
2. **Yellow Disease:** The foliar yellowing occurs which starts from the lower leaves and goes upwards. The plants wilt and dry up but do not fall on the ground in contrast to soft rot and bacterial wilt affected ginger. Rhizomes show creamy discolouration of the vascular system and cortical rot. The infected plants appear stunted with root rot and rhizome formation is affected. They show varying degrees of decay. For preventing this disease seed rhizomes are applied to the fungicide in the solution of Dithane M-45 (0.3 per cent) or benalate for two hours followed by two soil drenchings. One at the time of sowing and the second by 15th day of sowing.
3. **Pests Diseases:** This is an important pest of ginger. The grub bores into the shoot and kills the plant. This is occurred because of *Conogethes punctiferalis*. It is controlled by spraying 0.05 per cent endrin.

Medicinal and Economic Importance

- ☆ Rhizomes are used as spice and condiment, and are also used in medicines as carminative and digestive stimulant. Essential oil, obtained from rhizomes, is used for flavouring purposes.
- ☆ Ginger has a distinct spicy penetrating flavour and is largely used in the manufacture of ginger pill, ginger essence, ginger oleoresin or gingerin. Besides, starch from spent ginger, ginger soft drinks or nonalcoholic drinks, vitaminised effervescent ginger powder for use in soft drinks, ginger candy, lime ginger, pickles and ginger as a flavourant in some food products are the various products of ginger.
- ☆ It contains 0.25 to 3 per cent volatile oil of light yellow colour. Its manufacture for use in gingerale in USA and gingerbeer in England demand, import of considerable ground ginger annually.
- ☆ Ginger contains 1-2 per cent volatile oil and 5-8 per cent resinous matter, starch and mucilage. The oil of ginger is mixture of over 24 constituents, consisting of monoterpenes.
- ☆ The pungent component is gingerol, formed in the plant from phenylalnine, malonate and hexanoate. Minor constituents of an extract are gingreniols, methylgingediol, gingeryldiacetates and methyl gingediacetates.
- ☆ The main products from ginger are ginger powder, gingiber oil and crystallized ginger and candied ginger.

☆ Dry ginger is used for manufacture of several products such as ginger oil, ginger oleoresin, ginger essence, soft drink of non alcoholic beverages and vitaminised effervescent ginger powders used in soft drinks. It is used as flavourant in some food products like pies, cookes, cakes biscuits and spiced animal food.

References

Agharkar, S.P. (1953). Gazetteer of Bombay State. General-A, Botany Pt. I- Medicinal Plants, Bombay.

Anonymous (1966). Wealth of India, Raw materials, 7 vols, New Delhi

B.P.C.(1963). The British Pharmaceutical Codex, London.

Bhandari, Chandraraj (1957). Vanaushadhi Chandrodaya, Varanasi, 10 Parts (Hindi).

Bhatnagar, S.S., H. Santapau, J.D.H. Desa, A.C. Maniar, N.C. Ghadeally, M.J. Solomon, S. Yellore and T.N.S. Rao. (1961).Biological activity of Indian Medicinal Plants, *Ind. J. med. Res.* 49(5): 799-813.

Bhatnagar, S.S., Santapau, H. Fernandes, F. Kamat, V.N., Dastoor, N.J. and Rao, T.N.S. (1961). Biological activity of Indian Medicinal Plants. *J. Sci. and Industr. Res. Suppl.* 20A: 1-24.

Biswas, K. and Ghosh, E. (1950). Bhartiya Banaushadhi, Culcutta, 2 vols. (In Bengali)

Chopra, R.N. (1958). Chopra's Indigenous Drugs of India. Revised by R.N. Chopara, I.C. Chopara, K.L.Handa and L.D. Kapur, Culcutta.

Chopra, R.N., Bhadhwar, R.L. and Ghosh, S.. (1949). Poisonous plants of India, New Delhi.

Chopra, R.N. and Chopta, I.C. (1955). A review of work on Indian medicinal Plants, New Delhi.

Chopra, R.N., Chopra, I.C. and Varma, B.S.. (1969). Supplement to Glossary of Indian Medicinal Plants, New Delhi.

Chopra, R.N., Nayar, S.L. and Chopra, I.C. (1956). Glossary of Indian Medicinal Plants, New Delhi.

Dastur, J.F. (1951). Medicinal Plants of India and Pakistan, Bombay.

Dutt, U.C. (1877). The material medica of the Hindus, Calcutta.

Daniel, M. (2008). *Herbal technology: Concept and approaches*. Satish Serial Publishers, New Delhi, pp.580.

Daniel, M. and Nagar, P.S. (2010). A soujoun o he herbal treasures of MSU., M.S. University, Arihant offset Printers, Karelibaugh, Vadodara,.

Daniel, M. (2006). *Medicinal Plants- Chemistry and Properties*. Oxford and IBH Publishing Co. Pvt. Ltd., New Delhi, p. 250.

Gupta, R. (1972). " Vital drugs and essential oil bearing plants as future cash crops in India," *Indian Farming*, p.33.

Jadeja, B.A. and Nakar, R.N. (2010). Study on ethno-medico botany of weeds from saurashtra region, Gujarat, India. *Plant Archieves*. 10: 761-765.

Joshi, M.C. (1983). A floristic and phytochemical survey of some important South Gujarat forests with special reference to plants of medicinal and ethnobotanical interest, Baroda.

Jain, S.K. (1963). Studies in Indian ethnobotany-Plants used in Medicine by the tribals of Madhya Pradesh. *Bull. Region. Res. Lab. Jammu* 1(2): 126-128

Jain, S.K. (1965). On the prospects of some new or less known medicinal plan resources. *Ind. Med. J.* 59(12): 270-272.

Jain, S.K. (1965). Medicinal plant lore of the tribals of Bastar. *Economic Bot.* 19(3). 236-250.

Jain, S.K. (1981)(Ed.) Glimpses of Indian Ethnobotany, New Delhi.

Jain, S.K. and Tarafder, C.R.. (1963). Studies in Indian Ethnobotany-native plants remedies for snake bite among the advasis of Central India. *Indian Med. J.* 57(12):307-309.

Maity, S. and Geeta, K.A. (2007). Medicinal and Aromatic Plants in India. In: Horticulture-Ornamental, Medicinal and aromatic crops, National Research Centre for Medicinal and Aromatic Plants, pp 1-27

Mandavia, C.K., Dhaduk, H.L., Padaliya, P. and Nakar, R.N. (2010). *Issues Relating to Medicinal Plant Cultivation, Processing and Marketing*. Workshop souvenir. Department of Agricultural Botany, Junagadh Agricultural University, Junagadh.

Mhaskar, K.S. and Caius, J.F. (1931). Indian Plants remedies used in snake bite. *Indian Med. Res. Mem.*19.

Mukerji, B. (1953). The Indian Pharmaceutical Codex, New Delhi

Nadkarni, A.K. (1954). Dr. K.M. Nadkarni's Indian Materia Medica, Bombay. Revised. Ed.

Nagar, P.S. and Daniel, M. (2010). Floristic diversity of Gujarat. In National conference on Biodiversity conservation. M.S.University, Vadodara, pp. 63-77.

Nakar, R.N. and Jadeja, B.A. (2009). Study on morphology, ethnobotany and phenology of *Prosopis* in Girnar Forest Junagadh, Gujarat. *Proceedings of the*

National Symposium on Prosopis: ecological, economic significance and management challenges. Gujarat Institute of Desert Ecology, Bhuj, pp. 51-53.

Odedara, N.K. (2009). *Ethnobotany of Maher Tribe In Porbandar District, Gujarat, India,* PhD Thesis, Saurashtra University, Rajkot.

Planning Commission, (2000). Report of the Taskforce on Medicinal Plants in India. Planning Commission, Government of India, Yojana Bhawan, New Delhi, India

Punjani, B.L. (2002). Ethnobotanical aspects of some plants of Arravalli Hills in North Gujarat. *Ancient Life Sciencce.* 21: 268-280.

Punjani, B.L. (2010). Herbal folk medicines used for urinary complaints in tribal pockets of Northeast Gujarat. *Indian Journal of Traditional Knowledge*. 9: 126-130.

Santapau, H. (1953). The flora of Khadala on the Western Ghats of India. *Rec. Bot. Surv. India.* 16(1): 1-335.

Shah, N.C. (1981). Need of systematic cultivation of medicinal herbs used in indigenous systems and traditional medicine. Indian drugs. 18(6): 210-217.

State Forest Department (1998). Medicinal Plants of Madhya Pradesh, pp.127

Sunderraj, D. and Balasubramanyam, G. (1959). Guide to economic plants of South India, Madras.

Thakar, Jaikrisha Indraji (1926). Plants of Cutch and their utility, Bombay (In Gujarati).

Trivedi, Pandit Krishna Prasad. (1967). Dhanvantari, Vanaushadhi Visheshank, Aligarh, Parts 1-4 (in Hindi).

Uniyal, R.S., Uniyal, M.R. and Pushp Jain (2000). Cultivation of medicinal plants in India: A reference book. Traffic India/WWF India, pp.1-162

Upholf, J.C. Th (1959). Dictionary of Economic Plants, New York.

U.S.D. (1955). The united states dispensatory, Philadelphia.

World Health Organization (2002). Traditional Medicine Strategy 2002-2005.

Website: www.who.int/medicines/library/trm/trm_start_eng.pdf.

Appendix I:

GLOSSARY [BOTANICAL TERMS]

Achene: Single seeded, unicellular, dry, indehiscent fruit also called caryopsis

Acuminate: Long, pointed, gradually tapering towards apex

Amplexicaule: Encircling of the node by leaf bases

Apex: Tip, uppermost part

Apices (apex): Top

Aristate: Ending in bristle or awn

Articulate: Jointed

Bracteole: A small bract or leaf structures below perianth in aflower

Bulbils: Vegetative propagative spherical structure arising atthe leaf base (as in *Dioscorea bulbifera*)

Caducous: Falling soon

Campanulate: Bell-shaped

Capitulum: Head-shaped inflorescence, as in Asteraceae

Cauline: Arising from stem

Cladodes: Modification of dwarf branches into leaf-like structure

Clasping: Wrapping

Comose: With long, white bunch of hair

Cordate: Heart-shaped

Coriaceous: Thick, stiff

Corm: A stem modification, underground spherical inshape with reserve food material

Corona: Bundle of hair between corolla and stamens arisingfrom base of the corolla

Cuneate: Wedge-shaped, tapering towards base

Cuspidate: Tapering to long point at tip

Cyme: Arrangement of flowers with older flower on topand younger flower towards base

Diadelphous: Stamen divided into two groups 9+1 as in Fabaceae

Didymous: Two-sized (filament of stamens)

Dioceous: Male and female flowers on different plants

Discoid: Disc-shaped

Ellipsoid: Eclipse-shaped

Emarginate: Deeply and irregularly notched at apex

Entemophylous: Pollinated by insects

Entire: Even margin, complete margin, no cut or lobationon margin (of leaves)

Epigynous: Ovary seated above perianth

Exstipulate: Without stipule, a leafy structure at the base of leaf

Extrose: Facing outward

Fascicled: Clustered at one point

Fluted: Hollow

Follicle: Dry dehiscent fruit opening only by ventral suture

Fragrant: Emitting sweet smell

Gamo petalous: Petals united with each other

Gamo sepalous: Sepals united with each other

Gamo tepallus: Perianth united with each other

Glabrous: Without any hairy structure

Gregarious: Very long – robust, profuge

Gynaecium: Female part of flower having ovary, style, and stigma

Heart wood: Central hardest part of wood/trunk

Hypogynous: Ovary inferior, sepals, petals, and stamen above theovary

Imbricate: Arrangement of corolla with two outer, one inner,and two with one side outer other side inner

Imparipinnate: Leaflet in odd number on top

Lanceolate: Shape of convex lens

Latex: Oozing milky sap

Lenticellate: Slit-like raised cortical structure on the branches

Linear: Very narrow, like a line

Lomentum: Single seeded cell of pod, septate, and constrictedbetween two seeds

Moniliform: Beaded in a row like a garland

Monoecious: Unisexual, male and female flowers on the same plant

Mucronate: Small projection at the apex (acume)

Oblong: Longer than broad with narrowing margin towards base

Obpyramidal: Inverted pyramid shaped

Obsolete: Minute or wanting

Obtuse: Blunt top (apex)

Orbicular: Almost circular

Ovate: Egg shaped

Pedicel: Stalk of flower

Pedicillate: Stalked flowers

Peduncle: Stalk of inflorescence

Perianth: Vegetative covering of sexual organ in flower,sometime differentiated into calyx and corolla

Peripinnate: Leaflet in even number

Pesticide: An agent that kills unwanted plants and insects

Pinnate: Compound leaf with leaflets arranged on samerhachis at length

Polyhederal: Many faced, many angled

Procumbent: Creeping on ground then rising up

Pubescent: Carpeting of small soft hair

Radical: Arising from stem base (leaves or branches)

Reniform: Almost kidney-shaped

Reticulate: Weaved

Rhizome: Subterranean part between stem and root bearingbuds that may be used as a propagative part

Rhomboid: Quadrihedral with only opposite angles equal

Sarmentose: Growing among bushes, with long flexuous runners

Scandent: Weak plants that need support; climbing withoutany climbing organ, and so on

Serrate: Margin of leaf cut into saw-shaped structure,pointing upwards

Sessile: Without any stalk

Sinuate: Wavy margins

Spathulate: Service spoon shaped

Spike: Sessile flowers arranged on peduncle

Staminode: Barren stamen (infertile anthers)

Stellate: Star-shaped arrangement of short stiff hair (trichomes)

Stipitate: Stalked

Striate: Marked with vertical lines

Succulent: Thick, soft, and juicy

Suffruticose : A herb becoming perennial at base and herbaceousat apices

Syncarpous: Fused carpels

Terete: Lined

Terrestrial: Growing in soil

Tomentose: Dense, soft, layer of hair or cotton easily scraped off

Truncate: Flat topped

Tuber: A swollen, subterranean root containing reservefood material

Turbinate: Tube shaped

Variegated: Spotted with various colour

Villous: Long soft shaggy hair

Whorl: Arising more than two from one node (leaf or branches)

Zygomorphic: Asymmetrical plain of flowers not divisible intoequal halves

Appendix-II:

GLOSSARY [MEDICINAL TERMS]

Abortifacient: A drug that induces foetus expulsion

Alexiteric: Developing resistance against infectious diseases

Alterative: A drug that alters body condition by improvingmetabolism; used against long effect of a medicine

Amenorrhoea: Failure of menstruation

Analgesic: Pain killer

Anodyne: A drug used to allay pain

Antacid: To neutralize acidic effect in abdomen

Anthelmintic: A drug used to expel or destroy intestinal worms

Antibronchial: Working against respiratory track infection andcongestion

Antiemetic: A drug used to control vomiting

Antihistaminic: A drug used for controlling skin irritation and itching caused due to increase of blood histamine

Anti-inflammatory: A drug used to cure swellings

Antiperiodic: A drug that prevents recurrence of a disease. Used against malarial fever

Antiphlogestic: An agent used for reducing or subsiding inflammations

Antipyretic: A drug or a medicine used to lower bodytemperature in fever

Anti-rheumatic: A drug used against joint pain and swellings

Antirhinitis: Clearing of nasal mucous by subsiding nasalmembrane inflammation

Antiseptic: Prevention of putrefaction or sepsis of woundsand cuts

Antispasmodic: A medicine that releases nervous irritability and reduces spasm or convulsion

Antitussive: A drug controlling cough

Aperient: Mild laxative/cathartic

Aphrodisiac: Drug increasing the sexual desire and longevity

Appetizer: Increasing digestion and hunger

Aromatic: An agent that emits sweet smell

Astringent: A drug that contracts the muscular membrane

Bronchodiltaor: A drug that widens the trachea, thus easingcongestion

Carminative: A drug that releases intestinal gases or flatulence

Catarrhal: Mucous membrane inflammation with excessive secretion of mucous

Cathartic: Drastic purgative, totally expelling rectal stool

Cholagogue: A drug inducing excessive secretion of bile juice

Colic: Severe spasmodic and gripping pain in colon region

Demulcent: Soothing medicine for digestive function

Depurant: Purifier

Diaphoretic: Drug inducing perspiration

Diuretic: Increasing urination frequency

Dysmenorrhoea: Painful menstrual flow

Dyspepsia: Indigestion with gastric pain

Emmenagogue: A drug that restores regularity in menstrual cycle

Epilepsy: An affectation of the nervous system resulting fromexcessive or disordered discharge of cerebral neurons

Expectorant: A drug expelling phlegm from trachea

Febrifuge: A drug used to cure fever

Galactagogue: Increasing and activating mammary gland

Geriatric: Pertaining to old age

Gleet: Chronic discharge from vagina

Gonorrhoea: Inflammation of the gentio-urinary passage withpain and discharges

Gout: A purine metabolic disease with raised level ofserum uric acid (blood urea)

Gynaecological : Pertaining to female genital organ

Haemophilic: Loss of blood coagulation property in which bloodcontinues to flow on cuts

Haemoptypsies: Spitting of blood

Haemorrhage: Bleeding piles

Haemostatic: Blood coagulant, preventing bleeding

Hydrogogue: Promoting expulsion of water or serum

Hyper lipidemia: Reducing fat on joints

Hypoglycaemic: Lowering blood sugar

Hypotensive: Lowering blood pressure

Hysteria: Neurotic attack with unusual activities and symptoms

Insomnia: Sleeplessness

Lactagogue: Increasing milk secretion

Laxative: Smoothening rectal wall and loosening the stool

Leucorrhoea: White fluid discharge from vagina

Menorrhagia: Excessive menstrual flow

Neuropathy: Diseases related to CNS (central nervous system)

Oedima: Inflammation

Ophthalmic: Pertaining to eye diseases like conjunctivitis

Orchitis: Inflammation of testis with hypertrophy and pain

Paraplegia: Paralysis, loss of ability to move or feel in the lowerpart of body

Phlegm: Mucous secretion in respiratory track

Post-natal: After child birth

Pruritus: Skin itching

Purgative: Loosening stool to help exersion, thus curingconstipation

Refrigerant: Cooling effect

Rejuvinative: Antiageing, prolonging life

Resolvant: Causing resolution of a tumor or swelling

Rubifacient: Producing counter effect on external application

Scrofula: Tubercular cervical adenitis, with or without ulceration

Sedative: Central nervous system depressant in which a personis made calm or asleep

Stomachic: A drug used for improving digestion

Styptic: Blood purifier

Thermogenic: Producing heat offer metabolism

Thrombosis: A blockage preventing the flow of blood in the bodycaused by clot

Tranquilizer: A drug used to calm a person and reduce mental activity

Urticaria: Nettle rashes on skin

Vermifuge: Expelling or destroying intestinal worms

Vertigo: Dizziness, a feeling of spinning

Index

C

D

E

F

G

H

I

J

K

L

M

N

T

U

V

W

Z

www.ingramcontent.com/pod-product-compliance
Ingram Content Group UK Ltd.
Pitfield, Milton Keynes, MK11 3LW, UK
UKHW021436280726
14060UKWH00001BA/106

9 789351 309918